ML

Books are to be returned on or before
the last date below.

Principles of Broadband Switching and Networking

WILEY SERIES IN TELECOMMUNICATIONS AND SIGNAL PROCESSING

John G. Proakis, Editor
Northeastern University

A complete list of the titles in this series appears at the end of this volume.

Principles of Broadband Switching and Networking

Tony T. Lee and Soung C. Liew

A JOHN WILEY & SONS, INC., PUBLICATION

Copyright © 2010 by John Wiley & Sons, Inc. All rights reserved.

Published by John Wiley & Sons, Inc., Hoboken, New Jersey
Published simultaneously in Canada

No part of this publication may be reproduced, stored in a retrieval system, or transmitted in any form or by any means, electronic, mechanical, photocopying, recording, scanning, or otherwise, except as permitted under Section 107 or 108 of the 1976 United States Copyright Act, without either the prior written permission of the Publisher, or authorization through payment of the appropriate per-copy fee to the Copyright Clearance Center, Inc., 222 Rosewood Drive, Danvers, MA 01923, (978) 750-8400, fax (978)-750-4470, or on the web at www.copyright.com. Requests to the Publisher for permission should be addressed to the Permissions Department, John Wiley & Sons, Inc., 111 River Street, Hoboken, NJ 07030, (201) 748-6011, fax (201) 748-6008, or online at http://www.wiley.com/go/ permissions.

Limit of Liability/Disclaimer of Warranty: While the publisher and author have used their best efforts in preparing this book, they make no representations or warranties with respect to the accuracy or completeness of the contents of this book and specifically disclaim any implied warranties of merchantability or fitness for a particular purpose. No warranty may be created or extended by sales representatives or written sales materials. The advice and strategies contained herein may not be suitable for your situation. You should consult with a professional where appropriate. Neither the publisher nor author shall be liable for any loss of profit or any other commercial damages, including but not limited to special, incidental, consequential, or other damages.

For general information on our other products and services or technical support, please contact our Customer Care Department within the United States at (800) 762-2974, outside the United States at (317) 572-3993 or fax (317) 572-4002.

Wiley also publishes its books in a variety of electronic formats. Some content that appears in print may not be available in electronic books. For more information about Wiley products, visit our web site at www.wiley.com

ISBN: 978-0-471-13901-0

Library of Congress Cataloging-in-Publication Data is available.

Printed in the United States of America

*Dedicated to Professor Charles K. Kao for his guidance,
and to our wives, Alice and So Kuen, for their unwavering support.*

CONTENTS

Preface	xiii
About the Authors	xvii

1 Introduction and Overview   1
    1.1 Switching and Transmission   2
       1.1.1 Roles of Switching and Transmission   2
       1.1.2 Telephone Network Switching and Transmission Hierarchy   4
    1.2 Multiplexing and Concentration   5
    1.3 Timescales of Information Transfer   8
       1.3.1 Sessions and Circuits   9
       1.3.2 Messages   9
       1.3.3 Packets and Cells   9
    1.4 Broadband Integrated Services Network   10
       Problems   12

2 Circuit Switch Design Principles   15
    2.1 Space-Domain Circuit Switching   16
       2.1.1 Nonblocking Properties   16
       2.1.2 Complexity of Nonblocking Switches   18
       2.1.3 Clos Switching Network   20
       2.1.4 Benes Switching Network   28
       2.1.5 Baseline and Reverse Baseline Networks   31
       2.1.6 Cantor Switching Network   32
    2.2 Time-Domain and Time–Space–Time Circuit Switching   35
       2.2.1 Time-Domain Switching   35
       2.2.2 Time–Space–Time Switching   37
       Problems   39

3 Fundamental Principles of Packet Switch Design — 43

3.1 Packet Contention in Switches — 45
3.2 Fundamental Properties of Interconnection Networks — 48
    3.2.1 Definition of Banyan Networks — 49
    3.2.2 Simple Switches Based on Banyan Networks — 51
    3.2.3 Combinatoric Properties of Banyan Networks — 54
    3.2.4 Nonblocking Conditions for the Banyan Network — 54
3.3 Sorting Networks — 59
    3.3.1 Basic Concepts of Comparison Networks — 61
    3.3.2 Sorting Networks Based on Bitonic Sort — 64
    3.3.3 The Odd–Even Sorting Network — 70
    3.3.4 Switching and Contention Resolution in Sort-Banyan Network — 71
3.4 Nonblocking and Self-Routing Properties of Clos Networks — 75
    3.4.1 Nonblocking Route Assignment — 76
    3.4.2 Recursiveness Property — 79
    3.4.3 Basic Properties of Half-Clos Networks — 81
    3.4.4 Sort-Clos Principle — 89
Problems — 90

4 Switch Performance Analysis and Design Improvements — 95

4.1 Performance of Simple Switch Designs — 95
    4.1.1 Throughput of an Internally Nonblocking Loss System — 96
    4.1.2 Throughput of an Input-Buffered Switch — 96
    4.1.3 Delay of an Input-Buffered Switch — 103
    4.1.4 Delay of an Output-Buffered Switch — 112
4.2 Design Improvements for Input Queueing Switches — 113
    4.2.1 Look-Ahead Contention Resolution — 113
    4.2.2 Parallel Iterative Matching — 115
4.3 Design Improvements Based on Output Capacity Expansion — 119
    4.3.1 Speedup Principle — 119
    4.3.2 Channel-Grouping Principle — 121
    4.3.3 Knockout Principle — 131
    4.3.4 Replication Principle — 137
    4.3.5 Dilation Principle — 138
Problems — 144

5 Advanced Switch Design Principles — 151

5.1 Switch Design Principles Based on Deflection Routing — 151
    5.1.1 Tandem-Banyan Network — 151
    5.1.2 Shuffle-Exchange Network — 154

		5.1.3	Feedback Shuffle-Exchange Network	158
		5.1.4	Feedback Bidirectional Shuffle-Exchange Network	166
		5.1.5	Dual Shuffle-Exchange Network	175

5.2 Switching by Memory I/O — 184
5.3 Design Principles for Scalable Switches — 187
    5.3.1 Generalized Knockout Principle — 187
    5.3.2 Modular Architecture — 191
    Problems — 198

6 Switching Principles for Multicast, Multirate, and Multimedia Services — 205

6.1 Multicast Switching — 205
    6.1.1 Multicasting Based on Nonblocking Copy Networks — 208
    6.1.2 Performance Improvement of Copy Networks — 213
    6.1.3 Multicasting Algorithm for Arbitrary Network Topologies — 220
    6.1.4 Nonblocking Copy Networks Based on Broadcast Clos Networks — 228

6.2 Path Switching — 235
    6.2.1 Basic Concept of Path Switching — 237
    6.2.2 Capacity and Route Assignments for Multirate Traffic — 242
    6.2.3 Trade-Off Between Performance and Complexity — 249
    6.2.4 Multicasting in Path Switching — 254

6.A Appendix — 268
    6.A.1 A Formulation of Effective Bandwidth — 268
    6.A.2 Approximations of Effective Bandwidth Based on On–Off Source Model — 269
    Problems — 270

7 Basic Concepts of Broadband Communication Networks — 275

7.1 Synchronous Transfer Mode — 275
7.2 Delays in ATM Network — 280
7.3 Cell Size Consideration — 283
7.4 Cell Networking, Virtual Channels, and Virtual Paths — 285
    7.4.1 No Data Link Layer — 285
    7.4.2 Cell Sequence Preservation — 286
    7.4.3 Virtual-Circuit Hop-by-Hop Routing — 286
    7.4.4 Virtual Channels and Virtual Paths — 287
    7.4.5 Routing Using VCI and VPI — 289
    7.4.6 Motivations for VP/VC Two-Tier Hierarchy — 293

	7.5	ATM Layer, Adaptation Layer, and Service Class	295
	7.6	Transmission Interface	300
	7.7	Approaches Toward IP over ATM	300
		7.7.1 Classical IP over ATM	301
		7.7.2 Next Hop Resolution Protocol	302
		7.7.3 IP Switch and Cell Switch Router	303
		7.7.4 ARIS and Tag Switching	306
		7.7.5 Multiprotocol Label Switching	308
	Appendix 7.A ATM Cell Format		311
		7.A.1 ATM Layer	311
		7.A.2 Adaptation Layer	314
		Problems	319
8	**Network Traffic Control and Bandwidth Allocation**		**323**
	8.1	Fluid-Flow Model: Deterministic Discussion	326
	8.2	Fluid-Flow On–Off Source Model: Stochastic Treatment	332
	8.3	Traffic Shaping and Policing	348
	8.4	Open-Loop Flow Control and Scheduling	354
		8.4.1 First-Come-First-Serve Scheduling	355
		8.4.2 Fixed-Capacity Assignment	357
		8.4.3 Round-Robin Scheduling	358
		8.4.4 Weighted Fair Queueing	364
		8.4.5 Delay Bound in Weighted Fair Queueing with Leaky-Bucket Access Control	373
	8.5	Closed-Loop Flow Control	380
		Problems	381
9	**Packet Switching and Information Transmission**		**385**
	9.1	Duality of Switching and Transmission	386
	9.2	Parallel Characteristics of Contention and Noise	390
		9.2.1 Pseudo Signal-to-Noise Ratio of Packet Switch	390
		9.2.2 Clos Network with Random Routing as a Noisy Channel	393
	9.3	Clos Network with Deflection Routing	396
		9.3.1 Cascaded Clos Network	397
		9.3.2 Analysis of Deflection Clos Network	397
	9.4	Route Assignments and Error-Correcting Codes	402
		9.4.1 Complete Matching in Bipartite Graphs	402
		9.4.2 Graphical Codes	405
		9.4.3 Route Assignments of Benes Network	407

9.5	Clos Network as Noiseless Channel-Path Switching		410
	9.5.1 Capacity Allocation		411
	9.5.2 Capacity Matrix Decomposition		414
9.6	Scheduling and Source Coding		416
	9.6.1 Smoothness of Scheduling		417
	9.6.2 Comparison of Scheduling Algorithms		420
	9.6.3 Two-Dimensional Scheduling		424
9.7	Conclusion		430

Bibliography   **433**

PREFACE

The past few decades have seen the merging of many computer and communication applications. Enabled by the advancement of optical fiber, wireless communication, and very-large-scale integration (VLSI) technologies, modern telecommunication networks can be regarded as one of the most important inventions of the past century.

Before the emergence of Broadband Integrated Services Digital Network (B-ISDN), several separate communication networks already existed. They include the telephone network for voice communication, the computer network for data communication, and the television network for TV program broadcasting. These networks are designed with a specific application in mind and are typically not well suited for other applications. For example, the conventional telephone network cannot carry high-speed multimedia services, which require diverse quality-of-service (QoS) guarantees to support multirate and multicast connections. In addition, these heterogeneous networks often require expensive gateways equipped with different access interfaces running different protocols.

Meanwhile, the appeal of interactive video communication is on the rise in a society that is increasingly information-oriented. Images and facial expressions are more vivid and informative than text and audio for many types of human interactions. For example, video conferencing has made distant learning, medicine, and surgery possible, while 3D Internet games give rise to real-time interactions between remote players. All these applications are based on high-resolution video with large bandwidth demands. These developments led to the emergence of B-ISDN—the concept of an integrated network to support communication services of all kinds to achieve the most cost-effective sharing of resources was conceived in the late 1980s.

This book focuses on the design and analysis of switch architectures that are suitable for broadband integrated networks. In particular, the emphasis is on packet-switched interconnection networks with distributed routing algorithms. The focus is on the mathematical properties of these networks rather than specific implementation technologies. As such, although the pedagogical explanations in this book are in the context of switches, many of the fundamental principles are relevant to other communication networks with regular topologies. For example, the terminals in a multi-hop ad hoc wireless network could conceivably be interconnected together to form a logical topology that is regular. This could be enabled by the use of directional antennas, inexpensive multi-radio, and cognitive-radio technologies that can identify unused spectra. These technologies allow links to be formed among the terminals in a more flexible way, not necessarily based on proximity alone. There are two main advantages to regular network topologies: (1) very simple routing and scheduling are possible with their well-understood mathematical properties; and (2) performance and behavior are well understood and predictable. The performance and robustness of these ad hoc networks are by no means ad hoc.

The original content of this book was an outgrowth of an evening course offered at the Electrical Engineering Department of Columbia University, New York, in 1989. Since then, this course has been taught at Polytechnic Institute of New York University, Brooklyn, NY and the Chinese University of Hong Kong, Hong Kong. The target audience is senior undergraduate and first-year postgraduate students with solid background in probability theory. We found that many of our former students acquired an appreciation of the beauty of the mathematics associated with telecommunication networks after taking courses based on this book.

A general introduction and an overview of the entire book are given in Chapter 1, in which the roles of switching and transmission in the computer networks and telephone networks are discussed. The concept of the modern broadband integrated services network is explained and the reasons why this concept is necessary in modern society are also given in this chapter. The focus of Chapter 2 is on circuit switch design principles. Two types of the circuit switch design—space domain and time domain—are introduced in this chapter. Several classical nonblocking networks, including Clos network, Benes network, and Cantor network, are discussed.

Chapter 3 is devoted to fundamental principles of packet switch design, and Chapter 4 focuses on the throughput and delay analyses of both waiting and loss switching systems. The nonblocking and self-routing properties of packet switches are elaborated by the combination of sorting and Banyan networks. Throughput improvements are illustrated by some switch design variations such as speedup principle, channel-grouping principle, knockout principle, and dilation principle.

Chapter 5, following the previous chapter, explains some advanced switch design principles to alleviate the packet contention problem. Several networks based on the deflection routing principle such as tandem-banyan, shuffle-exchange, feedback shuffle-exchange, feedback bidirectional shuffle-exchange, and dual shuffle-exchange are introduced. Switch scalability is discussed, which provides some key principles to the construction of large switches out of modest-size switches, without sacrificing overall switch performance. Chapter 6, on switch design principles for broadband services, first presents several fundamental switch design principles for multicasting. Then we end the chapter by introducing the concept of path switching, which is a compromise of the dynamic and the static routing schemes.

Chapter 7 departs from switch designs and the focus moves to broadband communication networks that make use of such switches. The asynchronous transfer mode (ATM) being standardized worldwide is the technology that meets the requirements of the broadband communication networks. ATM is a switching technology that divides data into fixed-length packets called cells. Chapter 8, on network traffic control and bandwidth allocation, gives an introduction on how to allocate network resources and control the traffic to satisfy the quality-of-service (QoS) requirements of network users and to maximize network usage.

The content of Chapter 9 is an article "The mathematical parallels between packet switching and information transmission" originally posted at http://arxiv.org/abs/cs/0610050, which is included here as an epilogue. It is clear from the title that this is a philosophical discussion of analogies between switching and transmission. We show that transmission noise and packet contention actually have similar characteristics and can be tamed by comparable means to achieve reliable communication. From various comparisons, we conclude that packet switching systems are governed by mathematical laws that are similar to those of digital transmission systems as envisioned by Shannon in his seminal 1948 BSTJ paper "A Mathematical Theory of Communication."

We would like to thank many former students of Broadband Communication Laboratory at the Chinese University of Hong Kong, including Cheuk H. Lam, Philip To, Man Chi Chan, Cathy Chan, Soung-Yue Liew, Yun Deng, Manting Choy, Jianming Liu, Sichao Ruan, Li Pu, Dongjie Yin, and Pui King Wong, who participated in the discussions of the content of the book over the years. We are especially grateful for the delicate latex editing and figure drawing of the entire book by our student assistants Jiawei Chen and Yulin Deng. Our "family networks," though small, have given us the connectivity to many joys of life. We can never repay the debt of gratitude we owe to our families—our wives, Alice and So Kuen, and our children, Wynne and Edward Lee, and Vincent and Austin Liew—for their understanding, support, and patience while we wrote this book.

Tony T. Lee
Soung C. Liew

ABOUT THE AUTHORS

Professor Tony T. Lee received his BSEE degree from the National Cheng Kung University, Taiwan, in 1971, and his MS and PhD degrees in electrical engineering from the Polytechnic University in New York in 1976 and 1977, respectively. Currently, he is a Professor of Information Engineering at the Chinese University of Hong Kong and an Adjunct Professor at the Institute of Applied Mathematics of the Chinese Academy of Sciences. From 1991 to 1993, he was a Professor of Electrical Engineering at the Polytechnic Institute of University of New York, Brooklyn, NY. He was with AT&T Bell Laboratories, Holmdel, NJ, from 1977 to 1983, and Bellcore, currently Telcordia Technologies, Morristown, NJ, from 1983 to 1993. He is now serving as an Editor of the *IEEE Transactions on Communications*, and an Area Editor of *Journal of Communication Network*. He is a fellow of IEEE and HKIE. He is the recipient of many awards including the National Natural Science Award from China, the Leonard G. Abraham Prize Paper Award from the IEEE Communication Society, and the Outstanding Paper Award from IEICE.

Professor Soung Chang Liew received his SB, SM, EE, and PhD degrees from the Massachusetts Institute of Technology. From 1984 to 1988, he was at the MIT Laboratory for Information and Decision Systems, where he investigated fiber-optic communication networks. From March 1988 to July 1993, Professor Liew was at Bellcore (now Telcordia), New Jersey, where he was engaged in broadband network research. He is currently Professor and Chairman of the Department of Information Engineering, the Chinese University of Hong Kong. He is also an Adjunct Professor at the Southeast University, China. His current research interests include wireless networks, Internet protocols, multimedia communications, and packet switch design. Professor Liew and his students won the best paper awards at the 1st IEEE International Conference on Mobile Ad-Hoc and Sensor Systems (IEEE MASS 2004) and the 4th IEEE International Workshop on Wireless Local Network (IEEE WLN 2004). Separately, TCP Veno, a version of TCP to improve its performance over wireless networks proposed by Professor Liew and his students, has been incorporated into a recent release of Linux OS. In addition, Professor Liew initiated and built the first interuniversity ATM network testbed in Hong Kong in 1993.

Besides academic activities, Professor Liew is active in the industry. He cofounded two technology start-ups in Internet software and has been serving as consultant to many companies and industrial organizations. He is currently consultant for the

Hong Kong Applied Science and Technology Research Institute (ASTRI), providing technical advice as well as helping to formulate R&D directions and strategies in the areas of wireless internetworking, applications, and services. Professor Liew holds three U.S. patents and is a Fellow of IEE and HKIE. He is listed in Marquis Who's Who in Science and Engineering. He is the recipient of the first Vice-Chancellor Exemplary Teaching Award at the Chinese University of Hong Kong.

1

INTRODUCTION AND OVERVIEW

The past few decades have seen the merging of computer and communication technologies. Wide-area and local-area computer networks have been deployed to interconnect computers distributed throughout the world. This has led to a proliferation of many useful data communication services, such as electronic mail, remote file transfer, remote login, and web pages. Most of these services do not have very stringent "real-time" requirements in the sense that there is no urgency for the data to reach the receiver within a very short time, say, below 1s. At the other spectrum, the telephone network has been with us for a long time, and the information carried by the network has been primarily real-time telephone conversations. It is important for voice to reach the listener almost immediately for an intelligible and coherent conversation to take place.

With the emergence of multimedia services, real-time traffic will include not just voice, but also video, image, and computer data files. This has given rise to the vision of an integrated broadband network that is capable of carrying all kinds of information, real-time or non-real-time.

Many wide-area computer networks are implemented on top of telephone networks: transmission lines are leased from the telephone companies, and each of these lines interconnects two routers that perform data switching. Home computers are also linked to a gateway via telephone lines using modems. The gateway is in turn connected via telephone lines to other gateways or routers over the wide-area network. Thus, present-day computer networks are mostly networks overlaid on telephone networks. Strictly speaking, the telephone networks that are being used to carry computer data cannot be said to be integrated. The networks are designed with the intention that voice traffic will be carried, and their designs are optimized according to this assumption. A transmission line optimized for voice traffic is not necessarily optimal for other traffic types. The computer data are just "guests" to the telephone networks, and many components of the telephone network may not be optimized for the transport of non-voice services.

Principles of Broadband Switching and Networking, by Tony T. Lee and Soung C. Liew
Copyright © 2010 John Wiley & Sons, Inc.

2 INTRODUCTION AND OVERVIEW

The focus of this book is on future broadband integrated networks. Loosely, the terms "broadband" and "integration" imply that services with rates from below one kbps to hundreds of Mbps can be supported. Some of these services, such as video conferencing, are widely known and anticipated, whereas others may be unforeseen and created only when the broadband network becomes available. The broadband network must be flexible enough to accommodate these unanticipated services as well.

1.1 SWITCHING AND TRANSMISSION

At the fundamental level, a communication network is composed of switching and transmission resources that make it possible to transport information from one user to another. On top of the switching and transmission resources, we have the control functions, which could be implemented by either software or hardware, or both. Among other things, the control functions make it possible to automate the setting up of a connection between two users. At another level, they also ensure efficient usage of the switching and transmission resources. In a real network, the switching, transmission, and control facilities are typically distributed across many locations.

1.1.1 Roles of Switching and Transmission

When there are only two users, as shown in Fig. 1.1, information created by one user is always targeted to the other user: switching is not needed and only transmission is required. In essence, the transmission facilities serve to carry information directly from one end of the transmission medium, which could be a coaxial cable, an optical fiber, or the air space, to the other end.

As soon as we have a third user in our network, the question of who wants to communicate with whom, and at what time, arises. With reference to Fig. 1.2, user A may want to talk to user B at one time but to user C later. The switching function makes it possible to change the connectivity among users in a dynamic way. In this way, a user can communicate with different users at different times.

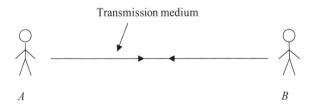

When there are only two users, information from A is by default destined for B, and vice versa

FIGURE 1.1 A two-user network; switching is not required.

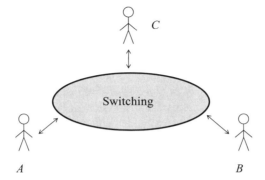

Information from *A* may be destined for *B* or *C*

FIGURE 1.2 A three-user network; switching is required.

It turns out that the locations of the switching facilities in a network have a significant impact on the amount of transmission facilities required in a network. Figure 1.3(a) depicts a telephone network in which the switching facilities are distributed and positioned at the N users' locations, and a user is connected to each of the other users via a unique line. Switching is performed when the user decides which of the N lines to use. When N is large, there will be many transmission lines and the transmission cost will be rather prohibitive.

In contrast, Fig. 1.3(b) shows a network in which each user has only one access line through which it can be connected to the other users. Switching is performed at

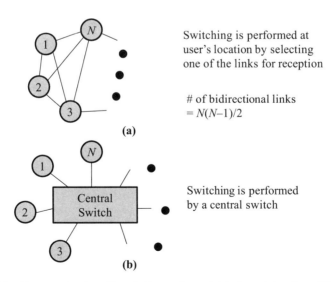

FIGURE 1.3 N-user networks with switching performed (a) at user's locations (b) by a central switch.

4 INTRODUCTION AND OVERVIEW

a central location. To the extent that a user does not need to speak to all the other users at the same time, this is a better solution because of the reduced number of transmission lines. Of course, if a user wants to be able to connect to more than one user simultaneously, multiple lines must still be installed between the user and the central switch.

In practice, a network typically consists of multiple switching centers at various locations that are interconnected via transmission lines. By locating the switching centers judiciously, transmission cost can be reduced.

1.1.2 Telephone Network Switching and Transmission Hierarchy

The switching and transmission facilities in a telephone network are organized in a hierarchical fashion. A simplified picture of a telephone network is given in Fig. 1.4. At the lower end of the hierarchy, we have subscribers' telephones located at business offices and households. Each telephone is connected to a switching facility, called a central office, via a subscriber loop. The switching center at this level is called the local office or the end office. If a subscriber wishes to speak to another subscriber linked to the same local office, the connection is set up by a switch at the local office.

Two local offices may be connected via either direct links or a higher level switching center, called a toll office. In the first case, there must be sufficient voice traffic between the two local central offices to justify the direct connection; otherwise, since the transmission capacities cannot be used by other local offices, they will be wasted. The second solution permits a higher degree of sharing of transmission resources. As illustrated in the figure, local offices A, B, and C are linked together via a toll office D. The transmission facilities between D and C are shared between A and B

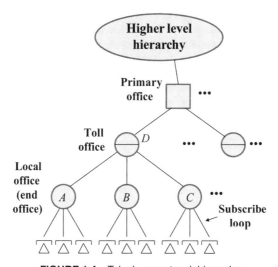

FIGURE 1.4 Telephone network hierarchy.

in the sense that both traffic between A and C and between B and C travel over them.

The toll offices are interconnected via an even higher level office, called the primary office. The primary offices are in turn connected by a yet even higher level office. Each level may move up to a higher level. In this way, a network hierarchy is formed.

The total amount of traffic reduces as we move up the hierarchy because of the so-called community-of-interest phenomenon. For instance, it is generally more likely for a user to make local phone calls than long-distance phone calls. The former may involve only a local switching office while the latter involves a series of switching offices at successive levels.

In short, an important objective achieved with the hierarchical network is the sharing of resources. The resources at the higher level are shared by a larger population of subscribers. The amount of resources can be reduced because it is statistically unlikely that all the subscribers will want to use the higher level resources at the same time.

Another advantage that comes with the hierarchical structure is the simplicity in finding a "route" for a connection between two subscribers. When subscriber i wants to connect to subscriber j, the local office of i first checks if j also belongs to the same office. If yes, switching is completed at the office. Otherwise, a connection is made between the local office and the next level toll office (assuming there are no direct links between the central offices of i and j). This procedure is repeated until an office with branches leading to both i and j is found.

1.2 MULTIPLEXING AND CONCENTRATION

Multiplexing and concentration are important concepts in reducing transmission cost. In both, a number of users share an underlying transmission medium (e.g., an optical fiber, a coaxial cable, or the air space).

As a multiplexing example, frequency-division multiplexing (FDM) is used to broadcast radio and TV programs on the air medium. In FDM, the capacity, or bandwidth, of the transmission medium is divided into different frequency bands, and each band is a logical channel for carrying information from a unique source. FDM can be used to subdivide the capacity of air medium, a coaxial cable, or any other transmission medium. Figure 1.5 depicts the transmission of digital information from a number of sources using FDM. Different carrier frequencies are used to transport different information streams. Receivers at the other end use bandpass filters to select the desired information stream.

Multiplexing can also be performed in the time domain. This is a more widely used multiplexing technique than FDM in telephone networks. Figure 1.6 illustrates a simple time-division multiplexing (TDM) scheme. The N sources take turns in a round-robin fashion to transmit on the transmission medium. Time is divided into frames, each having N time slots. Each source has a time slot dedicated to it in each frame. Thus, time slot 1 is assigned to source 1, time slot 2 to source 2, and so on. The slot positions i in successive frames all belong to the source i.

6 INTRODUCTION AND OVERVIEW

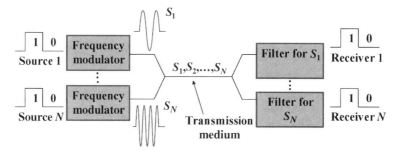

FIGURE 1.5 Frequency-division multiplexing.

In this book, we define switching as changing the connectivity between end users or equipments. The goal of a multiplexing system (consisting of the multiplexer, the transmission medium, and the demultiplexer) is not to perform switching; the goal is to partition a transmission medium into a number of logical channels, each of which can be used to interconnect a transmitter and a receiver. In the two scenarios above, each of the N multiplexed channels is dedicated exclusively to a transmitter–receiver pair, and which transmitter is connected to which receiver does not change over time. As an overall system, an input of the multiplexer is always connected to the same output of the demultiplexer. Thus, functionally, no switching occurs. Such is not the case with a concentrator.

Concentration achieves cost saving by making use of the fact that it is unlikely for all users to be active simultaneously at any given time. Therefore, transmission facilities need only be allocated to the active users. In the telephone network, for instance, it is unlikely that all the subscribers of the same local office want to use their phones at the same time. An $N \times M$ concentrator, as shown in Fig. 1.7, concentrates traffic from N sources onto M ($M < N$) outputs. A number of concentrators are usually placed at the "front end" of the local switching center to reduce the number

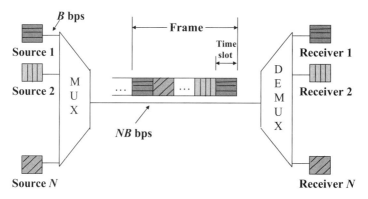

FIGURE 1.6 Time-division multiplexing.

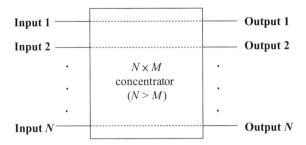

An active input is assigned to one of the outputs. It does not matter which output is assigned.

FIGURE 1.7 An $N \times M$ concentrator.

of ports of the switch. An output of a concentrator, hence an input to the switch, is allocated to the subscriber only when he picks up the phone. The connectivity (i.e., which input is connected to which output) of the concentrator changes in a dynamic manner over time. If more than M sources are active at the same time, then some of the sources may be "blocked." For telephone networks, M can usually be made considerably smaller than N to save cost without incurring a high likelihood for blocking.

Both multiplexers and concentrators achieve resource sharing, but in different ways. Let us refer to the sums of the capacities (bit rates) of the transmission lines connected to the inputs and outputs as the *total input and output capacities*, respectively. For a multiplexer, the total output capacity is equal to the total input capacity, whereas for a concentrator, the total output capacity is smaller than the total input capacity. The output capacity or bandwidth of the concentrator is said to be shared among the inputs, and that of the multiplexer is not. The concentrator outputs are allocated dynamically to the inputs based on need, and the allocation cannot be foretold in advance. In contrast, although a multiplexer allows the same transmission medium to be shared among several transmitter–receiver pairs, this is achieved by subdividing the capacity of the transmission medium and dedicating the resulting subchannels in an exclusive manner to individual pairs.

Statistical multiplexing is a packet-switching technique that can be considered as combining TDM with concentration. Consider a TDM scheme in which there are M ($M < N$) time slots in a frame. The output capacity is then smaller than the maximum possible total capacities of the inputs. A time slot is not always dedicated to a particular source. Slot 1 of frame 1 may be used by user 1, but user 1 may be idle in the next frame and slot 1 of frame 2 may be used by another user. The same slot positions of different frames can be used by different users, and therefore they can be targeted for different destinations in the overall communication network. To route the information in a slot to its final destination, we therefore need to encode the routing information in a "header" and attach the header to the information bits before putting both into

8 INTRODUCTION AND OVERVIEW

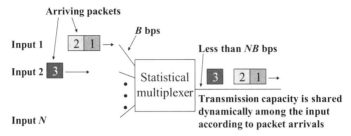

FIGURE 1.8 A statistical multiplexer.

the slot. Routing using header information is a basic operating principle in a packet-switched network. Each time slot is then said to contain a packet. The header in a packet generally contains either the explicit destination address or information from which the route to the final destination can be derived.

In circuit switching, the TDM scheme makes use of the fact that slots are assigned in a fixed way to derive the source and destination information implicitly. Since the same slot positions of successive frames are used by the same source and targeted for the same destination, they should be routed in a similar way. The route of these slots is determined during the call setup process (e.g., when setting up a voice connection); it will be used repeatedly in successive frames. In particular, no header is needed for routing purposes. In a packet network, the routing information must be incorporated into the header since the positions of a slot cannot be used to derive its route. To extend things even further, since the routing information is now contained in the header, we can even allow the time slot, or packet, to vary in length according to the amount of information transmitted by the source at each shot. The frame structure in TDM can be abandoned altogether in a packet network.

Note that unlike a time-division multiplexer, a statistical multiplexer performs switching by dynamically changing the user of its output (Fig. 1.8). We cannot foretell beforehand which source will use the output in advance. The users may also send data to the multiplexers in a random manner. It is therefore possible that there are more packets arriving at the statistical multiplexer than can be sent out on the output immediately. Thus, buffers or memories are required to store these outstanding packets until they can be cleared.

1.3 TIMESCALES OF INFORMATION TRANSFER

The above discussion of the concentrator and statistical multiplexer alluded to resource sharing at different timescales. In a telephone network, an output of a concentrator is assigned to a user only when the user picks up the phone and wants to make a call. The output is assigned for the duration of the call that typically lasts several minutes. In a packet-switched network, the output of a statistical multiplexer is assigned only for the duration of a packet, which typically is much less than 1s. A user may send out packets sporadically during a communication session. It is important to have a clear concept of the timescales of information transfer to appreciate the fact that resource sharing and network control can be achieved at different timescales.

1.3.1 Sessions and Circuits

Before two end users can send information to each other, a communication session needs to be established. A telephone call is a communication session. As another example, when we log onto a computer remotely, we establish a session between the local terminal and the remote computer.

Network resources are assigned to set up a circuit or connection for this session. Some of these resources, such as an output of a concentrator in a circuit-switched network, may be dedicated exclusively to this connection while it remains active. Some of these resources, such as the output of a statistical multiplexer in a packet network, may be used by other sessions concurrently. In the latter, the transmission bandwidth is not dedicated exclusively to the session and is shared among active sessions, and the associated circuit is sometimes called a virtual circuit.

Some packet networks are not connection-oriented. It is not necessary to preestablish a connection (hence a route from the source to the destination) before data are sent by a session. In fact, successive packets of the session may traverse different routes to reach the destination. Although the concept of a connection is absent within the network, the end users still need to set up a session before they start to communicate. The setup time, however, can be much shorter than in a connection-oriented network because the control functions inside the network need not be involved for connection setup.

1.3.2 Messages

Once a session is set up, the users can then send information in the form of messages in an on–off manner. For a remote login session, the typing of the carriage-return key by the end user may result in the sending of a line of text as a message. Files may also be sent as a message. So, messages tend to vary in length.

For a two-party telephone session, for example, it is known that a user speaks only 40% of the time. The activity of the user is said to alternate between idle period and busy period. The busy period is called a talkspurt, which can be viewed as a message in a voice session. A scheme called time assigned speech interpolation (TASI) is often used to statistically multiplex several voice sources onto the same satellite link on a talkspurt basis.

1.3.3 Packets and Cells

Messages are data units that have meaning to the end users and have a logical relationship with the associated service. It could be a talkspurt, a line of text, a file, and so on. Packets, on the other hand, are transport data units within the network.

In a packet network, a long message is often fragmented into smaller packets before it is transported across the network. One possible reason for fragmentation could be that the communication network does not support packets beyond a certain length. For instance, the Internet has a maximum packet size of 64 Kbytes.

Another reason for message fragmentation is that most computer networks are store-and-forward networks in which a switching node must receive the entire packet before it is forwarded to the next node. By fragmenting a long message into many smaller packets, the end-to-end delay of the message can be made smaller, especially when the network is lightly loaded (see Problem 1.4).

Yet another motivation for fragmentation is to prevent a long packet from hogging a communication channel. Consider the output of a statistical multiplexer. While a long packet is being transmitted, newly arriving packets from other sources must wait until the completion of its transmission, which may take an excessively long time if there were no limit on its length. On the other hand, if the long packet has been cut into many smaller packets, the newly arriving packets from other sources have a chance to jump ahead and access the output channel after a short wait for the transmission of the current packet to complete.

Packet length can be variable or fixed. One advantage of the fixed packet-length scheme is that more efficient packet switches can be implemented. For instance, by time aligning the boundaries of the packets across the inputs of a packet switch, higher throughput can be achieved with the fixed packet-length scheme than with the variable packet-length scheme.

The fixed packet-length scheme has a disadvantage when the messages to be sent are much shorter than the packet length. In this case, only a small part of each packet contains the useful message, and the rest is stuffed with dummy bits to make up the whole packet. The observation suggests that small packets are preferable. However, the length of the packet header is largely independent of the overall packet length (e.g., the destination address length in the header is independent of the packet length). Hence, the header overhead (ratio of header length to packet length) tends to be larger for smaller packets. Thus, too small a packet can lead to high inefficiency as well.

In general, the determination of packet size is a complicated issue involving considerations from many different angles, not the least the characteristics of the underlying network traffic. The ITU (International Telecommunication Union), an international standard body, has chosen the asynchronous transfer mode (ATM) to be the information transport mechanism for the future broadband integrated network. An essence of the ATM scheme is that the basic information data unit is a 53-byte fixed-length packet called *cells*. The details of ATM and the motivations for the small-size cell will be covered in the later chapters.

1.4 BROADBAND INTEGRATED SERVICES NETWORK

The discussion up to this point forms the backdrop of the focus of this book—broadband integrated services networks. As the name suggests, an integrated network must be capable of supporting many different kinds of services.

Traditionally, services are segregated and each communication network supports only one type of service. Some examples of these networks are the telephone, television, airline reservation, and computer data networks. There are many advantages to having a single integrated network for the support of all these services. For instance, efficient resource sharing may be achievable. Take the telephone service. More phone calls are made during business hours than during the evening. The television service, on the other hand, is in high demand during the evening. By integrating these services on the same network, the same resources can be assigned to different services at different times.

Traditionally, whenever a new service is introduced, a new communication network may need to be designed and set up. Carrying the information traffic of this service on a network designed for another service may not be very efficient because of the dissimilar characteristics of the traffic. If an integrated network is designed with the forethought that some unknown services may need to be introduced in the future, then these services can be accommodated more easily.

The design of an integrated network taking into account the above concern is by no means easy. Figure 1.9 shows some services with widely varying traffic characteristics; the holding times (durations of sessions) and bit rate requirements of different services may differ by several orders of magnitude. Furthermore, some services, such as computer data transfer, tend to generate traffic in a bursty manner during a session (Fig. 1.10). Other services such as telephony and video conferencing generate traffic in a more continuous fashion.

The delay requirements may also be different. Real-time services are highly sensitive to network delay. For example, if real-time video data do not arrive at the display monitor at the receiver within certain time, they might as well be considered as lost.

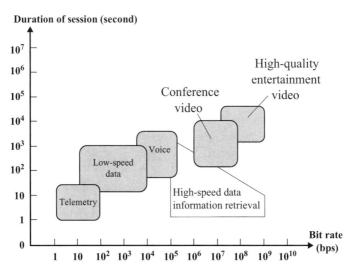

FIGURE 1.9 Holding times and bit rates of various services.

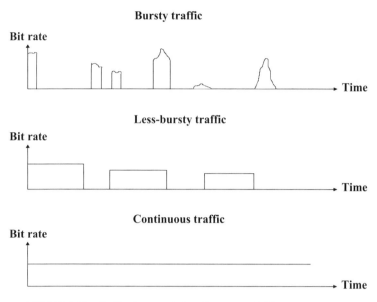

FIGURE 1.10 Traffic characteristics of sources of different burstiness.

How to control the traffic of these services to satisfy their diverse requirements is a nontrivial challenge that is still being worked on actively by researchers in this field.

As mentioned above, part of the challenge is to design the network to be flexible enough for the support of new services and services whose requirements have changed over time. As an example of the latter, advances in video and speech coding algorithms may well influence the characteristics of traffic generated by some services and thus change the service requirements. Finally, the integrated network must also be cost-effective and efficient for it to be successful.

PROBLEMS

1.1 This is a simplified problem related to resource sharing. You are among the 10 people sharing four telephones. At any time, a person is using or attempting to use a telephone with probability 0.2. What is the probability that all the telephones are being used when you want to make a call to your girl/boy friend? What if there are 100 people sharing 40 phones? Is it better to have a larger number of people sharing a pool of resources?

1.2 Does TDM perform the switching or transmission function?

1.3 Each telephone conversation requires 64 kbps of transmission capacity. Consider a geographical region with 6 million people, each with a telephone. We want to multiplex all the telephone traffic onto a number of optical fibers, each with capacity of 2.4 Gbps. How many fibers are needed?

1.4 Consider the problem of fragmenting a message of 1000 bytes into packets of x bytes each. Each packet has a header (overhead) of 10 bytes. The message passes through five links of the same transmission rate in a store-and-forward network on its way to destination. There is no other traffic in the network. What is the optimal packet size x in order to minimize end-to-end delay?

1.5 Consider 100 active phone conversations being multiplexed onto 60 lines by TASI. Assume a person speaks only 40% of the time when using a telephone. Talkspurts are clipped when all 60 lines have been used. What is the probability of a talkspurt being clipped?

2

CIRCUIT SWITCH DESIGN PRINCIPLES

The telephone network is the most widespread and far-reaching among all communication networks. It is difficult to imagine a modern society without it. In fact, the telephone network is one of the largest and most complex engineering systems created by mankind. A large portion of the Internet is built upon the telephone network infrastructure. Much thought and work have gone into the design of the telephone network.

The telephone network is largely a circuit-switched network in which resource sharing is achieved at the session level. Transmission and switching facilities are dedicated to a session only when a telephone call is initiated. Using these resources, a circuit, or a connection, is formed between two end users. These facilities will be released upon the termination of the call so that other sessions may use them. While the session is ongoing, however, the resources assigned to the circuit are not shared by other circuits and can be used to carry the traffic of the circuit only. In other words, once the resources are assigned to a circuit, they are guaranteed. Of course, there is a no guarantee that a newly initiated session will get the resources that it requests. In this case, the call is said to be "blocked." One important design issue of the circuit-switched network is how to minimize the blocking probability in a low-cost and efficient manner.

An example of this problem is the determination of the minimum number of "trunks" or transmission lines between two switching centers in order to meet a certain blocking probability requirement. In general, more trunks will be needed for smaller blocking probability. One of the exercises of this chapter goes over this in more detail. A call can also be blocked because of the design of the switch even if there are enough trunks. This chapter focuses on the fundamental design principles of circuit switches; in particular, how switches can be designed to be nonblocking,

---

Principles of Broadband Switching and Networking, by Tony T. Lee and Soung C. Liew
Copyright © 2010 John Wiley & Sons, Inc.

16 CIRCUIT SWITCH DESIGN PRINCIPLES

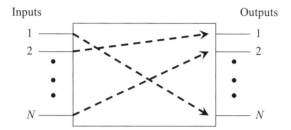

FIGURE 2.1 An $N \times N$ switch used to interconnect N inputs and N outputs.

in which case blocking will be due entirely to number of calls exceeding number of trunks.

2.1 SPACE-DOMAIN CIRCUIT SWITCHING

An $N \times N$ space-division switch can be used to connect N incoming links to N outgoing links, as illustrated in Fig. 2.1. Switches are often constructed using 2×2 switching elements, called crosspoints, shown in Fig. 2.2. Each crosspoint has two states. In the *bar* state, the upper input is connected to the upper output and the lower input is connected to the lower output. In the *cross* state, the upper input is connected to the lower output and the lower input connected to the upper output.

Figure 2.3 shows two ways of constructing 4×4 switches out of the 2×2 elements. For the crossbar switch, the elements are arranged in a square grid, and by setting the individual elements in bar and cross states, any input can be connected to any output without internal blocking. The baseline switch, although requires fewer number of switch elements, is blocking. As illustrated, if input 1 is already connected to output 1, then input 2 cannot be connected to output 2. In general, there is a trade-off between the number of crosspoints needed and the blocking level of the switch.

2.1.1 Nonblocking Properties

Circuit switches are often classified according to their nonblocking properties.

Definition 2.1 (Strictly Nonblocking). A switch is strictly nonblocking if a connection can always be set up between any idle (or free) input and output without the need to rearrange the paths of the existing connections.

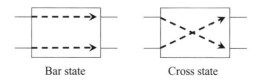

FIGURE 2.2 Bar and cross states of 2×2 switching elements.

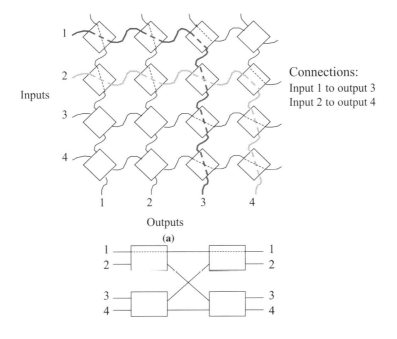

FIGURE 2.3 (a) Crossbar switch; (b) baseline switch.

The reader can verify that the crossbar switch shown in Fig. 2.3(a) is strictly nonblocking. By setting selected switch elements to bar or cross states, you can always find a path from an idle input to an idle output without the need to rearrange the paths taken by the existing connections. The number of switch elements required in a crossbar switch is N^2, where N is the number of input or output ports.

Definition 2.2 (Rearrangeably Nonblocking). A switch is rearrangeably nonblocking if a connection can always be set up between any idle input and output, although it may be necessary to rearrange the existing connections.

A 4×4 rearrangeably nonblocking switch is shown in Fig. 2.4(a). Figure 2.4(b) depicts a situation where input 2 is connected to output 2 and input 3 to output 3 with all the switch elements in the bar states. If a new connection arrives requesting connection from input 4 to output 1, it will be blocked by the two current connections. However, if we rearrange the connection between input 2 and output 2 as in Fig. 2.4(c), we find that the new request can now be accommodated.

Definition 2.3 (Wide-Sense Nonblocking). A switch is wide-sense nonblocking if a route selection policy exists for setting connections in such a way that a new connection can always be set up between any idle input and output without the need to rearrange the paths of the existing connections.

18 CIRCUIT SWITCH DESIGN PRINCIPLES

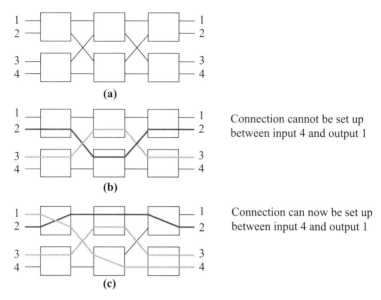

FIGURE 2.4 (a) A 4 × 4 rearrangeably nonblocking switch; (b) a connection request from input 4 to output 1 is blocked; (c) same connection request can be satisfied by rearranging the existing connection from input 2 to output 2.

Thus, associated with the wide-sense nonblocking is an algorithm for setting the internal paths of the switch. The study and the proof of wide-sense nonblocking property is generally not easy, since not only the arrivals of connection requests must be considered, but the departures (terminations) of existing connections must also be considered.

From the above definitions, it is easily seen that the strictly nonblocking property poses the most stringent requirements, followed by wide-sense nonblocking property and then rearrangeably nonblocking. A switch that is strictly nonblocking is also nonblocking by the other two definitions, but not vice versa. If a switch is wide-sense nonblocking, it is also rearrangeably nonblocking. To see this, consider a new request arriving at a wide-sense nonblocking switch. Suppose we have not been following the connection setup algorithm required for it to be nonblocking and the new request is blocked. But we can rearrange the existing connections! In the worst case, we can disconnect all the existing connections and then reconnect them one by one following the wide-sense nonblocking algorithm. The new request is guaranteed to be satisfiable after the rearrangement. Note that a rearrangeably nonblocking switch, however, is not necessarily wide-sense nonblocking.

2.1.2 Complexity of Nonblocking Switches

An interesting question is what is the minimum number of crosspoints needed to build an $N \times N$ nonblocking switch. It turns out that an order of $N \log N$ crosspoints are

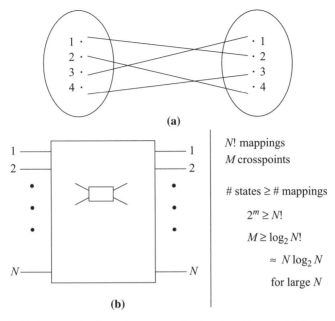

FIGURE 2.5 (a) An example of one-to-one mapping from input to output; (b) number of crosspoints needed for nonblocking switch.

needed. This can be seen by relating the number of possible input–output mappings to the number of states of the switch.

A one-to-one mapping from inputs to outputs is a unique connection of inputs to outputs in a one-to-one fashion, an example of which is given in Fig. 2.5(a). Clearly, there are $N!$ possible mappings if all inputs and outputs are busy. In practice, there may be some idle inputs and outputs at any given time. However, since there is no signal (or information flow) on these idle ports, it does not matter how they are connected as long as they are not connected to a busy port. Thus, the mappings with idle inputs and outputs can be subsumed under one of the $N!$ one-to-one mappings. That is, the realization of one of the $N!$ mappings can also be used to realize a mapping with idle ports, and we need only be concerned about the $N!$ mappings.

We must set the states of the individual crosspoints within the switch to realize the $N!$ mappings. Let there be M crosspoints in the overall switch. Since each crosspoint has two states and the state of the overall switch is defined by the combination of the states of the crosspoints, there are 2^M states for the overall switch. Each of these states can realize one and only one of the $N!$ mapping. Thus, to realize all the $N!$ mappings, we must have

$$2^M \geq N! \tag{2.1}$$

The inequality is due to the fact that two states may realize the same mapping. The reader can easily verify this by experimenting with the switch in Fig. 2.4.

20 CIRCUIT SWITCH DESIGN PRINCIPLES

The Stirling's formula is

$$N! = N^N e^{-N} \sqrt{2\pi N}(1 + \epsilon(N)),$$

where $\epsilon > 0$ is a decreasing function of N. Substituting this into Eq. (2.1) and taking the log on both sides, we have

$$M \geq N \log_2 N - N \log_2 e + 0.5 \log_2(2\pi N) + \log_2(1 + \epsilon(N)). \quad (2.2)$$

The dominant term on the right-hand side of inequality (2.2) is $N \log N$ for large N. Thus, asymptotically, the number of crosspoints required must be at least $N \log_2 N$.

Note that the bound applies to all three nonblocking properties defined previously. The derivation is nonconstructive in that it does not tell us how to construct a switch that achieves the $N \log_2 N$ bound. It only tells us that if such a switch exists, its order of complexity cannot be lower than $N \log_2 N$. Thus, the question of whether there are switches satisfying this bound arises. We now turn our attention to some specific switch constructions.

2.1.3 Clos Switching Network

An important issue in switch design is the construction of a large switch out of smaller switch modules. A three-stage Clos switching network is shown in Fig. 2.6. Switch modules are arranged in three stages and any module is interconnected with any module in the adjacent stage via a unique link. The modules are nonblocking and could be, for example, the crossbar switches described previously. To motivate the study of a three-stage network rather than a two-stage network, Problem 2.11 argues that two-stage networks are not suitable for constructing nonblocking switches.

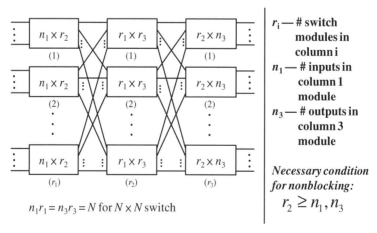

FIGURE 2.6 A three-stage Clos switch architecture.

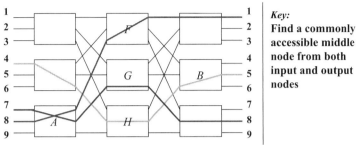

A request for connection from input 9 to output 4 is blocked
S_A = set of middle-stage nodes used by A
  = { F, G }
S_B = set of middle-stage nodes used by B
  = { H }

FIGURE 2.7 An example of blocking in a three-stage switch.

There are five independent parameters in the Clos architecture. We have r_1, r_2, and r_3 modules in stages 1, 2, and 3, respectively. The dimensions of the modules in the stages 1, 2, and 3 are $n_1 \times r_2$, $r_1 \times r_3$, and $r_2 \times n_3$, respectively. If the overall switch has equal number of inputs and outputs, then the number of inputs or outputs is $N = n_1 r_1 = n_3 r_3$, and there are only four independent parameters.

Figure 2.7 shows a Clos switch with $n_1 = r_1 = r_2 = r_3 = n_3 = 3$. This particular switch structure is not strictly nonblocking. We want to derive the relationship among the parameters that will guarantee nonblocking operation. A crucial point is the number of middle-stage modules r_2. By making r_2 larger, there are more alternative paths between stage-1 and stage-3 modules, and therefore we should expect the likelihood of blocking to be smaller. In fact, if r_2 is made large enough, blocking can be eliminated altogether. On the other hand, the switch becomes more complex in terms of both the number of stage-2 modules and the dimensions of stage-1 and stage-3 modules.

It is easy to see that for the switch to be nonblocking, we must have

$$r_2 \geq n_1, n_3. \tag{2.3}$$

Otherwise, if all of the inputs (outputs) of a stage-1 (stage-2) module are active, some of the connections cannot be set up. Inequalities (2.3) are necessary conditions. We shall see that they are also sufficient for achieving the rearrangeably nonblocking property but not the strictly nonblocking property.

We need to develop some notational tools before we proceed. Figure 2.8 shows that the connection state of the switch can be represented in the form of a connection matrix. Row i corresponds to first-stage module i and column j corresponds to third-stage module j. Entry (i, j) is associated with the middle-stage modules. As shown, if there are connections between module A and module B through modules F, G, and

22 CIRCUIT SWITCH DESIGN PRINCIPLES

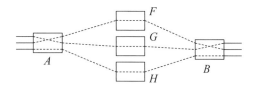

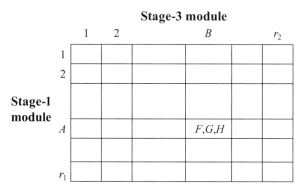

FIGURE 2.8 The connection matrix of the three-stage network.

H, then entry $(A, B) = \{F, G, H\}$. In other words, entry (i, j) contains the symbols or labels of the middle-stage modules that are used to connect calls from module i in stage 1 to module j in stage 3.

Let S_A and S_B be the sets of symbols in any row A and column B, respectively. There are three conditions that must be satisfied by a legitimate connection matrix:

1. Each row can have at most n_1 symbols:

$$|S_A| \leq n_1. \tag{2.4}$$

This is because each first-stage module has n_1 inputs and can have at most n_1 connections. Each connection needs to go through one and only one middle-stage module.

2. Each column can have at most n_3 symbols:

$$|S_B| \leq n_3. \tag{2.5}$$

3. The symbols in each row or each column must be distinct: this is because each first- or third- stage module is connected to each middle-stage module by one and only one link. There can be at most r_2 symbols in each row or column. Thus,

$$|S_A|, |S_B| \leq r_2. \tag{2.6}$$

Theorem 2.1. Assuming the underlying switch modules are individually strictly nonblocking, a three-stage $N \times N$ Clos network is strictly nonblocking if and only if

$$r_2 \geq \min\{n_1 + n_3 - 1, N\}. \tag{2.7}$$

Proof. The trivial case when $N \leq n_1 + n_3 - 1$ is easy to see: since there can be no more than N calls in progress at the same time in the overall switch, r_2 need not be more than N. For $n_1 + n_3 - 1 < N$, suppose there is a new connection request from an input in module A to an output in module B. Consider the worst-case situation in which all other inputs and outputs of A and B are busy. Then,

$$|S_A| = n_1 - 1,$$
$$|S_B| = n_3 - 1. \tag{2.8}$$

Furthermore,
$$|S_A \cup S_B| = |S_A| + |S_B| - |S_A \cap S_B|$$
$$\leq |S_A| + |S_B| = n_1 + n_3 - 2. \tag{2.9}$$

The above inequality is satisfied with equality if $|S_A \cap S_B| = 0$; that is, the middle-stage modules used by connections from A and B are disjoint. If $r_2 \geq n_1 + n_3 - 1$, there must be at least one symbol not in either S_A or S_B. This is a symbol corresponding to a middle-stage module as yet unused by the existing connections of A and B, and it can be used to set up the new connection request. □

The value of r_2 can be made smaller if we only require the switch to be rearrangeably nonblocking. In fact, r_2 needs only be no less than $\max(n_1, n_3)$. Certainly, by the previous theorem, the switch is not strictly nonblocking with this value of r_2. Therefore, we must find a way to rearrange the existing circuits to accommodate a new request whenever it is blocked.

Substituting a symbol of an entry in the connection matrix with another symbol corresponds physically to rearranging an existing connection. Specifically, the connection is disconnected and reestablished over the middle-stage module represented by the new symbol. Certainly, we cannot simply substitute the old symbol with any arbitrary symbol. The new symbol must not have already occurred in the row and column of the entry in order not to violate the legitimacy of the matrix.

With reference to Fig. 2.9, suppose we want to establish a new connection between first-stage module A and third-stage module B. Suppose that all symbols except D occur in row A and all symbols except C occur in column B. Then, the connection is blocked because we could not find a symbol that have neither occurred in row A nor in column B for entry (A, B).

Since C is not currently in column B, we might try to change the symbol D in column B to symbol C so that we can put D in entry (A, B). But if there were a C already in the row occupied by the D in column B (see Fig. 2.9), we could not change the D to C without violating the matrix constraint.

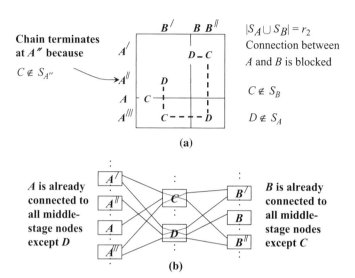

FIGURE 2.9 (a) A chain of C and D originating from B; (b) physical connections corresponding to the chain.

In other words, we may need to rearrange more than just one existing connection. To see which connections need to be rearranged, let us introduce the concept of the rearrangement chain. A *chain* of two symbols, C and D, with *end points* at column B and row A'' is shown in Fig. 2.9(a). An end point must lie in a row or column containing either C or D but not both. For example, column B contains the end point with symbol D and it does not have a symbol C. The row and column of an intermediate point each contains one C and one D. To identify the chain, we search across the rows and columns alternatively for C and D. Thus, if we find a C during a row search, we then search for D across the column containing C, and so on. The process stops when a row search fails to find a C or a column search fails to find a D. The current C or D then forms the other end point of the chain. Note that the chain has a finite number of points and a loop is impossible if we start with an end point; otherwise, C or D will occur more than once in some row or column, as illustrated in Fig. 2.10.

The physical interpretation of the chain is shown in Fig. 2.9(b). Specifically, the chain corresponds to a set of first-stage and third-stage modules with connections across middle-stage modules C and D.

The chain is said to be rearranged if we switch the symbols C and D in the chain. This corresponds to rearranging the associated connections, as illustrated in Fig. 2.11. Connections that used to be established across C are now established across D, and vice versa. Note that rearranging the chain as such will not lead to a violation of the rule that a symbol can occur at most once in each row or column. As shown, by rearranging the chain starting in B as in Fig. 2.11, we can put D in entry (A, B).

Now, a problem could arise if the chain from column B ends in the symbol C in row A, as indicated in Fig. 2.12(b), since the rearrangement would have switched the

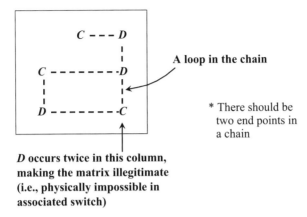

FIGURE 2.10 Illustration showing loops in chains are not permitted in legitimate connection matrix.

C in row A to D and we cannot then put D in the entry (A, B) for the new connection without violating the constraint that a symbol can occur at most once in each row. Fortunately, as indicated in Fig. 2.12(b), it is impossible for the chain from column B to end in row A. To see this, note that for the search starting from column B, each row search always attempts to find a C and each column search always attempts to find a D: having the chain connected to the C in row A leads to the contradiction that a column search finds the C in row A.

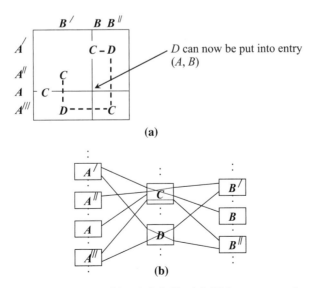

FIGURE 2.11 (a) Rearrangement of the chain in Fig. 2.9; (b) the corresponding rearrangement of connections.

26 CIRCUIT SWITCH DESIGN PRINCIPLES

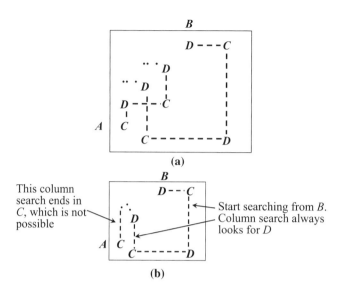

FIGURE 2.12 (a) Two chains, one originates from B and one from A; (b) illustration that the two chains in (a) cannot be connected.

The example assumes that originally there was a symbol C in S_A but not in S_B and a symbol D in S_B but not in S_A. The following theorem states that this is always true given that $r_2 \geq \max(n_1, n_3)$.

Theorem 2.2. Assuming the underlying switch modules are individually rearrangeably nonblocking, a three-stage $N \times N$ Clos network is rearrangeably nonblocking if and only if

$$r_2 \geq \max(n_1, n_3). \qquad (2.10)$$

Proof. We first consider the more specific case where the underlying modules are strictly nonblocking. The "only if" part is a trivial corollary of condition (2.3). For the "if" part, suppose $r_2 \geq \max(n_1, n_3)$, and we want to set up a new connection between A and B. For the existing connections, we have

$$|S_A| \leq n_1 - 1,$$
$$|S_B| \leq n_3 - 1. \qquad (2.11)$$

There are two cases: (i) $|S_A \cup S_B| < r_2$ and (ii) $|S_A \cup S_B| = r_2$. In case (i), there is a symbol not in A and B and the corresponding middle-stage module can be used to set up the connection. In case (ii), inequalities (2.11) yield

$$|S_A - S_B| = |S_A \cup S_B| - |S_B| \geq r_2 - n_3 + 1$$
$$\geq n_3 - n_3 + 1 = 1 \qquad (2.12)$$

and

$$|S_B - S_A| = |S_A \cup S_B| - |S_A| \geq r_2 - n_1 + 1$$
$$\geq n_1 - n_1 + 1 = 1. \qquad (2.13)$$

Thus, there is a symbol $C \in S_A - S_B$ and a symbol $D \in S_B - S_A$. That is, C is in row A but not column B, and D is in column B but not row A. This situation corresponds to the one in Fig. 2.9, and we can use the rearrangement as depicted in Fig. 2.11 to accommodate the new connection.

If the underlying switch modules are only rearrangeably nonblocking but not strictly nonblocking, the existence of C and D can still be proved as above. The situation now is slightly different in that in addition to the "external rearrangements" of the connectivity switch modules of adjacent stages, rearrangements within the switch modules may also be necessary since the external arrangements may necessitate setting up of new connections in the switch modules. But the setting up of these new connections is always possible by rearranging the existing connections in the switch modules since they are by assumption rearrangeably nonblocking. This observation completes the proof. □

Suppose that the switch modules are strictly nonblocking. A question is how many existing connections need to be rearranged. Suppose we begin traversing the chain starting from the end point in B. Each time a point is included, a new row or column is covered. Since there is no loop in the chain, it takes at most $r_1 + r_3 - 2$ moves before all the other rows and columns (in addition to the initial row and column of the B end point) are covered. Thus, there can be at most $r_1 + r_3 - 1$ points in the chain, which is also the number of arrangements needed. This bound can be improved to $2\min(r_1, r_3) - 2$ if one considers a more careful argument (see Problem 2.9).

The preceding paragraph considers only the chain that starts from B. There is another chain that starts from A, as illustrated in Fig. 2.12(a). These two chains cannot be connected and therefore cannot be the same. This fact has already been argued and is depicted in Fig. 2.12(b).

Theorem 2.3. Assuming the switch modules are individually strictly nonblocking, the number of rearrangements needed in a rearrangeably nonblocking three-stage Clos switch is at most $\min(r_1, r_3) - 1$ if we choose the shorter of the two chains.

Proof. Suppose that we start searching simultaneously from A and B. A composite move includes a move in each chain. Whenever a chain cannot be extended further, we choose that chain for rearrangement.

28 CIRCUIT SWITCH DESIGN PRINCIPLES

In each composite move, a new row will be traversed by one of the chains while a new column will be traversed by the other chain. Since the chains are disjoint, the same row or column cannot be covered by both chains. There can be at most $r_1 - 2$ composite moves before all rows are included (2 is subtracted from r_1 because the initial two end points occupy two rows). Therefore, the number of points in the shorter chain is no more than $r_1 - 1$. Similarly, by considering the columns, the number of points in the shorter chain is no more than $r_3 - 1$. Thus, the number of points in the shorter chain is at most $\min(r_1, r_3) - 1$. □

2.1.4 Benes Switching Network

The modules in the Clos switch architecture are usually relatively large switches compared to the 2×2 crosspoints. However, this does not preclude the use of the theory developed above for 2×2 switch elements. Figure 2.13 shows a symmetric three-stage network in which $n_1 = n_3 = 2$. Each of the first- and third-stage modules are 2×2 switching elements. It can be seen that the problem of constructing an $N \times N$ switch has been broken down to the problem of constructing two $N/2 \times N/2$ switches in the middle. By Theorem 2.2, the $N \times N$ switch is rearrangeably nonblocking if the $N/2 \times N/2$ switches are rearrangeably nonblocking.

To construct the $N/2 \times N/2$ rearrangeably nonblocking modules, we can use the same decomposition in a recursive manner. That is, each $N/2 \times N/2$ module can be broken down into three stages consisting of 2×2 elements in the first and third stages and two $N/4 \times N/4$ modules in the middle. Repeating this, recursively, only 2×2 elements will remain in the end. An 8×8 switch constructed this way is shown in Fig. 2.14. This architecture is called the Benes network.

A question is how many crosspoints are there in an $N \times N$ Benes network. Let us assume that $N = 2^n$; that is, it is a power of 2. Let the number of stages in a $k \times k$

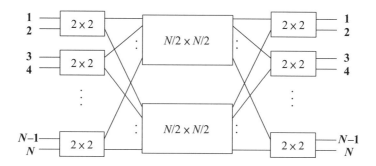

The $N \times N$ network is rearrangeably nonblocking if
the $N/2 \times N/2$ networks are rearrangeably nonblocking.

FIGURE 2.13 Recursive decomposition of a rearrangeably nonblocking network.

SPACE-DOMAIN CIRCUIT SWITCHING

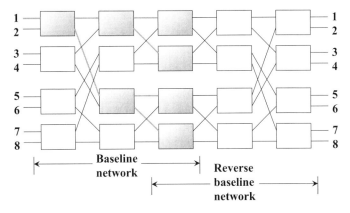

FIGURE 2.14 An 8 × 8 Benes network.

Benes switch be denoted by $f(k)$. By the recursive construction, we have

$$f(N) = f\left(\frac{N}{2}\right) + 2. \qquad (2.14)$$

Applying the above as below yields a closed-form expression of $f(N)$:

$$\begin{aligned} f(2^n) &= f(2^{n-1}) + 2 \\ &= f(2^{n-2}) + 4 \\ &\;\;\vdots \\ &= f(2^{n-j}) + 2j \\ &\;\;\vdots \\ &= f(2) + 2(n-1) \\ &= 1 + 2(n-1) \\ &= 2n - 1. \end{aligned} \qquad (2.15)$$

Since each stage has $N/2$ crosspoints, the total number of crosspoints is

$$\frac{N}{2}(2n-1) = N \log_2 N - \frac{N}{2}. \qquad (2.16)$$

Notice that asymptotically (i.e., for large N), the Benes network satisfies the $N \log N$ lower bound on the number of crosspoints required in a nonblocking switch.

Although the number of crosspoints in the Benes network is of order $N \log N$, lacking is a *fast* control algorithm that can set up paths between the inputs and outputs.

30 CIRCUIT SWITCH DESIGN PRINCIPLES

This makes it less appealing for fast- packet switching in which the paths taken by packets must be determined in a very short time.

Let us consider a looping algorithm. As illustrated in Fig. 2.13, the Benes network consists of two subnetworks sandwiched between the first and last columns of 2×2 switch elements. Therefore, one way to set up the path for an input–output pair is to first determine whether it should go through the upper or lower subnetwork. The constraints that must be satisfied are that the paths of the inputs (outputs) sharing the same 2×2 elements (e.g., inputs (outputs) 1 and 2) must go through different subnetworks since there is only one link from a 2×2 element to a subnetwork. Once the set of disjoint paths has been determined at this level, we can then go down to the next level of setting up paths within each of the two $N/2 \times N/2$ subnetworks. The same algorithm can be applied in a recursive manner until the complete paths are determined for all input–output pairs.

Let us illustrate the algorithm with an example. With reference to Fig. 2.15, in which we wish to set up paths for the following input–output pairs: (1,4), (2,5), (3,6), (4,3), (5,7), (6,8), (7,1), (8,2), we start out by routing the first path (the one from input 1) through the upper subnetwork. This path reaches output 4. We next satisfy the constraint generated at the output (i.e., the path to output 3, which shares the same switch element as output 4) by routing the path from input 4 to output 3 through the lower subnetwork. This generates the constraint that the path from input 3 must go through the upper subnetwork. Performing this iteratively by satisfying the constraints created at the output and input alternatively, eventually, we will close the "loop" by establishing a path from input 2, which shares an element with input 1, through the lower subnetwork. This is the situation as depicted in Fig. 2.15. Notice that since this is a Clos network, we can use the connection matrix already discussed to represent the

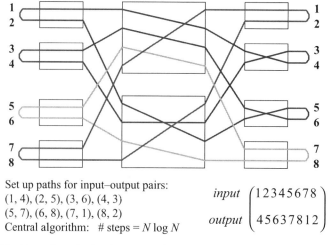

Set up paths for input–output pairs:
(1, 4), (2, 5), (3, 6), (4, 3)
(5, 7), (6, 8), (7, 1), (8, 2)
Central algorithm: # steps = $N \log N$

$$\text{input} \begin{pmatrix} 1 & 2 & 3 & 4 & 5 & 6 & 7 & 8 \\ 4 & 5 & 6 & 3 & 7 & 8 & 1 & 2 \end{pmatrix} \text{output}$$

FIGURE 2.15 Illustration of a looping connection setup algorithm; we want to set up paths for input–output pairs: (1, 4), (2, 5), (3, 6), (4, 3), (5, 7), (6, 8), (7, 1), (8, 2).

connection state. Each loop here corresponds to a chain with a loop in the connection matrix.

The loop may not involve all inputs and outputs. If some inputs and outputs remain to be connected, then we start the path setup procedure as above again starting from one of these inputs or outputs. In the example in Fig. 2.15, the connection from input 5 has not been considered in the first loop. Therefore, we may start off by setting the path from input 5 through the upper subnetwork. In this way, the connections from all input–output pairs will be established in the end. But this is only the end of the "first-level" connection setup, which in turn imposes certain input-output connection requirements on each of the $N/2 \times N/2$ middle switch modules. The same looping algorithm can be applied when setting up these "second-level" connections in the each $N/2 \times N/2$ switch modules, leaving us with "third-level" connections in four $N/4 \times N/4$ switch modules. Applying the looping algorithm recursively solves the connection problem of the overall switch.

The number of steps required in the first-level path setup is N. Furthermore, these steps cannot be executed concurrently because each step depends on the previous step. At the next level when we set up paths within a $N/2 \times N/2$ subnetwork, $N/2$ steps are required. Taking into consideration the two subnetworks, a total of N steps are required. The path setup at the two subnetworks, however, can be executed together. It is easy to see that a total of $N \log_2 N$ steps are needed since there are $\log_2 N$ levels. If we parallelize the path setup procedures in separate subnetworks, the time complexity is of order

$$N + N/2 + N/4 + \cdots + 2 = 2(N-1).$$

Notice that the algorithm is to be executed by a central controller and that the central controller is also responsible for setting the states of the 2×2 elements at the end of the algorithm. Lacking is a self-routing algorithm in which each 2×2 element determines its own state based on the local information at its inputs and outputs. Self-routing networks will be discussed in the next two chapters.

2.1.5 Baseline and Reverse Baseline Networks

The Benes network can be considered as being formed by two subnetworks, as indicated in Fig. 2.14. The middle stage and the stages before it form a network structure called the baseline network. The middle stage and the stages after it are a mirror image of the baseline network called the reverse baseline network.

Both the baseline and reverse baseline networks have the interesting property that within each network there is a unique path from any input to any output. In other words, there is one and only one way of connecting an input to an output in these networks. The unique paths from any input to all outputs form a binary tree. To see this, consider the baseline network. We note that an input is connected to only one node (i.e., crosspoint) at stage 1. From this node, the signal from the input can branch out in two directions to two nodes in stage 2, and then four nodes in stage 3. Thus, the input can reach 2^{j-1} nodes at stage j. The last stage is stage $n = \log_2 N$, and the

32 CIRCUIT SWITCH DESIGN PRINCIPLES

input can reach all the $N/2$ nodes in the last stage, from which it can reach all the N outputs. An input can only reach a subset of the nodes in the stages before stage n, and the path to each of the node in the subset is clearly unique from the above explanation. In general, we observe the following:

> *Reachability of nodes in baseline/reverse baseline networks:* A node in stage i can be reached by 2^i inputs and can reach 2^{n-i+1} outputs.

2.1.6 Cantor Switching Network

The Benes network is only rearrangeably nonblocking. However, the above observations about the two subnetworks of the Benes network can be used to construct a parallel Benes network, called the Cantor network, which is strictly nonblocking. Figure 2.16 shows a Cantor network. It is made up of $\log_2 N$ Benes networks arranged in a parallel fashion. Input i of the overall switch is connected to inputs i of all the $\log_2 N$ Benes switches via a front-end $1 \times \log_2 N$ switch. Similarly, output j of the overall switch is connected to outputs j of all the $\log_2 N$ Benes switch via a back-end $\log_2 N \times 1$ switch. Basically, the function of the front-end and back-end switches is to increase the number of alternative paths for any input–output pair so as to increase the chance of a successful connection. It just turns out that $\log_2 N$ Benes networks are sufficient to always allow connection between any idle input–output pair.

Theorem 2.4. *The Cantor network is strictly nonblocking.*

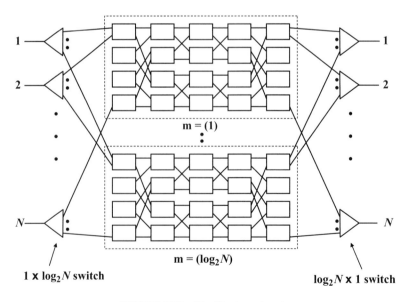

FIGURE 2.16 The Cantor network.

Proof. For each Benes network, an input can be connected to any of the middle-stage nodes if there are no existing connections. However, if there are already some existing connections in the network, it may not be possible to connect the input to some of the middle-stage nodes. We say that an input can *access* a node at or before the middle stage only if none of the links in the unique path from the input to that node is already used by an existing connection. For example, with respect to Fig. 2.4(b), which shows the connection state of a 4×4 Benes Switch made up of 2×2 nodes, the lower node in the middle stage cannot be accessed by input 1. We define accessibility from an output to those nodes at or after the middle stage in a similar fashion.

The essence of our argument is to show that there is a node in the middle stage in one of the Benes networks that can be accessed by both the input and the output of the new request. The $1 \times \log_2 N$ and $\log_2 N \times 1$ switches can then be used to connect the input and output to that Benes switch and the connection is established thereof.

Let m be the number of Benes networks required to make the Cantor network strictly nonblocking. Suppose we want to set up a new connection between input a and output b. We shall first examine accessibility from input a. In each Benes network, the paths leading from input a to the $N/2$ middle-stage nodes form a binary tree, as illustrated in the example of Fig. 2.17, in which $a = 3$. There are altogether m binary trees, one in each Benes network, and they contain the paths from input a to the $Nm/2$ middle-stage nodes in the Cantor network.

Consider the worst-case situation in which all the other $N - 1$ inputs and output are currently busy. Each of the $N - 1$ connections has an associated path from its input to its output. Each path must intersect or meet with one of the m binary trees at some node at or before the middle stage. We say that the path *meets* the binary trees at stage i if the path intersects with one of the trees *for the first time* at stage i. Note that the rest of the path must also intersect with the tree thereof.

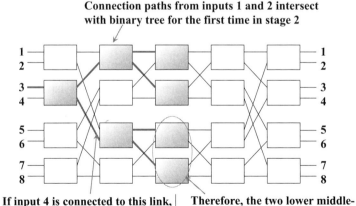

FIGURE 2.17 Binary tree extended from an input to all middle-stage nodes.

By the construction of the Benes network, we can verify that out of the $N-1$ paths, one path meets the binary trees at stage 1. Consider Fig. 2.16 and suppose that input a is input 3. Then, the path of the existing connection from input 4 meets a binary tree of input 3 at a stage-1 node. Next, two paths meet the binary trees at stage 2, four paths meet the binary trees at stage 3, and so on. In general, the number of paths meeting the binary trees at stage i is

$$A_i = 2^{i-1}.$$

For verification, the reader can check that

$$\sum_{i=1}^{\log_2 N} A_i = N - 1.$$

For each of the paths meeting with the trees before the middle stage, say at stage i, the path will exit on one of the outgoing links of the intersecting node. The subtree extending out of this link cannot be used by input a for connection (see the example in Fig. 2.16). This subtree has

$$B_i = \frac{N}{2^{i+1}}$$

nodes in the middle stage, and they are not accessible by input a.

A total of

$$C \leq \sum_{i=1}^{\log_2 N - 1} A_i B_i \qquad (2.17)$$

middle-stage nodes will be eliminated from accessibility by input a. The inequality is satisfied with equality only in the worst case when all the subtrees generated by the $N-1$ paths are disjoint and therefore the middle-stage nodes eliminated are disjoint. For the strictly nonblocking property, what is important is the worst-case situation. Notice, however, that in general some of the subtrees may overlap: for instance, by the rearrangeably nonblocking property of the Benes network, it is possible for all the $N-1$ paths to concentrate on one of the m Benes networks, eliminating only $N/2 - 1$ middle-stage nodes.

Substituting A_i and B_i into the above yields

$$C \leq (\log_2 N - 1)N/4. \qquad (2.18)$$

Thus, the number of middle-stage nodes that can be accessed by input a is

$$|I| = Nm/2 - C \geq Nm/2 - (\log_2 N - 1)N/4, \qquad (2.19)$$

where I is the set of middle-stage nodes accessible by input a. By symmetry,

$$|O| \geq Nm/2 - (\log_2 N - 1)N/4, \qquad (2.20)$$

where O is the set of middle-stage nodes accessible by output b. But

$$Nm/2 \geq |I \cup O| = |I| + |O| - |I \cap O|, \qquad (2.21)$$

the rearrangement of which yields

$$|I \cap O| \geq |I| + |O| - Nm/2 \geq Nm/2 - (\log_2 N - 1)N/2. \qquad (2.22)$$

There is a middle-stage node accessible by both input a and output b if $|I \cap O| > 0$, which is the case if

$$Nm/2 - (\log_2 N - 1)N/2 > 0. \qquad (2.23)$$

The above simplifies to

$$m > \log_2 N - 1. \qquad (2.24)$$

Thus, if $m = \log_2 N$, the connection between input a and output b can be set up. □

2.2 TIME-DOMAIN AND TIME–SPACE–TIME CIRCUIT SWITCHING

The preceding section concerns switching in the space domain. That is, a set of physically separate inputs are to be connected to a set of physically separate outputs. In telephone networks, most digital information streams are time-division multiplexed (TDM) at the higher network hierarchy. It is often less costly and more convenient to switch this traffic directly in the time domain.

2.2.1 Time-Domain Switching

Recall from Fig. 1.6 that the time domain of a TDM transmission medium is divided into frames. In each frame, there are N time slots, and the ith time slot is dedicated for the transmission of information from source i to receiver i. A question that arises is what if the sources want to send information to different receivers at different times.

Figure 2.18 shows a possible switching scheme. There is a switching center in between the sources and the receivers. At the switching center, the TDM traffic is first demultiplexed into N physically separate streams. The N streams are then switched using an $N \times N$ space-division switch, which could be one of those discussed in the previous section. After switching, the N output streams are then multiplexed back into one physical stream. Note that the time slots occupied by the sources could have been interchanged at this point. At the receiver, when this stream is demultiplexed,

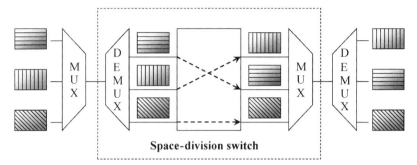

FIGURE 2.18 Performing time-slot interchange using space-division switch.

the information received by receiver i depends on switching performed by the space-division switch.

A more direct way to achieve the same function is to use a time-slot interchanger (TSI) in the middle, as shown in Fig. 2.19. As the name suggests, a TSI interchanges the time slots occupied by the logical channels within a TDM stream. The TSI is usually implemented using random access memory (RAM). After a whole frame is written into the RAM, the time slots in the frame can then be read out in an order dictated by the switching (or time-slot interchange) to be performed. Thus, if the data on time slot N are to be switched to time slot 1, these data will be read out first on the output frame. In this way, switching is performed directly in the time domain. Logically, the TSI is equivalent to an $N \times N$ space-division switch.

Notice that there must be a delay equal to one frame time in the TSI. After a whole frame is written, it is read. Meanwhile the next frame is being written. Two data frames of memory are needed.

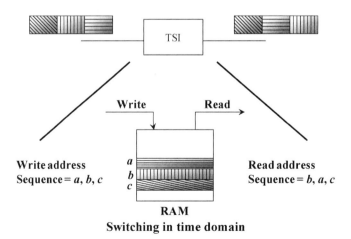

FIGURE 2.19 Direct time-slot interchange using random access memory (switching in the time domain).

Since the TSI is implemented by RAM, its operation is limited by the memory access speed, considering the fact that data must be written into and read out of the memory as fast as they arrive. As an example, suppose $N = 24$, the bit rate per source is 64 kbps (the data rate for a telephone conversation), and there is a byte to each time slot. Then,

$$\text{Arrival rate} = \frac{24 \times 64{,}000}{8} = 192{,}000 \text{ slots/s}.$$

A read and a write are required per time slot. Thus,

$$\text{Memory access time} = \frac{1}{192{,}000 \times 2} \text{ s} \approx 2.6 \text{ μs}.$$

2.2.2 Time–Space–Time Switching

There are generally many TDM lines connected to a switching center at the higher network hierarchy. In this situation, a combination of time and space switching is often performed. A switching structure that is often used in practice is the so-called time–space–time switch architecture shown in Fig. 2.20.

It turns out that this switch is logically equivalent to the three-stage Clos switch discussed in the previous section. To see this, let us refer back to the general Clos switch architecture depicted in Fig. 2.6 and consider the simple symmetric case in which $n_1 = n_3 = n$, $r_1 = r_3 = r$, and $r_2 = m$. Now, for the time–space–time switch architecture in Fig. 2.20, let n be the number of input or output time slots per frame, r be the number of input or output ports of the space-division switch, and m be the internal number of time slots per frame. If $m \neq n$, then the number of time slots per frame inside the switch is different from that at the external lines. Otherwise, for the input TSI, the output bit rate is m/n times the inputs bit rate because of the different numbers of time slots in a frame; for the output TSI, the input bit rate is m/n times the output bit rate. Each time slot in an input frame can be mapped onto any time

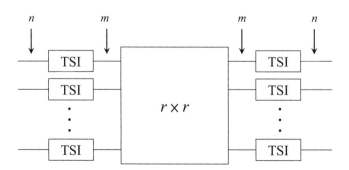

n, m: number of time slots per frame at various points

FIGURE 2.20 A time–space–time switch.

38 CIRCUIT SWITCH DESIGN PRINCIPLES

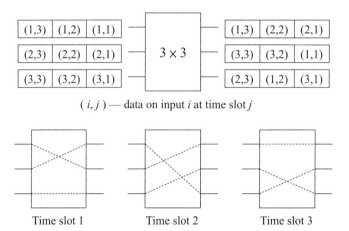

FIGURE 2.21 Input–output mapping changes from slot to slot in space-division switch in time–space–time switching.

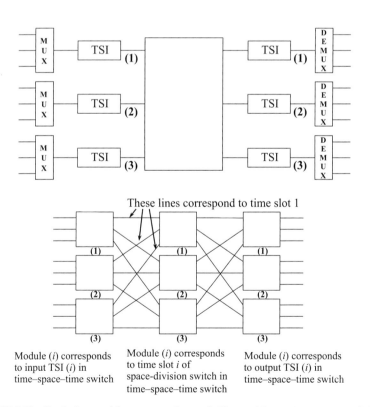

FIGURE 2.22 Equivalence of time–space–time switching and three-stage space switching.

slot in an output frame. Thus, logically, the input TSI is like an $n \times m$ switch and the output TSI is like an $m \times n$ switch.

The operation of the center space-division switch can be considered as being divided into m time slots. In the first time slot, it switches the information from time slot 1 of the inputs to time slot 1 of the outputs. In the second time slot, a different input–output mapping is used to switch information from time slot 2 of inputs to time slot 2 of outputs. Thus, the input–output mapping realized by the switch changes very dynamically from time slot to time slot, and as illustrated in Fig. 2.21, it is as if there were m copies of space-division switches, one for each time slot. In fact, if we were to put a time-division multiplexer before each input TSI, and a time-division demultiplexer after each output, as in Fig. 2.22, the switch will be functionally identical to the three-stage Clos architecture. Thus, all the nonblocking conditions we derived previously for the Clos switch also apply here. For instance, if $m = 2n - 1$, the switch will be strictly nonblocking, and if $m = n$, the switch will be rearrangeably nonblocking. Of course, in the latter case the way time slots are interchanged in the TSI may need to be rearranged.

PROBLEMS

2.1 A formula that is often used to engineer the trunk facilities between two switching centers is the Erlang B formula given by

$$B = \frac{a^n/n!}{\sum_{i=0}^{n} a^i/i!},$$

where B is the blocking probability, n is the number of trunks, and $a = \lambda/\mu$ is the product of the arrival rate of new connections λ and the average holding time of a connection $1/\mu$. Suppose new connections arrive at the rate of 20 calls /min and the average holding time of a call is 5 min. Find the minimum number of trunks needed to make the blocking probability 10^{-3}.

2.2 Explain why a rearrangeably nonblocking circuit switch is not necessary wide-sense nonblocking.

2.3 True or false? Constructing a wide-sense nonblocking switch requires a larger number of crosspoints than constructing a strictly nonblocking switch. Explain.

2.4 Section 2.1.2 mentions that two different switch states may realize the same input–output mappings. Give an example illustrating this.

2.5 In Section 2.1.2, we examined the lower bound on the number of crosspoints required by a nonblocking switch. Derive a lower bound for an $N \times N'$ asymmetric switch where $N' = N/2$ using the same method.

2.6 Consider a strictly nonblocking symmetric Clos switch with $n_1 = n_3 = n$ and $r_1 = r_3 = r$ made up of three stages of crossbar switch modules. Express the number of crosspoints in the overall switch in terms of n and r. For $N = 1000$, find the values of n and r that minimize the number of crosspoints. (*Note:* If it is necessary to have an N that is slightly larger than 1000 in order to minimize the crosspoint count, the larger N should be adopted since we can always use just 1000 of the N ports.)

2.7 Give an example of a strictly nonblocking $N \times N$ Clos network in which $r_2 = N < n_1 + n_3 - 1$. Explain why in practice no one would construct a three-stage switch this way by showing that one of the stages can be eliminated. In other words, show that a simpler two-stage nonblocking switch can be constructed if $N < n_1 + n_3 - 1$.

2.8 Consider a modified three-stage Clos switching network in which there are *two* links connecting every switch module in first and third stages to every module in the middle stage. Each first-stage switch module is $n \times 2n$, each third-stage module is $2n \times n$, and the numbers of modules in all the stages are equal. Is this switch strictly nonblocking? If yes, prove it. If no, give a counterexample.

2.9 The text proves that the number of rearrangements needed in a rearrangeably Clos nonblocking switch is bounded above by $r_1 + r_3 - 1$ by considering one chain, where r_1 and r_3 are the number of modules in the first and third stages, respectively. Improve the bound to $2\min(r_1, r_3) - 2$, still considering only one chain. That is, you choose one of the two chains before starting to identify the points in the chain, but you do not know which chain is the shorter one. (*Hint:* You may choose the chain based on your knowledge of r_1 and r_3, but there is no guarantee that the one you choose will be the shorter one. After choosing the chain, first consider the columns covered and then the rows covered.)

2.10 Show how to construct $N \times 1$ and $1 \times N$ switches using the minimum number of crosspoints. Suppose your designs are used in the construction of the Cantor network. How many crosspoints are needed in the Cantor network?

2.11 Instead of a three-stage network, let us consider a two-stage network as shown in Fig. 2.23, in which there is one link interconnecting any module at the first stage and any module at the second stage.
  (a) Give an example showing that this network is blocking.
  (b) Argue that this network could be highly blocking.
  (c) Instead of just one link interconnecting any module at the first module and any module at the second stage, suppose that there is a group of m links interconnecting them, as shown in Fig. 2.24. How large should m be to make this network nonblocking?
  (d) Suppose we want to explore a seldom blocking network in which the value of m is less than the above value. Assume that $n = 8$, all inputs and outputs are active, and the inputs are equally likely to be destined for any of the outputs. What is the probability that there are more than m connections wanting to go through the same group of m links? How large should m be to make this probability less than 10^{-3}.

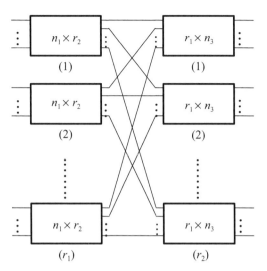

FIGURE 2.23 Two-stage network.

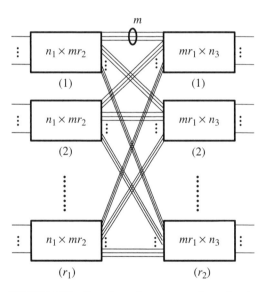

FIGURE 2.24 Two-stage channel-grouping network.

2.12 The looping algorithm for setting connections in a Benes network alternatively considers the constraints generated at the inputs and outputs. Suppose we consider the scheme of things only from the inputs by routing the connections from input 1 via upper middle switch, input 2 via the lower middle switch, input 3 via the upper switch, and input 4 via the lower middle switch. Give an example showing that this does not work.

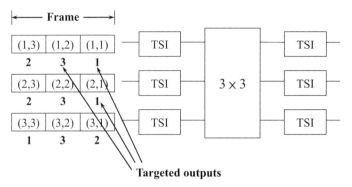

FIGURE 2.25 A time–space–time switch.

2.13 Benes network is a rearrangeably nonblocking network with a recursive structure. Suppose we want to construct a strictly nonblocking structure using a similar recursion, but with 2×3 switch modules at the first stage, 3×2 switch modules at the third stage, and three $N/2 \times N/2$ modules in the middle stage.
  (a) Constructing the switch recursively this way, how many 2×3, 3×2, and 2×2 switch modules are needed altogether, assuming N is a power of 2?
  (b) The ratio of the number of internal links at each stage to the number of inputs is called the bandwidth expansion factor. What is the maximum bandwidth expansion factor for this switch?
  (c) Compare this switch with the Cantor network.

2.14 Consider an 8×8 Cantor network. The number of Benes networks needed to make it strictly nonblocking is $\log_2 8 = 3$. Give an example showing that if the number of Benes networks is only 2, the Cantor network is not strictly nonblocking.

2.15 The text explains time–space–time switching in terms of the Clos switch architecture. This problem attempts to look at time–space–time switching from a different viewpoint. With reference to Fig. 2.25, let (i, j) denote the data in time slot j of input 1. In the example the data on different inputs may be destined for the same output in the same time slot. Show how the input TSIs can be used to interchange the time slots occupied by the data to avoid output conflict. Explain why the output TSIs are needed after the data arrive at the outputs.

3

FUNDAMENTAL PRINCIPLES OF PACKET SWITCH DESIGN

The next few chapters focus primarily on fixed-length packet (or cell) switching. To switch variable-length packets using a fixed-length switch, one could first fragment the variable-length packets at the inputs into small fixed-length packets, switch the packets, and then reassemble the packets back into the original variable-length packets at the outputs. Generally, it is possible for the packet boundaries of different inputs to be unaligned (see Fig. 3.1). Because of the fixed packet length, we can deliberately delay the early arriving packets (e.g., store them into a buffer) in order to align the boundaries. Thus, as far as the operation of the switch is concerned, we may assume that time is divided into fixed slots of one-packet duration. In the beginning of each time slot, new packets arrive at all the inputs simultaneously.

One of the exercises in the preceding chapter examines time–space–time circuit switching from the viewpoint of contention resolution. Specifically, the information on different inputs may be destined for the same output of the space-division switch at any given time slot (see Fig. 2.23). The input TSI rearranges the time slots occupied by the information so that output conflicts do not occur.

In circuit switching, the time slots occupying the same relative position in successive frames of an input are assigned to the same circuit, and they contain information destined for the same output. In other words, the output link to which the information should be forwarded is known implicitly *a priori*. In packet switching, the output links to which a packet should be forwarded is not known before their arrivals. Unlike in time-division multiplexing, there is no telling exactly how many packets from an input will be destined for each output over a given period of time. Transmission capacity may not be dedicated to each communication session in an exclusive manner. This and the lack of a transmission frame structure make it impossible to use a TSI to

---

Principles of Broadband Switching and Networking, by Tony T. Lee and Soung C. Liew
Copyright © 2010 John Wiley & Sons, Inc.

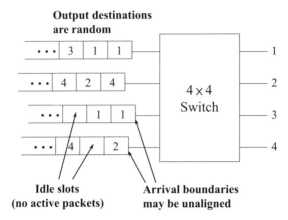

FIGURE 3.1 Packet arrivals in a 4 × 4 packet switch.

eliminate blocking the way it is done in circuit switching.[1] Other methods are needed to deal with the contention problem.

The output destination of a packet must be contained in a header or derivable from it. In virtual-circuit networks, a virtual circuit is set up for each session and the route of the virtual circuit is determined during the call setup time. Once the route is set up, all packets of the virtual circuit will follow the same route while traveling to the destination. The packet header does not contain the final output destination explicitly. Instead, it contains a number called the virtual-circuit identifier (VCI). At the switch input, this number is translated into a local address associated with an output of the switch. In addition, the input VCI is also mapped onto an output VCI. The input VCI in the header will be replaced by the output VCI, and the packet will be switched to the output. The output is in turn connected to an outgoing link that may be connected to an input of a subsequent switch. In this case, the output VCI becomes the input VCI of the subsequent switch, and it will be used to identify the output and output VCI of the packet at the subsequent switch. In general, the switch output to which an input VCI on an input should be mapped forms part of the end-to-end route that was determined during the setting up of the virtual circuit. Once the mapping is determined, it is stored in a memory.

Figure 3.2 shows a schematic diagram of the header translation process. An incoming packet is first disassembled into the information part and the header part. The header contains the input VCI of the packet. The header processor, in addition to performing other functions, uses the input VCI as an index to locate from the memory the switch output to which the packet should be forwarded. The output address, along with the bits of the original header, is then attached to the packet by the assembler.

---

[1]The reader should be aware that the point here is not to argue that circuit switching is superior to packet switching. It is to indicate that because of the different ways in which transmission resources are used in these two schemes, the switching requirements are different.

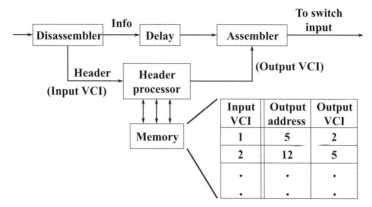

FIGURE 3.2 Input packet processor.

The original header and the information bits will not be examined while the packet is in the switch; only the output address will be used for switching purposes. Generally, the output-address header is attached before the original header. In the next few chapters when we describe about switch operation, the term "header" refers to the output-address header (and other bits used for routing packets within the switch). It is the information that will be examined inside the switch and it does not include the original header bits. After the packet is switched to its desired output, this will be stripped before the packet is sent to the next switching node.

3.1 PACKET CONTENTION IN SWITCHES

Fundamental to the design of a packet switch is the resolution of output contention. To see what it takes to eliminate contention entirely in packet switching, consider Fig. 3.3(a). The figure shows three packets destined for the same output. By speeding up the operation of the switch three times with respect to the input rate, contention among the three packets can be eliminated. This is because each switch cycle then takes up only one-third time slot so that before the next batch of packets arrive at the inputs, all three packets will have been switched to their target output. But generally, it is possible, though not probable, in an $N \times N$ switch that all the N input packets in a time slot are targeted for the same output. When the switch dimensions are large, switching mechanisms that let N packets reach the same destination increase design complexity and become impractical. As long as only fewer than N packets can reach the same output address, the potential contention problem remains, and it must be dealt with in other ways.

There are only two alternative solutions: the switch can drop excess packets that cannot be switched, or it can buffer them for output access in the next time slot. Accordingly, switches based on interconnection networks can be classified as either *loss systems* or *waiting systems*, depending on how contention is resolved.

46 FUNDAMENTAL PRINCIPLES OF PACKET SWITCH DESIGN

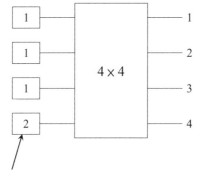
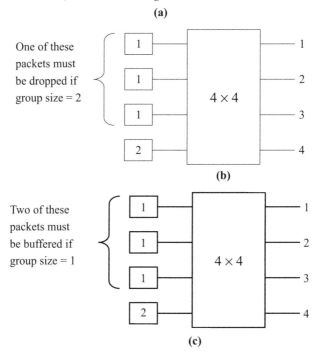

FIGURE 3.3 Packet contention dealt with by (a) speeding up switch operation by N times; (b) dropping packets that cannot be switched; (c) queueing packets that cannot be switched.

In a loss system, packets are either dropped or switched to their outputs immediately. By providing sufficient paths from inputs to outputs, packet loss probability can be made as small as desired. The maximum number of packets that can be received by an output in one time slot will be called the *group size*. Since switch complexity generally increases with group size, one would look for the minimum group size to meet the packet loss probability requirement in practice. In addition, if the group size is more than 1, packet buffers are needed at outputs, since the output port may not be able to transmit all arriving packets at once.

As an example, Fig. 3.3(b) shows a 4×4 switch that has a group size of 2. If three packets arrive for the same output, one of them will be dropped. As long as the group size is less than N, there is the likelihood that a packet will be dropped. There is a trade-off between switch complexity and packet-loss probability. In some switches, even though there is no output contention, it is possible that there is internal contention due to their blocking structures. An example is the baseline switch discussed in the preceding chapter. If these switches are designed as a loss system, packets may be dropped internally.

For waiting systems, excess packets are buffered whenever contention arises so that attempts can be made to switch them in subsequent switch cycles. These packets may be buffered at the inputs or internally in the switch. Figure 3.3(c) shows a scenario in which two of the three packets destined for the same output are buffered at the inputs. We shall see that the throughput of a waiting-system switch is generally less than 100%. This limitation can be relaxed by increasing the group size so that more packets can access their destination outputs simultaneously.

To summarize, switches can be classified into two classes with the following characteristics:

- Loss system
    - The switch fabric has no input or internal buffers; packets already reaching their outputs may be queued at outputs waiting to be transmitted if the group size is more than 1.
    - Packets may be dropped internally or at outputs due to contention. The loss probability can be made arbitrarily small by adjusting the group size or some related switch design parameters.
- Waiting system
    - Output conflicts are resolved by some contention-resolution mechanism to select the packets that will be switched. Packets that have lost contention will be buffered so that they can be switched in one of the later time slots. Packets already reaching their outputs may be queued at outputs waiting to be transmitted if the group size is more than 1.
    - The throughput of the switch can be made arbitrarily close to 100% by increasing the group size or other design parameters.

In the switching literature, the term "input-buffered switch" is often used to refer to a waiting system with a group size of one that buffers packets only at the inputs. Packets are not queued at the outputs because no more than one packet may arrive at an output in a time slot, and by the time the next packet arrives, the current packet at the output, if any, would have been transmitted. The term "output-buffered switch" is often used to refer to an ideal switch with a group size of N. Since all packets can be switched to their desired outputs in the same time slot, there is no contention and buffers at the inputs are then not needed. However, buffers are needed at the outputs since the output link may not be able to transmit all simultaneously arriving packets due to its limited transmission rate. In the output-buffered switch, there is neither throughput limitation nor packet loss in the switch. Although the names "input-buffered switch" and "output-buffered switch" are rather descriptive, they could also be misleading in that the superior performance of the output-buffered switch improvement is often attributed either explicitly or implicitly to buffering packets at the outputs. In actuality, the performance improvement of the output-buffered switch over the input-buffered switch is really an outcome of having a large group size. As we shall see in the next chapter, this is not without cost and the switch necessarily becomes more complex. That buffers are needed at the outputs is really a consequence of this switch design principle rather than the cause of the performance improvement: simply putting buffers at the outputs without increasing the group size will not lead to any throughput improvement.

3.2 FUNDAMENTAL PROPERTIES OF INTERCONNECTION NETWORKS

Many packet switches proposed to date are based on interconnection networks, originally intended for multiprocessor interconnect in highly parallel computer systems. These switches make use of many small switch elements in their overall architectures. An attractive feature of these switches is their regular topological interconnection pattern, that can be easily implemented by VLSI technology.

Some of the circuit switches discussed in the preceding chapter are also made up of small switch elements. However, the control algorithms used for establishing the circuits are central algorithms. One can imagine that there is a central controller that is fully aware of which inputs are to be mapped onto which outputs. It then sets the switch elements into bar or cross state to establish the connections in a nonconflicting manner.

A packet switch is said to be internally nonblocking if packets with nonoverlapping targeted outputs can be routed through it in a nonconflicting manner. Centrally controlled switches that are rearrangeably nonblocking in the circuit switching sense are also internally nonblocking in the packet-switching sense. To see the equivalence, consider a set of packets arriving to a rearrangeably nonblocking switch at the beginning of a time slot. Before they can be launched into the switch, a set of nonconflicting paths within the switch must be determined by the central controller. The controller can decide the paths in a one-by-one manner and "rearrange" the paths decided earlier should conflicts arise during the computation. Notice that these rearrangements do not change the paths *already used* by packets. Rather, the rearrangements change the paths *tentatively assigned* to packets earlier in the computation: after all, no packet has been launched into the switch yet. By the definition of rearrangeability, if all packets have non-overlapping outputs, a set of nonconflicting paths can be determined in the end. Of course, there may be algorithms that are more clever than others that require less computation time.

FUNDAMENTAL PROPERTIES OF INTERCONNECTION NETWORKS   49

The practical difference between rearrangeably nonblocking circuit switches and internally nonblocking packet switch is the speed at which the path setup algorithm must be executed. In packet switching, to avoid large overhead, the execution must be completed within a small fraction of a time slot. When the switch size N is large, this becomes difficult with a central algorithm. An advantage of switches constructed using interconnection networks is their distributed routing algorithm that can be scaled more easily. These networks to be discussed in this chapter are sometimes called self-routing networks in the sense that the switch elements make use of only the destination labels in the packets to perform switching and that they do not need to know the states of the other switch elements. There is no direct coordination or communication among the switch elements and a central controller is not needed. This is an important property, especially for a large switch with packets of small size, since many packets need to be switched within a short time.

It is also important to point out that unlike a store-and-forward packet communication network in which packets are received in their entirety before they are forwarded, the switch elements in an interconnection network need to process only the header in order to set their states. Thus, a switch element may start to forward a packet before the rest of it has arrived. Generally, there are at most a few bits of delay at each switch element; in many designs, there is only one bit of delay.

3.2.1 Definition of Banyan Networks

Let us first consider a fundamental class of interconnection networks that are *not* internally nonblocking called the Banyan networks. Figure 3.4 shows four networks

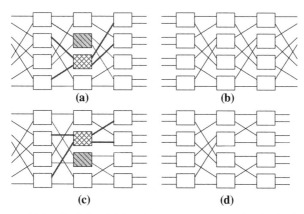

Networks **(a)** and **(c)** are isomorphic: one can be obtained
from the other by interchanging the shaded elements

FIGURE 3.4 Four different Banyan networks: (a) shuffle-exchange (omega) network; (b) reverse shuffle-exchange network; (c) banyan network; (d) baseline network.

belonging to this class: the shuffle-exchange network (also called the omega network), the inverse shuffle-exchange network, the banyan network, and the baseline network. A point of confusion is that the term "banyan network" is sometimes also used to refer to the network in (c) and sometimes used to refer to any of the networks. To avoid confusion and to accommodate the common usage, when referring to the specific network in (c), we shall use the term "banyan network" with a lowercase "b", and when referring to any network in the class, we shall use the term "Banyan network" with an uppercase "B". Thus, a banyan network and a baseline network are both Banyan networks, but the baseline network is *not* a banyan network. For our purpose a network is considered a Banyan network if it has these two properties:

1. There is a unique path from any input to any output.
2. There are $\log_2 N$ columns, each with $N/2$ 2×2 switch elements.

These networks may have different properties besides the two common properties above and it is important to distinguish them. On the other hand, characteristics that depend only on the two properties above are common to all networks within the class.

Different Banyan networks may be isomorphic with each other—that is there are different ways of drawing the same network. For example, by interchanging the positions of the two shaded switch elements of network (a) (without changing the adjacent switch elements to which they are connected), we get network (c). If we enclose each of these two networks in a black box, there is no way of telling them apart just by their properties. Network (b) is not isomorphic to network (a), because one network cannot be obtained from the other by just rearranging the drawing. In fact, the input–output mappings that can be realized in one network may not be realizable in the other network, and vice versa.[2]

Let us now examine how a packet can be routed from an input to its desired output destination. Suppose the output destination is labeled as a string of $n = \log_2 N$ bits, $b_1 \cdots b_n$. This output address is encoded into the header of the packet. In the first column (or stage), the most significant bit b_1 is examined. If it is 0, the packet will be forwarded to the upper outgoing link; if it is 1, the packet will be forwarded to the lower outgoing link. In the next stage, the next most significant bit b_2 will be examined and the routing performed likewise. The example in Fig. 3.5 illustrates that after $\log_2 N$ stages when all the destination bits have been used, the packet reaches the desired destination. It is easy to understand why this strategy works. After each stage, the packet enters a smaller subnetwork, and the number of outputs that can be reached is halved. The subsets of reachable outputs as the packet progresses through the stages always include the desired output. After $\log_2 N$ stages, only the desired output

---

[2] The four networks, however, are "internally isomorphic" in the sense that if we ignore the external connections of inputs and outputs to the switch elements at the first and last stages, we may obtain one network from the other by rearranging the positions of switch elements. This means that by rearranging the connections of inputs and outputs to the first and last stages, all the four networks can be made isomorphic to each other, and as a result they realize the same input–output mappings.

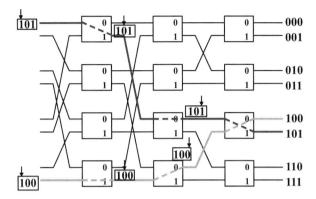

Destination addresses are in binary form. The $\log_2 N$-bit address is used as the routing bits for the packet: bit i is used in stage i

FIGURE 3.5 Routing in the banyan network.

remains. Notice that the routing algorithm is *distributed*. Each switching element makes its decision based on only one bit, and a central controller is not needed.

Since the shuffle-exchange network in (a) is isomorphic to this network, the same routing strategy will also work for the shuffle-exchange network. We shall explain why the same routing network works for the shuffle-exchange network from a different viewpoint in one of the subsequent chapters. This routing strategy, however, will not work for the reverse shuffle-exchange network in (c). The reverse shuffle-exchange network is a mirror image of the shuffle-exchange network. In this network, one should route a packet starting from the least significant bit of the output address and proceeding to the most significant bit.

3.2.2 Simple Switches Based on Banyan Networks

So far we have examined the routing of one packet without considering contention among packets while they are being routed. As discussed before, packets may contend for the same output. For Banyan networks, packets with different output addresses may also contend with each other internally for the same outgoing link of a switch element, as illustrated in Fig. 3.6.

Let us consider the throughput and the loss probability of the Banyan network when it is operated as a loss system. Assume a uniform-traffic situation in which each packet is equally likely to be destined for any of the outputs. Because of the unique-path property, a packet is also equally likely to be destined for either of the two outgoing links at each switch element. Let

$$P_m = \Pr[\text{there is a packet at an input link at stage } m + 1].$$

The input load of the network is $P_0 = \rho_o$. Note that P_m decreases as m increases as more and more packets are dropped. To express P_{m+1} in terms of P_m, consider

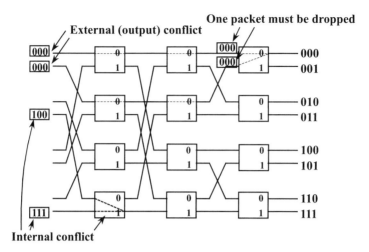

FIGURE 3.6 Internal and external conflicts when routing packets in a banyan network.

an outgoing link of a switch element in stage $m+1$, as shown in Fig. 3.7. There is a packet on this link only if at least one incoming packet is destined for it. The probability that there is no packet destined for it is $(1 - P_m/2)^2$; taht is, neither of the inputs has a packet destined for it. Thus,

$$P_{m+1} = 1 - (1 - P_m/2)^2 = P_m - P_m^2/4. \tag{3.1}$$

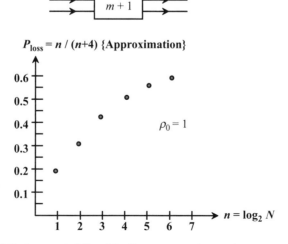

FIGURE 3.7 Loss probability of the Banyan network operating as a loss system.

From the above, we can calculate $P_{\log_2 N}$ in a recursive manner and obtain the loss probability from

$$P_{loss} = \frac{P_0 - P_{\log_2 N}}{P_0}. \tag{3.2}$$

However, a closed-form expression for P_{loss} cannot be obtained this way, since $P_{\log_2 N}$ is not in closed form. As an approximation, suppose we treat m as a "continuous" variable and expand P_{m+1} as a Taylor series:

$$P_{m+1} = P_m + \frac{dP_m}{dm} + \frac{1}{2!}\frac{d^2 P_m}{dm^2} + \cdots$$

If $d^n P_m/dm^n$ is small for $n \geq 2$, then (3.1) can be viewed as the recursion relation for the following differential equation:

$$\frac{dP_m}{dm} = -\frac{P_m^2}{4}. \tag{3.3}$$

From this, we have

$$\frac{dP_m}{P_m^2} = -\frac{dm}{4}.$$

Integrating from $m = 0$ and substituting $P_0 = \rho_o$ yields

$$P_m = \frac{4\rho_o}{m\rho_o + 4}. \tag{3.4}$$

The overall packet loss probability is then given by

$$P_{loss} = \frac{\rho_o - P_n}{\rho_o} = \frac{n\rho_o}{n\rho_o + 4}, \tag{3.5}$$

where $n = \log_2 N$. The maximum throughput is obtained when $\rho_o = 1$, and it is

$$\rho^* = 1 - P_{loss} = \frac{4}{n+4}. \tag{3.6}$$

Figure 3.7 plots the loss probability as a function of n for $\rho_o = 1$. The loss probability increases quite rapidly with n. Even for a small 4×4 switch, the loss probability is already greater than 0.5. Thus, this switch is not suitable for a communication network that must provide good quality of service. The problem with the Banyan network is that there is one and only one path from an input to an output. One must provide more paths from inputs to outputs in order to reduce the loss probability. We shall examine in the next chapter how switch parameters, such as the group size, can be engineered to improve performance.

54 FUNDAMENTAL PRINCIPLES OF PACKET SWITCH DESIGN

As an alternative, we can also operate the Banyan network as a waiting system. Packets that have lost contention can be queued either at the inputs or at the switch elements where they lost contention. Although there is no packet lost due to contention, the switch throughput will still be severely limited. For example, if $\rho_o > \rho^*$ in (3.6), then no matter how much buffer we have for the packet queues, they will overflow, since the Banyan network will not be able to clear the packets as fast as they arrive.

3.2.3 Combinatoric Properties of Banyan Networks

To look at the limitations of the Banyan network from a more fundamental angle, let us consider how many input–output mappings can be realized by the Banyan network. There are altogether $\frac{N}{2} \log_2 N$ switch elements in the Banyan network. Thus, there are $2^{\frac{N}{2} \log_2 N} = N^{N/2}$ states. By the unique-path property of the network, each of these states corresponds to a unique input–output mapping.

Since each incoming packet can be destined for any output regardless of the other packets, there are N^N possible input–output mappings. Thus, the fraction of realizable input–output mappings is

$$\frac{N^{N/2}}{N^N} = \frac{1}{N^{N/2}},$$

which approaches zero very quickly as N becomes large.

3.2.4 Nonblocking Conditions for the Banyan Network

It turns out that if the output addresses of the packets are sorted either in an ascending or in a descending order, the banyan network (*note*: not all the networks in the class of Banyan networks), and therefore its isomorphic shuffle-exchange network, will be internally nonblocking; Fig. 3.8 illustrates this with an example. Thus, if the banyan network is preceded by a network that sorts the packets according to their output destinations, the overall sort-banyan network will be internally nonblocking.

Theorem 3.1. *The banyan network is nonblocking if the active inputs (inputs with arriving packets) $x_1, \ldots, x_m$ ($x_j > x_i$ if $j > i$) and their corresponding output destinations $y_1, \ldots, y_m$ satisfy the following:*

1. *Distinct and monotonic outputs*: $y_1 < y_2 < \cdots < y_m$ or $y_1 > y_2 > \cdots > y_m$.
2. *Concentrated inputs*: Any input between two active inputs is also active. That is, $x_i \leq w \leq x_j$ implies input w is active.

Proof. We know that as a packet is routed in the banyan network, it enters a smaller and smaller subnetwork, as illustrated in Fig. 3.9. The entrance of stage k of an n-stage banyan network sees 2^{k-1} subnetworks from top to bottom. Each node in stage k can be uniquely represented by two binary numbers $(a_{n-k} \cdots a_1, b_1 \cdots b_{k-1})$. Intuitively,

FUNDAMENTAL PROPERTIES OF INTERCONNECTION NETWORKS 55

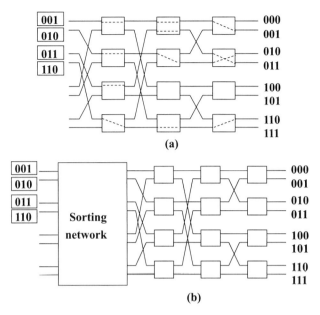

FIGURE 3.8 (a) An example showing the banyan network is nonblocking for sorted inputs; (b) nonblocking sort-banyan network.

$b_1 \cdots b_{k-1}$ is the subnetwork label and $a_{n-k} \cdots a_1$ is the relative position of the node numbered from the top within the subnetwork.

Node $(a_{n-k} \cdots a_1, b_1 \cdots b_{k-1})$ in stage k is connected by output link b_k (0 or 1) to node $(a_{n-k-1} \cdots a_1, b_1 \cdots b_{k-1} b_k)$. Thus, the path from an input $x = a_n \cdots a_1$ to an output $y = b_1 \cdots b_n$, denoted by $\langle x, y \rangle$, consists of the following sequence of nodes:

$$(a_{n-1} \cdots a_1, \emptyset) \xrightarrow{b_1} (a_{n-2} \cdots a_1, b_1) \cdots (a_{n-k} \cdots a_1, b_1 \cdots b_{k-1}) \cdots \xrightarrow{b_{n-1}}$$
$$(\emptyset, b_1 \cdots b_{n-1}).$$

Suppose that two packets, one from input $x = a_n \cdots a_1$ to output $y = b_1 \cdots b_n$ and the other from input $x' = a'_n \cdots a'_1$ to output $y' = b'_1 \cdots b'_n$, collide in stage k. That is, the two paths $\langle x, y \rangle$ and $\langle x', y' \rangle$ merge at the same node $(a_{n-k} \cdots a_1, b_1 \cdots b_{k-1}) = (a'_{n-k} \cdots a'_1, b'_1 \cdots b'_{k-1})$ and share the same outgoing link $b_k = b'_k$. Then, we have

$$a_{n-k} \cdots a_1 = a'_{n-k} \cdots a'_1 \tag{3.7}$$

and

$$b_1 \cdots b_k = b'_1 \cdots b'_k. \tag{3.8}$$

Since the packets are concentrated, the total number of packets between inputs x and x', inclusively, is

$$|x' - x| + 1. \tag{3.9}$$

56 FUNDAMENTAL PRINCIPLES OF PACKET SWITCH DESIGN

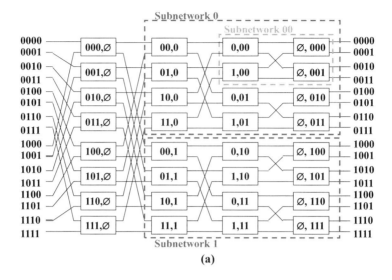

FIGURE 3.9 (a) Labeling of nodes in the banyan network; (b) sequence of nodes traversed by a packet from input $a_n \cdots a_1$ to output $b_1 \cdots b_n$.

By condition 1, all these packets are destined for different outputs, and therefore there must be $|x' - x| + 1$ distinct output addresses among them. Considering the monotone condition, the largest and the smallest addresses must be y and y', respectively, or y' and y, respectively. Hence, $|y - y'| + 1 \geq$ number of distinct output addresses $= |x' - x| + 1$, or

$$|x' - x| \leq |y' - y|. \tag{3.10}$$

According to (3.7) and (3.8),

$$\begin{aligned} |x' - x| &= |a'_n \cdots a'_1 - a_n \cdots a_1| \\ &= |a'_n \cdots a'_{n-k+1} 0 \cdots 0 - a_n \cdots a_{n-k+1} 0 \cdots 0| \\ &= 2^{n-k} |a'_n \cdots a'_{n-k+1} - a_n \cdots a_{n-k+1}| \\ &\geq 2^{n-k} \end{aligned} \tag{3.11}$$

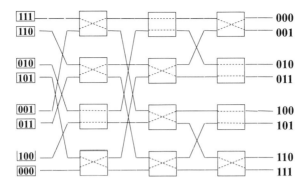

FIGURE 3.10 An example of unsorted packets having no conflict in the banyan network.

and

$$|y' - y| = |b'_1 \cdots b'_n - b_1 \cdots b_n|$$
$$= |b'_{k+1} \cdots b'_n - b_{k+1} \cdots b_n|$$
$$\leq 2^{n-k} - 1. \qquad (3.12)$$

But (3.11) and (3.12) contradict (3.10). Therefore, the two paths $\langle x, y \rangle$ and $\langle x', y' \rangle$ must be link-independent, and the theorem is proved. □

As long as the two conditions above are satisfied, the banyan network is nonblocking. On the other hand, when the two conditions are not satisfied, the banyan network is not necessarily blocking. An example in which the conditions are not satisfied but the banyan network is still nonblocking is shown in Fig. 3.10. One can conjure up many other examples.

For instance, if a set of packets can be routed without internal conflicts then we can shift (cyclically) the inputs occupied by the packets by a constant amount without creating internal conflict (see Fig. 3.11 for an example).

Theorem 3.2. Let the input–output pair of packet i be denoted by (x_i, y_i). If the packets can be routed through the banyan network without conflicts, so can the set of packets $(x_i + z \,(\mathrm{mod}\, N), y_i)$.

Proof. By contradiction, suppose that two packets, one from input $x + z \,(\mathrm{mod}\, N) = a_n \cdots a_1$ to output $y = b_1 \cdots b_n$ and the other from input $x' + z \,(\mathrm{mod}\, N) = a'_n \cdots a'_1$ to output $y' = b'_1 \cdots b'_n$, collide in stage k. Then, using the same argument as in the proof of the preceding theorem, we have

$$a_{n-k} \cdots a_1 = a'_{n-k} \cdots a'_1 \qquad (3.13)$$

58 FUNDAMENTAL PRINCIPLES OF PACKET SWITCH DESIGN

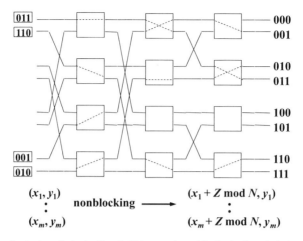

FIGURE 3.11 Sorted packets in Fig. 3.8(a) remain unblocked after their inputs are shifted (mod 8) by 6.

and

$$b_1 \cdots b_k = b'_1 \cdots b'_k. \tag{3.14}$$

Also,

$$x = (x+z) - z = (x+z) + (N-z) \pmod{N},$$
$$x' = (x'+z) - z = (x'+z) + (N-z) \pmod{N}. \tag{3.15}$$

Let $N - z = c_n \cdots c_1$. Then,

$$x = a_n \cdots a_1 + c_n \cdots c_1 \pmod{N},$$
$$x' = a'_n \cdots a'_1 + c_n \cdots c_1 \pmod{N}. \tag{3.16}$$

In other words, x and x' are the addition of $c_n \cdots c_1$ to $a_n \cdots a_1$ and $a'_n \cdots a'_1$, respectively, with only the n least significant bits retained. If $a_{n-k} \cdots a_1 = a'_{n-k} \cdots a'_1$, then

$$a_{n-k} \cdots a_1 + c_{n-k} \cdots c_1 = a'_{n-k} \cdots a'_1 + c_{n-k} \cdots c_1. \tag{3.17}$$

The above expressions may have more than $n - k$ bits. The $n - k$ least significant bits must be equal for the whole expressions to be equal. Thus, the $n - k$ least significant bits of x and x' are equal. Therefore, if the two packets with input–output pairs $(x + z \pmod{N}, y)$ and $(x' + z \pmod{N}, y')$ collide in stage k, so will the two packets with input–output pairs (x, y) and (x', y'). □

The properties summarized in the two theorems let us implement many useful functions using the banyan and reverse banyan networks. We already see that an internally nonblocking network can be constructed by cascading a sorting network and a banyan network. We can also build a packet concentrator using a reverse banyan network. That is, suppose not all N inputs are always active and we want to concentrate the incoming packets to $M < N$ links. Unlike for switching, it is not necessary for the concentrator to route a packet to a specific output—any one of the M outputs would do. To concentrate packets using the reverse banyan network in a nonblocking fashion, we can assign the first packet (from top to bottom) the output address 0, the second packet the output address 1, and so on. By routing from the least significant bit of the address to the most significant bit, Theorem 3.1 says that the packets would not have internal conflict—of course, we have to interpret the inputs as outputs and outputs as inputs.

We can also build a *shifter* using a reverse banyan network. Suppose that we want to distribute the incoming packets in a cyclic fashion to the outputs. Say, in the previous time slot, a batch of packets arrive and they are assigned to adjacent outputs with output i being the last (from top to bottom) having a packet assignment. In the current time slots, the batch of incoming packets is to be distributed to outputs $i + 1 \pmod N$, $i + 2 \pmod N$, and so on. Instead of assigning the output addresses starting from 0, as in the ordinary concentrator, we offset the assignment by $i + 1 \pmod N$. Thus, the first packet is assigned to output $i + 1 \pmod N$, the second packet to output $i + 2 \pmod N$, and so on. By keeping track of the next output to be assigned, we can distribute packets to the outputs in a cyclic and evenly manner. Theorem 3.1 together with Theorem 3.2 says that assigning output addresses this way does not cause internal conflict.

One can implement the concentration and shifting functions together in a reverse banyan network. Suppose that the number of outputs is $M < N$, with M a power of 2. Then, instead of offsetting output addresses using mod N arithmetic, we can use mod M arithmetic. The reader can easily check that the shifting property in Theorem 3.2 is still valid. Both the concentrator and the shifter are useful building blocks for constructing other switching and computation architectures, as will be seen in the next chapter.

3.3 SORTING NETWORKS

A sorting network is a nonblocking network by itself if

1. All inputs are active.
2. No two packets are destined for the same output.

This is illustrated in Fig. 3.12(a). Let us first examine condition 1 assuming condition 2 is satisfied. When some of the inputs are inactive, the packets may arrive at the wrong destination, as illustrated in Fig. 3.12(b). This is the reason why we need to

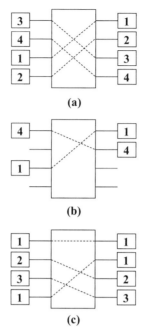

FIGURE 3.12 Examples showing that the sorting network (a) switches correctly when all inputs are active and have no common outputs; (b) switches incorrectly when some inputs are inactive; (c) switches incorrectly when some inputs have common outputs.

cascade a banyan network with the sorting network in order to build an internally nonblocking network based on a distributed routing algorithm.

The qualifier "based on a distributed routing algorithm" is an important one. With a central controller, the sorting network by itself can be easily made nonblocking as follows. On the inactive inputs, the central controller may introduce dummy (artificial) packets that are destined for the idle outputs in order to make the sorting network nonblocking by itself (see Fig. 3.13). These dummy packets can be discarded at the

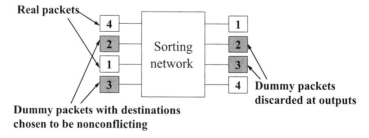

FIGURE 3.13 An example showing that dummy packets with nonconflicting destinations may be introduced to make the sorting network switch correctly when not all inputs are active; this requires knowledge of the destinations of active inputs.

outputs. But this requires the knowledge of the destinations of all the "real packets" in order that the destinations of the dummy packets can be assigned in a nonconflicting manner, hence the need for a central agent that gathers this information. The requirement of this global knowledge defeats the purpose of the self-routing operation.

Condition 2 is often not satisfied in packet switching. When packets contend for the same output, the contention needs to be resolved one way or another. This has been discussed at a preliminary level earlier in this chapter and will be dealt with at a more detailed level later.

3.3.1 Basic Concepts of Comparison Networks

A *comparison network* is constructed of 2×2 comparators shown in Fig. 3.14(a). A comparator takes two input numbers and places the smaller number on the upper output and the larger number on the lower output. A comparator can sometimes be more conveniently represented by drawing a vertical line with inputs on the left and outputs on the right, as shown in Fig. 3.14(b).

A comparison network is an interconnection comparator. Not all comparison networks are sorting networks. Sorting networks are those that can take N input numbers and transform them into a sorted sequence at the outputs. Figure 3.15 shows a sorting network.

A comparison network consists of several stages of comparators. The comparators at stage d take the outputs of the comparators at stages before d (i.e., $d-1$, $d-2$, ..., 1) as their inputs. For a unique representation, we also insist that a comparator at stage d must have at least one of its inputs taken from stage $d-1$; otherwise it will be placed at a stage before d. The structure of a comparison network corresponds to an algorithm or procedure that specifies how comparisons are to occur.

The zero–one principle states that if a sorting network sorts those inputs consisting of only 0's and 1's correctly, then it sorts arbitrary numbers correctly. This is a very powerful principle. It allows us to concentrate on inputs drawn from $\{0, 1\}$ in constructing a working sorting network. Once we have proved that the sorting network works for 0's and 1's, we can appeal to the zero–one principle to show that it works properly for numbers of arbitrary values. Before proving the zero–one principle, let us establish the order-preserving property:

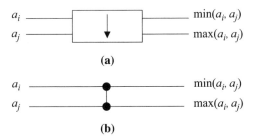

FIGURE 3.14 (a) A comparator; (b) a compact way of representing a comparator.

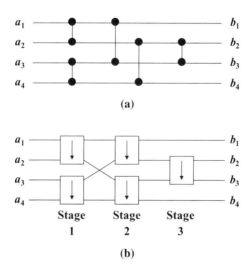

FIGURE 3.15 A 4×4 sorting network (a) is the compact representation; (b) is the full representation.

Order-Preserving Property. Suppose a comparison network transforms the input sequence $a = \langle a_1, a_2, \ldots, a_N \rangle$ into the output sequence $b = \langle b_1, b_2, \ldots, b_N \rangle$, then for any monotonically increasing function f, the network transforms the inputs sequence $f(a) = \langle f(a_1), f(a_2), \ldots, f(a_N) \rangle$ into the output sequence $f(b) = \langle f(b_1), f(b_2), \ldots, f(b_N) \rangle$.

Proof. With reference to Fig. 3.16, consider a comparator with inputs x and y and upper output $\min(x, y)$ and lower output $\max(x, y)$. Without loss of generality, let $x \leq y$. Since f is monotonically increasing, this implies $f(x) \leq f(y)$. Thus, if the

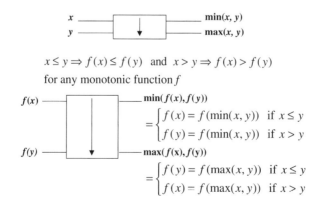

FIGURE 3.16 Illustration that a comparator has the order-preserving property.

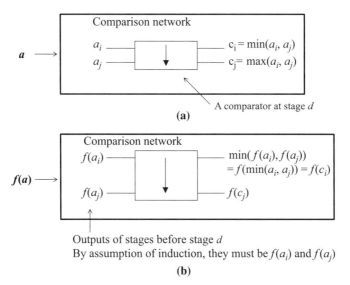

FIGURE 3.17 (a) The inputs and outputs of a comparator at stage d when input sequence is a; (b) the inputs and outputs of the same comparator when input sequence is $f(a)$.

inputs were $f(x)$ and $f(y)$, the upper and lower outputs would be, respectively,

$$\min(f(x), f(y)) = f(x) = f(\min(x, y)),$$
$$\max(f(x), f(y)) = f(y) = f(\max(x, y)). \quad (3.18)$$

Thus, a comparator has the order-preserving property. For a general comparison network, we can show by induction the correctness of this statement: If an output of a comparator at stage d assumes the value of c_i when the input sequence to the overall network is a, it assumes a value of $f(c_i)$ when the input sequence is $f(a)$. Note that since stage d can be any stage, including the last stage, the correctness of the above statement implies the order-preserving property.

By (3.18), the statement is true for $d = 1$, since numbers at the outputs of stage 1 have passed through at most one comparator. By induction, assume that the statement is true for stages $d - 1$ and below, we want to show that it is also true for stage d. The inputs to a comparator at stage d are outputs from the stages $d - 1$ and below. Thus, if the inputs assume the value of a_i and a_j when the input sequence is a, they must assume the values $f(a_i)$ and $f(a_j)$ (see Fig. 3.17) when the input sequence is $f(a)$. When the input sequence is a, the upper and lower outputs of the comparator are $c_i = \min(a_i, a_j)$ and $c_j = \max(a_i, a_j)$, respectively. Therefore, by (3.18), the upper and lower outputs are $f(\min(a_i, a_j)) = f(c_i)$ and $f(\max(a_i, a_j)) = f(c_j)$, respectively, when the input sequence is $f(a)$. □

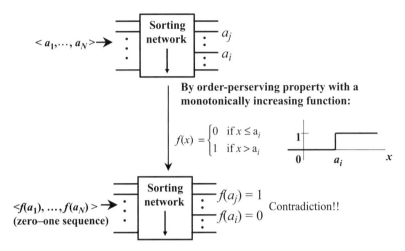

FIGURE 3.18 Illustration of the proof of the zero–one principle.

With the order-preserving property, it is easy to prove the zero–one principle.

Theorem 3.3. If a sorting network with N inputs sorts all the 2^N possible sequences of 0's and 1's correctly, then it sorts all sequences of arbitrary input numbers correctly.

Proof. Consider a network that sorts input sequence of 0's and 1's correctly. By contradiction, suppose that it does not sort input sequences of arbitrary numbers correctly. That is, there is an input sequence $\langle a_1, a_2, \ldots, a_N \rangle$ containing two elements a_i and a_j such that $a_i < a_j$, but the network places a_j before a_i. Define a monotonically increasing function

$$f(x) = \begin{cases} 0, & \text{if } x \leq a_i, \\ 1, & \text{if } x > a_i. \end{cases}$$

According to the order-preserving property, since the network places a_j before a_i when the input sequence is $\langle a_1, a_2, \ldots, a_N \rangle$, it places $f(a_j) = 1$ before $f(a_i) = 0$ when the input sequence is $\langle f(a_1), f(a_2), \ldots, f(a_N) \rangle$. But this input sequence consists of only 0's and 1's, and yet the network does not sort it correctly, leading to a contradiction. □

3.3.2 Sorting Networks Based on Bitonic Sort

There are many ways to construct sorting networks. Merging is a divide-and-conquer technique for sorting. A k-merger takes two sorted input sequences and merges them

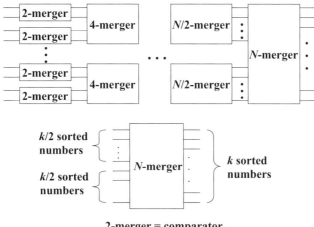

FIGURE 3.19 Sorting based on merging; successive shorter sorted sequences are merged into longer sorted sequences.

into one sorted sequence of k elements. Intuitively, merging is easier than sorting in general because the inputs are already partially sorted. Suppose we have mergers of different sizes; Fig. 3.19 shows how these mergers can be interconnected together to sort an arbitrary input sequence. We start with N unsorted numbers, and merge them using 2-mergers into $N/2$ sorted sequences, each with two numbers. The $N/2$ sequences are then merged into $N/4$ sequences, each with four numbers, and so on, until we are left with one sorted sequence. One way to construct the mergers is to use the bitonic sorting algorithm invented by Batcher.

A bitonic sequence is a sequence that either increases monotonically and then decreases monotonically, or decreases monotonically and then increases monotonically. In other words, it is a concatenation of two sequences sorted in opposing directions. Sequences that are either increasing monotonically or decreasing monotonically are also considered bitonic. For example, the sequences $\langle 0, 1, 4, 6, 7, 7, 5, 4\rangle$, $\langle 9, 8, 7, 3, 3, 2, 4, 6\rangle$, $\langle 0, 1, 4, 7, 7, 7, 8, 8\rangle$, and $\langle 8, 7, 3, 3, 2, 1, 1, 0\rangle$ are all bitonic. A *bitonic sorter* is a merger that takes a bitonic sequence (i.e., two concatenated sorted sequences, one ascending and one descending) and sorts it into a monotonic sequence, as depicted in Fig. 3.20.

To explain the sorting network based on bitonic sorters, let us only consider inputs consisting of only 0's and 1's. If the resulting sorting network can sort the 0's and 1's, by the zero–one principle, it can sort arbitrary numbers. The zero–one principle does not, however, imply that the bitonic sorter that works for 0's and 1's will work for arbitrary numbers, even though that is true. It only implies that the overall sorting network constructed out of many bitonic sorters will work for any arbitrary numbers. One of the exercises asks you to use the order-preserving property to show the analog of the zero–one principle that applies to bitonic sorters specifically.

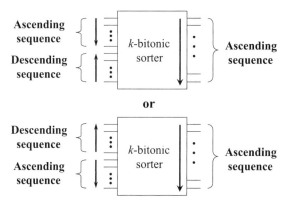

*The two input sequences do not have to be of the same length
*The two input sequences are of opposing directions

FIGURE 3.20 Bitonic Sorters.

A bitonic sequence consisting of only 0's and 1's is either of the form $0^i 1^j 0^k$ or of the form $1^i 0^j 1^k$, for some $i, j, k \geq 0$, where the notation x^i means i successive x values. A bitonic sequence a is said to be no less (greater) than another bitonic sequence if none of the element in a is less (greater) than any of the elements in b; symbolically, we write $a \geq b$ if a is not less than b and $a \leq b$ if a is not greater than b. For example,

$$\langle 001100 \rangle \geq \langle 000000 \rangle,$$

$$\langle 111100 \rangle \leq \langle 111111 \rangle.$$

Two sequences do not necessarily have an ordering relationship. For example, $\langle 001100 \rangle$ is neither not less than nor not greater than $\langle 111100 \rangle$. Two zero–one sequences have an ordering relationship if and only if at least one of them is consisting of all 0's or all 1's. This is one reason why it is easier to focus on zero–one sequences in constructing a comparison network.

The following theorem shows how a bitonic sequence a can be decomposed into two bitonic subsequences a' and a'' with $a' \leq a''$ using only one stage of comparators. The same decomposition method can then be applied to a' and a'' to produce four bitonic subsequences with each subsequence not greater than the next. Applying this decomposition in a recursive manner allows us to sort the original bitonic sequence into a monotonic sequence (when all the subsequences have only one element).

Theorem 3.4. If a zero–one sequence of $2n$ elements $a = \langle a_1, a_2, \ldots, a_{2n} \rangle$ is bitonic, then the two n-element sequences

$$a' = \langle \min(a_1, a_{n+1}), \min(a_2, a_{n+2}), \ldots, \min(a_n, a_{2n}) \rangle$$

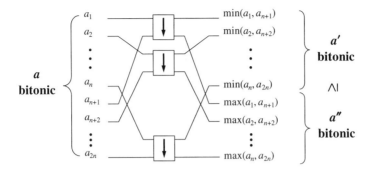

If *a* is a zero–one sequence, either a' is all 0's or a'' is all 1's, or both

FIGURE 3.21 A half-cleaner.

and

$$a'' = \langle \max(a_1, a_{n+1}), \max(a_2, a_{n+2}), \ldots, \max(a_n, a_{2n}) \rangle$$

have two properties:

1. They are both bitonic.
2. $a' \leq a''$.

Comment. The above decomposition can be implemented by a single stage of n comparators, called a *half-cleaner*, as depicted in Fig. 3.21. As shown, elements in the first half and the second half of the input sequence are paired and compared. The larger elements and smaller elements of the comparisons form two subsequences with the above properties. Since a' and a'' have an ordering relationship and they consist of only 0's and 1's, at least one of them must be all 0's or all 1's. A sequence consisting of all 0's or all 1's is said to be *clean*, hence the name *half-cleaner* for the comparison network.

Proof. Without loss of generality, suppose that the bitonic sequence is of the form $0^i 1^j 0^k$. Proving the case $1^i 0^j 1^k$ is similar to the argument presented here. There are only four possible cases depending on where the midpoint falls, as shown in Fig. 3.22. Cases (i) and (ii) have midpoints falling on a block of 0's. From the figure, it can be seen that the theorem holds for both cases. Cases (iii) and (iv) have midpoints falling on the block of 1's. In case (iii), there are more 0's than 1's in the sequence a, and in case (iv), there are at least as many 1's as 0's. Again, as shown in the figure, the theorem holds for both cases. □

68 FUNDAMENTAL PRINCIPLES OF PACKET SWITCH DESIGN

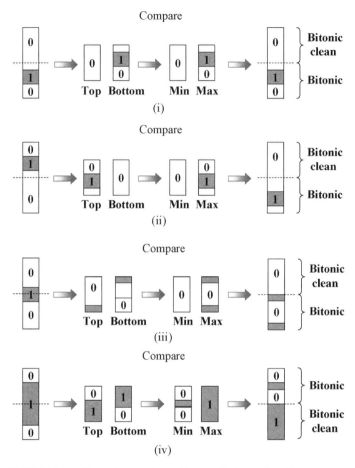

FIGURE 3.22 Operations performed by a half-cleaner for different cases.

Figure 3.23 shows the decomposition of a k-bitonic sorter (one that sorts k elements) into a k-half cleaner followed by two $k/2$-bitonic sorters. The above theorem states that the inputs to the lower $k/2$-bitonic sorter are not less than the inputs to the upper $k/2$-bitonic sorter. Furthermore, each of the input sequences is bitonic. The $k/2$-bitonic

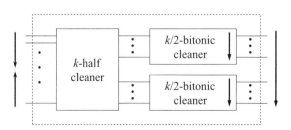

FIGURE 3.23 Recursive construction of a k-bitonic sorter.

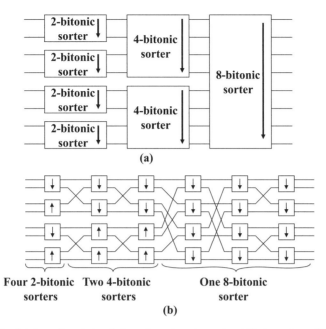

FIGURE 3.24 (a) A sorting network based on merging using bitonic sorters; (b) the same network broken down into comparators.

sorters can be further broken down into two $k/2$-half cleaners and four $k/4$-bitonic sorters using the same decomposition. Applying the decomposition recursively, only 2-bitonic sorters are left in the end, and 2-bitonic sorters are simply comparators. Notice that the structure of the bitonic sorter is the same as that of the banyan network, and it consists of $\log_2 N$ stages of comparators.

Now that we know how to build bitonic sorters, we can use them as the mergers in Fig. 3.19 to build an overall sorting network that sorts arbitrary zero–one sequences (i.e., sequences that are not necessarily bitonic). Figure 3.24 shows the construction of an 8×8 Batcher sorting network based on the bitonic sort.

To compute the number of comparators in a Batcher sorting network, note that the number of stages in a k-bitonic sorter is

$$f(k) = \log_2 k. \qquad (3.19)$$

Figure 3.24(a) shows that an input element passes through one 2-bitonic sorter, one 2^2-bitonic sorter, one 2^3-bitonic sorter, and so on, on its way to the output of the overall sorting network. Therefore, the total number of stages in the Batcher network is

$$\sum_{i=1}^{\log_2 N} f(2^i) = \sum_{i=1}^{\log_2 N} i = \frac{\log_2 N(\log_2 N + 1)}{2}. \qquad (3.20)$$

70 FUNDAMENTAL PRINCIPLES OF PACKET SWITCH DESIGN

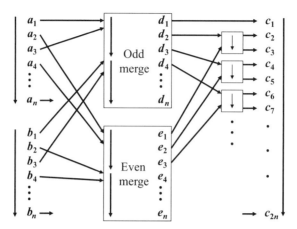

FIGURE 3.25 Recursion for odd–even merging networks.

Since each stage in the sorting network has $N/2$ comparators, the total number of comparator is

$$\frac{N \log_2 N(\log_2 N + 1)}{4}. \tag{3.21}$$

The number of elements in the sorting network is therefore of order $N \log_2^2 N$.

3.3.3 The Odd–Even Sorting Network

Batcher also invented another order $N \log_2^2 N$ sorting network. As in the bitonic sorting algorithm, sorting by merging is used. Instead of using bitonic sorters for merging, odd–even mergers are used. The basic idea of the recursive odd–even merger is illustrated in Fig. 3.25. The essence is contained in the following theorem.

Theorem 3.5. Consider two sorted input sequences $a = \langle a_1, \ldots, a_n \rangle$ and $b = \langle b_1, \ldots, b_n \rangle$. The odd-indexed elements of a and b, $a' = \langle a_1, a_3, \ldots \rangle$ and $b' = \langle b_1, b_3, \ldots \rangle$, and the even-indexed elements, $a'' = \langle a_2, a_4, \ldots \rangle$ and $b'' = \langle b_2, b_4, \ldots \rangle$, are all sorted subsequences. Suppose that the odd-indexed subsequences a' and b' are merged into a sorted sequence $d = \langle d_1, d_2, \ldots \rangle$, and the even-indexed subsequences a'' and b'' are merged into $e = \langle e_1, e_2 \rangle$. Then, it can be shown that

$$e_{i-2} \leq d_i \leq e_i. \tag{3.22}$$

Comment. The above merging of a' and b', and a'' and b'', is depicted in Fig. 3.25. The theorem implies that $c = \langle d_1, \min(d_2, e_1), \max(d_2, e_1), \ldots, \min(d_i, e_{i-1}), \max(d_i, e_{i-1}), \ldots, \rangle$ must be a sorted sequence: that each element is at least as large as the preceding element can be easily verified using (3.22). The sequence c is produced at the last stage of the odd–even merger in Fig. 3.25. The smaller mergers in the middle are implemented by the same algorithm in a recursive manner. Note the difference between the bitonic recursion and the odd–even recursion: the smaller mergers in the former are positioned after the one-stage comparison whereas the latter is the other way round.

Proof. Let us assume n is even. The proof for odd n is similar. We appeal to the zero–one principle and shall assume a and b to consist of only 0's and 1's. Let $f(x) = (n_x, m_x)$ be a tuple denoting the number of 0's, n_x, and the number of 1's, m_x, in the sequence x. Then, for even n, it can be easily seen that

$$f(a') = \left(\left\lceil \frac{n_a}{2} \right\rceil, \left\lfloor \frac{m_a}{2} \right\rfloor\right),$$

$$f(a'') = \left(\left\lfloor \frac{n_a}{2} \right\rfloor, \left\lceil \frac{m_a}{2} \right\rceil\right),$$

$$f(b') = \left(\left\lceil \frac{n_b}{2} \right\rceil, \left\lfloor \frac{m_b}{2} \right\rfloor\right),$$

$$f(b'') = \left(\left\lfloor \frac{n_b}{2} \right\rfloor, \left\lceil \frac{m_b}{2} \right\rceil\right).$$

Therefore, since d and e are sorted sequences resulting from merging a' and b', and a'' and b'', respectively, we have

$$f(d) = \left(\left\lceil \frac{n_a}{2} \right\rceil + \left\lceil \frac{n_b}{2} \right\rceil, \left\lfloor \frac{m_a}{2} \right\rfloor + \left\lfloor \frac{m_b}{2} \right\rfloor\right),$$

$$f(e) = \left(\left\lfloor \frac{n_a}{2} \right\rfloor + \left\lfloor \frac{n_b}{2} \right\rfloor, \left\lceil \frac{m_a}{2} \right\rceil + \left\lceil \frac{m_b}{2} \right\rceil\right).$$

The number of elements in d and e is the same. The number of 0's in d is no smaller than that in e, implying $d_i \leq e_i$. The number of 0's in d can be more than that in e by at most 2, implying $e_{i-2} \leq d_i$ for $i = 3, 4, \ldots$. These observations complete the proof. □

3.3.4 Switching and Contention Resolution in Sort-Banyan Network

Recall that in packet switching, the output address of a packet is contained in the header. When using the Batcher sorting network for switching, the comparator only looks at the header and passes on the information bits without examining them. Only a delay of one-bit duration is experienced by packets passing through a comparator.

Figure 3.26 illustrates the operation of a comparator used for packet switching. Suppose that the output address of the upper packet is 0010 and that of the lower packet is 0001, and suppose that the comparator is in the bar state initially. Each time a bit of the output address has arrived on both inputs, the comparator does a comparison. If both bits are 0's or 1's, as in the first and second bits in our example,

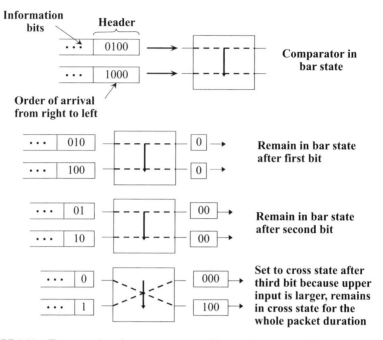

FIGURE 3.26 The operation of a comparator used in a sorting network for packet switching.

the comparator remains in the bar state and the bits are forwarded to the outputs. This comparison of the bits introduces a one-bit delay. Note that if both bits are the same, it does not matter whether the comparator is in the bar or cross state. Upon the arrival of the first pair of bits that differ, the comparator sets its state so that the input with bit 0 is connected to the upper output and the input with bit 1 is connected to the lower output. In our example, the two output addresses 0010 and 0001 start to differ in the third bits, and when the third bits on both inputs have arrived, the comparator sets itself into the cross state, since the upper bit is 1 and the lower bit is 0. The comparator remains in this state until all the bits of the packets, including the information bits, have passed through and the next batch of two packets arrives, at which time it starts to compare headers again.

Not all inputs of the sorting network have packets all the time. We can assign an additional bit to the header called the activity bit. This bit will be treated as the most significant bit and is the first bit of the packet to arrive at a comparator. For the active inputs, the activity bit is set to 0. For the idle inputs, a dummy packet is constructed and its activity bit is set to 1; it does not matter what the rest of the bits of the dummy packet are. With the activity bits thus set, the sorting network will "push" all the dummy packets to the lower end of the outputs. Therefore, the output packets of the sorting network are concentrated and have output addresses that are monotone. These packets can then be routed in the banyan network without conflict if there is no output conflict.

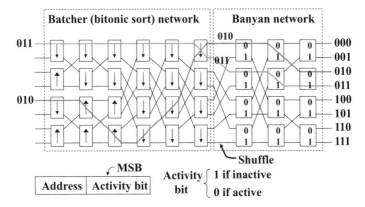

FIGURE 3.27 An 8 × 8 Batcher–banyan network.

A complete Batcher banyan network is shown in Fig. 3.27. The Batcher network is based on the bitonic sorting algorithm. In packet switching, we cannot guarantee that no two arriving packets are destined for the same output. Each input, without knowing the output addresses of the other input, does not know whether its packet conflicts with the others' packets. A contention-resolution scheme is needed so that only a subset of nonconflicting packets can enter the Batcher–banyan network. We shall focus on a three-phase switching scheme. The first two phases are for contention resolution and third phase is for switching packets that have won contention.

In the *probe* phase, only the headers of packets enter the sorting network. Packets having the same output address will be adjacent to each other at the output of the sorting network. Output $j + 1$ then checks with output j to see if their packet addresses are the same. If yes, we can let output j be the winning packet and output $j + 1$ be the losing packet. In this way, we will only choose one winning packet for each output destination. This comparison operation is local and therefore consistent with the self-routing requirement.

The next phase is the *acknowledgment* phase. If the sorting network is designed with back-propagating paths, acknowledgments can be sent to the inputs of the winning packets through the backpropagating paths. To implement the backpropagating paths, each node in the Batcher network can have a companion backward 2 × 2 switch element in addition to the forward sorting element. The companion elements of adjacent stages are interconnected in the same way as the forward sorting cell. Furthermore, the state of the companion switch element is set in the same way as the sorting element: in other words, if the sorting cell is in the bar (cross) state, so is the backward switch element. As depicted in Fig. 3.28, the backpropagating paths travel through the same nodes as the forward paths set by the sorting network.

The third phase is the *send* phase. The inputs that receive an acknowledgment can now send their packets into the sort-banyan network without internal conflict. The inputs that have lost contention buffer their packets in a buffer so that they can attempt for output access in the next cycle.

74 FUNDAMENTAL PRINCIPLES OF PACKET SWITCH DESIGN

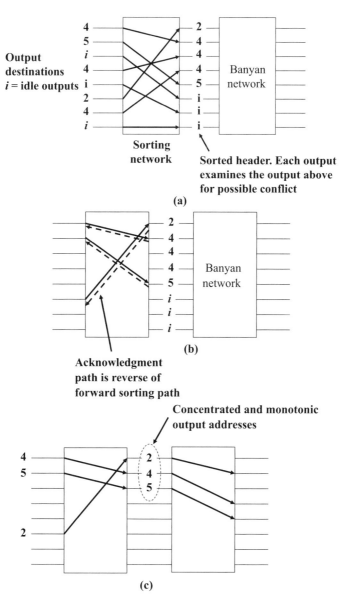

FIGURE 3.28 Three-phase scheme for sort-banyan network (a) is for probing for conflict; (b) is for acknowledgment of winning packets; (c) is for routing winning packets.

There are many possible variations to the above basic scheme. The main idea, however, is to use the sort-banyan network as the contention arbiter as well as the switching network. The initial two phases are for contention resolution. While the first two phases are ongoing, no packets are being routed to their outputs, and this introduces a certain amount of contention-resolution overhead, reducing the throughput of the switch to below 100%.

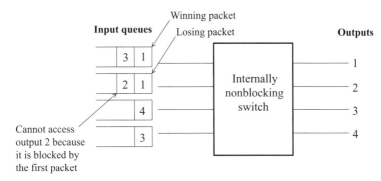

FIGURE 3.29 Illustration of head-of-line blocking.

The so-called head-of-line (HOL) blocking further reduces the throughput of the switch. Since the packets losing contention are buffered at their inputs, the sort-banyan network as described above is a waiting system. While the packets are being buffered, new packets may arrive, forming queues at the inputs. As illustrated in Fig. 3.29, the second packet of an input queue whose first packet has lost contention may be destined for an idle output. If the packets at the inputs access their outputs in a first-in-first-out fashion, the second packet cannot access its output until the first packet has accessed its own output. It is said to be blocked by the first (or HOL) packet. The throughput limitation due to HOL blocking will be derived in the next chapter.

3.4 NONBLOCKING AND SELF-ROUTING PROPERTIES OF CLOS NETWORKS

In the previous sections, we have introduced the fundamental properties of Banyan networks and demonstrated the way to construct a nonblocking packet switch based on them. Here, we turn our concern to the Clos networks. We shall focus on their nonblocking conditions and self-routing properties.

Before proceeding, we need to present a scheme to assign addresses to the inputs and outputs of Clos networks. Let us consider a three-stage Clos network as shown in Fig. 3.30, which is characterized by two integer parameters p and q. Here, the number of inputs and outputs N is given by $N = pq$. There are two ways to address the inputs and outputs. One is to number them in a consecutive top-to-bottom manner from 0 to $N - 1$. The other is to use a 2-tuple addressing scheme. An input is numbered (a_i, b_i) if it is the $(b_i + 1)$th input port on switch module a_i. Similarly, an output is numbered (x_i, y_i) if it is the $(y_i + 1)$th output port on switch module x_i. The 2-tuple addresses $\tilde{s}_i$ and $\tilde{d}_i$ can be obtained from their corresponding consecutive addresses s_i and d_i by

$$\tilde{s}_i = \left(\left\lfloor \frac{s_i}{q} \right\rfloor, [s_i]_q \right) \quad \text{and} \quad \tilde{d}_i = \left(\left\lfloor \frac{d_i}{q} \right\rfloor, [d_i]_q \right), \quad (3.23)$$

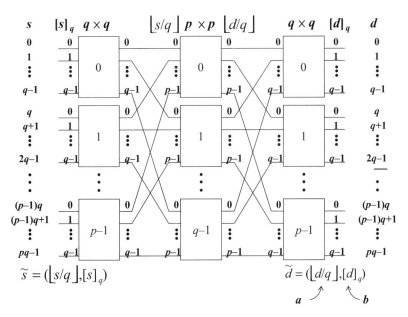

FIGURE 3.30 A three-stage Clos network with address numbering scheme.

where $\lfloor \frac{a}{b} \rfloor$ is the largest integer smaller than a/b and $[a]_b$ is the reminder of a/b. For simplicity, we will denote the address (a_i, b_i) by $a_i b_i$ whenever there is no ambiguity.

Actually, the 2-tuple addressing scheme can be used for routing. For instance, to go from any middle-stage module to an output $x_i y_i$, a packet should go to the $(x_i + 1)$th output in the middle-stage module and to the $(y_i + 1)$th output in the output-stage module. Since each middle-stage module here is of size $p \times p$ and each output-stage module is $q \times q$, x_i is of base p ($0 \leq x_i \leq p-1$) and y_i is of base q ($0 \leq y_i \leq q-1$).

3.4.1 Nonblocking Route Assignment

For nonblocking routing of packets across a Clos network, we need a route assignment algorithm to assign a middle-stage module for each connection request. An assignment is nonblocking if none of the internal links is shared by two paths. Consider a three-stage Clos network with parameters p and q. There are q middle-stage modules, which provide totally q alternate paths for each input–output pair. Let $C = \{0, 1, \ldots, q-1\}$ be the set of middle-stage modules and $\pi = \{(s_0, d_0), \ldots, (s_{n-1}, d_{n-1})\}$ the input–output permutation (which may be partial). A sufficient condition for an assignment $f : \pi \to C$ to be nonblocking is

$$f(s_i, d_i) = f(s_j, d_j) \Rightarrow |s_i - s_j| \geq q \text{ and } |d_i - d_j| \geq q, \tag{3.24}$$

where $d_i = \pi(s_i)$ and $d_j = \pi(s_j)$. That is, if two connections (s_i, d_i) and (s_j, d_j) go through the same middle-stage module, the condition $|s_i - s_j| \geq q$ guarantees that the inputs are on different switch modules and the condition $|d_i - d_j| \geq q$ guarantees that the outputs are on different switch modules. With this, no more than one input or output on the same module will be assigned to the same middle-stage module and hence the assignment is nonblocking. Note that (3.24) is a strong condition in the sense that we only require s_i and s_j to be on different input-stage modules and d_i and d_j to be on different output-stage modules for the assignment to be nonblocking. They do not need to be differed by q.

For an assignment f, the routing tag for the connection (s_i, d_i) is given by

$$\left(f(s_i, d_i), \left\lfloor \frac{d_i}{q} \right\rfloor, [d_i]_q \right). \tag{3.25}$$

The first component in the routing tag is the middle-stage module to which the connection request is assigned. The second and third components are derived from the address of the output destination. Thus, once the first component is determined by the route assignment algorithm, the routing tag can be determined immediately.

Consider the reverse path from output d_i to input s_i. In order that s_i and d_i to be connected, they must be assigned the same middle-stage module. Therefore, to go from d_i to s_i, the routing tag for the reverse path is

$$\left(f(s_i, d_i), \left\lfloor \frac{s_i}{q} \right\rfloor, [s_i]_q \right). \tag{3.26}$$

The set of connection requests π is monotonic if the active inputs $s_0, s_1, \ldots, s_{n-1}$ and their corresponding outputs $d_0, d_1, \ldots, d_{n-1}$ satisfy either one of the following conditions:

$$s_i < s_j \Rightarrow d_i < d_j \quad \text{for all } i, j; \text{ or} \tag{3.27}$$

$$s_i < s_j \Rightarrow d_i > d_j \quad \text{for all } i, j.$$

Equivalently, the set of output destinations corresponding to the set of active inputs is either strictly increasing or strictly decreasing. The reason why we consider monotonic connection requests is because they exhibit a simple nonblocking self-routing scheme due to the following fundamental lemma.

Lemma 3.1. Let $x_0, x_1, \ldots, x_{n-1}$ be a strict monotonic sequence of integers and define $g(x_i)$ as follows:

$$g(x_i) = [m + i]_q \quad \text{for all } i,$$

where m and q are constant integers. Then $g(x_i) = g(x_j)$ implies $|x_i - x_j| \geq q$ for $i \neq j$.

78 FUNDAMENTAL PRINCIPLES OF PACKET SWITCH DESIGN

Proof. Without loss of generality, assume that the sequence is increasing and let $i < j$. It follows that $g(x_i) = g(x_j)$, which implies $i + lq = j$, where $l \geq 1$. Thus,

$$x_j - x_i = (x_j - x_{j-1}) + (x_{j-1} - x_{j-2}) + \cdots + (x_{i+1} - x_i)$$
$$\geq \underbrace{1 + \cdots + 1}_{j-i} = j - i = lq \geq q.$$

□

Based on this lemma, we obtain a route assignment algorithm for ordered connection requests, which routes the calls according to their "ranks" and is the so-called the *rank-based assignment algorithm*.

Theorem 3.6 (Rank-Based Assignment Algorithm). Let the set of connection requests $\pi = \{(s_0, d_0), \ldots, (s_{n-1}, d_{n-1})\}$ be monotonic. The assignment

$$f(s_i, d_i) = [m + i]_q, \tag{3.28}$$

where m is a constant integer and i is the rank of the connection request $(s_i, d_i = \pi(s_i))$, is nonblocking.

Proof. Since the set of connection requests $\pi = \{(s_0, d_0), \ldots, (s_{n-1}, d_{n-1})\}$ is a monotonic input–output permutation, both sequences $(s_0, \ldots, s_{n-1})$ and $(d_0, \ldots, d_{n-1})$ are monotonic. Let

$$g_1(s_i) = f(s_i, \pi(s_i)) = f(s_i, d_i) = [m + i]_q \quad \text{and}$$
$$g_2(d_i) = f(\pi^{-1}(d_i), d_i) = f(s_i, d_i) = [m + i]_q$$

for all i. Thus, $f(s_i, d_i) = f(s_j, d_j)$ implies $g_1(s_i) = g_1(s_j)$ and $g_2(d_i) = g_2(d_j)$. It follows from Lemma 3.1 that

$$\mid s_i - s_j \mid \geq q \quad \text{and} \quad \mid d_i - d_j \mid \geq q.$$

Hence, the assignment is nonblocking according to the condition given in (3.24). □

As an example, consider a three-stage Clos network with $q = 3$ and $p = 4$ and a set of monotonic connection requests given by

$$\pi = \begin{pmatrix} 1 & 3 & 5 & 6 & 9 & 11 \\ 2 & 4 & 7 & 8 & 9 & 11 \end{pmatrix}.$$

Using the rank-based assignment algorithm, we obtain a nonblocking route assignment as shown in Table 3.1. In this case, $m = 0$. The resulting route assignment is shown graphically in Fig. 3.31.

TABLE 3.1 Route Assignment for π

s_i	1	3	5	6	9	11
d_i	2	4	7	8	9	11
d_i	(0,2)	(1,1)	(2,1)	(2,2)	(3,0)	(3,2)
$m+i$	0	1	2	3	4	5
$f(s_i, d_i) = [m+i]_q$	0	1	2	0	1	2
Routing tag	(0,0,2)	(1,1,1)	(2,2,1)	(0,2,2)	(1,3,0)	(2,3,2)

3.4.2 Recursiveness Property

In the preceding chapter, we have demonstrated that Clos networks can be recursively constructed. To determine a nonblocking route assignment for such Clos networks, we establish the following order-preserving property, which states that in a Clos network with monotonic connection requests, with the use of the rank-based assignment algorithm, each middle-stage module is itself a Clos network with monotonic connection requests. That is, we can apply the rank-based assignment algorithm in a recursive manner to obtain the route assignment of each subnetwork.

Theorem 3.7 (Order-Preserving Property of Clos Networks). If the connection requests to a three-stage Clos network is ordered, and their routes are assigned by the rank-based assignment algorithm, then the set of connection requests to each middle-stage switch module is also monotonic. That is, the order of their ranks is preserved.

Proof. Given a Clos network with monotonic connection requests, consider a middle-stage module M and consider two inputs s_i and s_j assigned to M. Let their corresponding outputs be d_i and d_j respectively. Without loss of generality, assume that $s_i < s_j$ and $d_i < d_j$. Due to the rank-based assignment algorithm, we have $s_j - s_i \geq q$ and $d_j - d_i \geq q$. Refer to Fig. 3.32. Let the ports on M to which s_i and s_j are connected be m_i and m_j, respectively, and let the ports to which d_i and d_j are connected be n_i

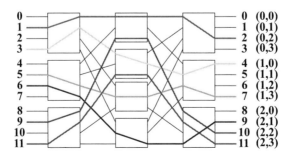

FIGURE 3.31 Route assignment for π.

80 FUNDAMENTAL PRINCIPLES OF PACKET SWITCH DESIGN

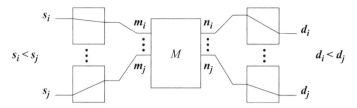

FIGURE 3.32 Order is preserved in a middle-stage module M.

and n_j, respectively. Now, from (3.25) and (3.26), we have

$$m_j = \left\lfloor \frac{s_j}{q} \right\rfloor \geq \left\lfloor \frac{s_i + q}{q} \right\rfloor = \left\lfloor \frac{s_i}{q} \right\rfloor + 1 = m_i + 1 > m_i,$$

$$n_j = \left\lfloor \frac{d_j}{q} \right\rfloor \geq \left\lfloor \frac{d_i + q}{q} \right\rfloor = \left\lfloor \frac{d_i}{q} \right\rfloor + 1 = n_i + 1 > n_i.$$

(3.29)

The rank order is preserved in each middle-stage module. The active inputs and their corresponding outputs on M are therefore monotonic. □

As an example, suppose we have an $N \times N$ Clos network with monotonic connection requests. Assume that N can be factorized into $N = pqr$. As shown in Fig. 3.33,

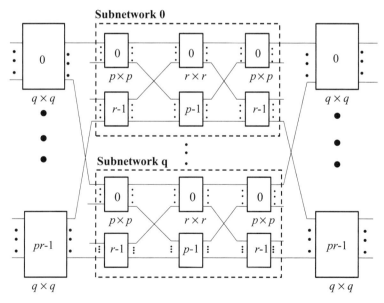

FIGURE 3.33 Recursive construction of Clos network.

each switch module in the input stage and in the output stage is of size $q \times q$. There are q middle-stage modules, each of size $pr \times pr$. By Theorem 3.7, each middle-stage module is a Clos network with monotonic connection requests and can be further decomposed. Suppose a connection request (s_i, d_i) is assigned to a middle-stage module M. The rank of the corresponding connection request on M is given by $\lfloor i/q \rfloor$. To see this, recall that a connection request (s_i, d_i) is assigned to a middle-stage module M if $f(s_i, d_i) = [m+i]_q = M$. Without loss of generality, suppose $m = 0$ and $f(s_i, d_i) = [i]_q$. Those connection requests with rank $M, M+q, M+2q, \ldots$ are assigned to M. Taking the integer division of their ranks by q therefore gives the rank of their corresponding connections on M. In this example, there are p middle-stage modules in each smaller Clos network. The route assignment within each smaller Clos network can then be determined recursively and the routing tag for this five-stage Clos network is

$$\left([r(s_i)]_q, \left[\left\lfloor \frac{r(s_i)}{q} \right\rfloor\right]_p, \left\lfloor \frac{\lfloor \frac{d_i}{q} \rfloor}{p} \right\rfloor, \left[\left\lfloor \frac{d_i}{q} \right\rfloor\right]_p, [d_i]_q\right). \qquad (3.30)$$

The first two components in this routing tag are due to the rank-based assignment and indicate which middle-stage module a call should be routed to. The last three components are used to route the packet correctly to destination starting from the middle-stage module.

3.4.3 Basic Properties of Half-Clos Networks

A Clos network can be considered as the cascade combination of two symmetric subnetworks, referred to as half-Clos networks. One well-known example is the formation of the Benes network by combining the baseline network and the reverse baseline network. In these half-Clos networks, there is a unique path from any input to any output that can be completely determined by the output destination address. Thus, we can see that the half-Clos networks possess the nonblocking self-routing property and all the multistage interconnection networks (MINs) previously described belong to this class. When two half-Clos networks are combined to form a Clos network, multiple alternative paths exist for each input–output pair. These paths are the products of the unique paths in the half-Clos networks.

Figure 3.34 shows that when a Clos network is divided into two parts, the omega network and the reverse omega network are formed. Without loss of generality, we shall study the properties of half-Clos networks based on them and explain the fundamental principle behind the rank-based assignment algorithm by combining them together. This leads us to the observation that any MIN can be combined with its reverse network to form a general Clos-type network taht also possesses the same nonblocking self-routing property as that of the classical Clos networks.

3.4.3.1 Omega Network
An omega network is the right half of a Clos network. An example of omega network, together with the address numbering scheme, is shown

82 FUNDAMENTAL PRINCIPLES OF PACKET SWITCH DESIGN

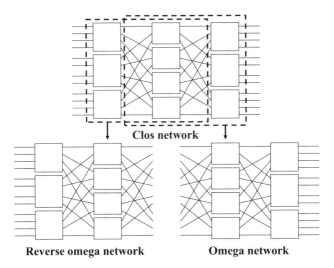

FIGURE 3.34 Dividing a Clos network into two half-Clos networks.

in Fig. 3.35. There are three modules in the input stage and four in the output stage. Using the numbering scheme introduced earlier, we can number each input port and each output port by a 2-tuple address. The shuffle stage at the inputs is required to ensure that the same numbering scheme is used in the input and the output ports. Suppose there is no shuffle stage. Denote an input address by $a_i b_i$ and an output

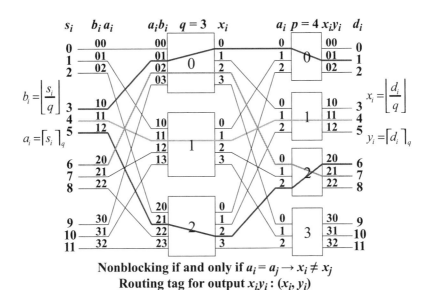

Nonblocking if and only if $a_i = a_j \rightarrow x_i \neq x_j$
Routing tag for output $x_i y_i$: (x_i, y_i)

FIGURE 3.35 An omega network.

address by $x_i y_i$. It is easy to see that a_i is of base $q = 3$ and b_i is of base $p = 4$ while x_i is of base $p = 4$ and y_i is of base $q = 3$. The numbering of inputs and outputs is therefore inconsistent. With the shuffle stage, the input addresses become $b_i a_i$. Each tuple in the input address then has the same base as the corresponding tuple in the output address.

Denote an input s_i by $b_i a_i$ and an output d_i by $x_i y_i$. For the omega network to be externally nonblocking (i.e., no output port conflict) we must have

$$x_i = x_j \quad \Rightarrow \quad y_i \neq y_j, \tag{3.31}$$

and it is internally nonblocking if and only if the active inputs satisfy

$$a_i = a_j \quad \Rightarrow \quad x_i \neq x_j. \tag{3.32}$$

The following theorem gives the sufficient conditions for an omega network to be internally nonblocking.

Theorem 3.8 (Nonblocking Conditions for Omega Networks). The omega network is internally nonblocking if the active inputs $s_0, \ldots, s_{n-1}$ ($s_j > s_i$ if $j > i$) and their corresponding output destinations $d_0, \ldots, d_{n-1}$ satisfy the following:

1. *Concentrated inputs*: Any input between two active inputs is also active. That is, if s_i and s_j are active and $s_i \leq w \leq s_j$, then w is active.
2. *Monotonic outputs*: $d_0 < d_1 < \cdots < d_{n-1}$ or $d_0 > d_1 > \cdots > d_{n-1}$.

Proof. Using the same 2-tuple numbering scheme, denote two active inputs s_i and s_j ($s_i < s_j$) by $b_i a_i$ and $b_j a_j$, respectively. Similarly denote their corresponding outputs d_i and d_j by $x_i y_i$ and $x_j y_j$, respectively. Without loss of generality, assume that $d_i < d_j$. By the concentrated inputs and monotonic outputs conditions, we have

$$x_j y_j - x_i y_i \geq b_j a_j - b_i a_i.$$

We are going to show that the omega network is nonblocking by contradiction. If $a_i = a_j$, we have

$$x_j y_j - x_i y_i \geq b_j a_j - b_i a_i = (b_j - b_i) \cdot q \geq q \tag{3.33}$$

because a_i is of base q. However, $x_i = x_j$ implies

$$x_j y_j - x_i y_i = y_j - y_i \leq q - 1. \tag{3.34}$$

Since (3.33) and (3.34) cannot hold simultaneously, the omega network must be internally nonblocking with concentrated and ordered connection requests according to condition (3.32). □

84   FUNDAMENTAL PRINCIPLES OF PACKET SWITCH DESIGN

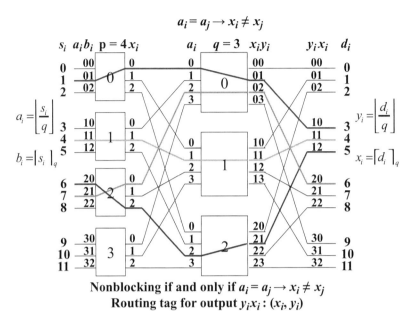

FIGURE 3.36 A reverse omega network.

In fact, the set of active inputs needs only be concentrated in a modulo fashion. We can shift a set of concentrated inputs by a constant amount, take modulo N, and the network is still nonblocking. That is, the set of active inputs can be cyclically concentrated and "wrap around" the inputs. To see this, consider two arbitrary active inputs $b_i a_i$ and $b_j a_j$ and their corresponding outputs $x_i y_i$ and $x_j y_j$. If the network is nonblocking, we have by (3.32) that $a_i = a_j$ implies $x_i \neq x_j$. Suppose the inputs are shifted by m and become $b'_i a'_i$ and $b'_j a'_j$, respectively. Since a_i is of base q, we have $a'_i = a_i + [m]_q$ and $a'_j = a_j + [m]_q$. It is easy to see that $a'_i = a'_j$. Therefore, for this shifted set of active inputs, we still have $a'_i = a'_j \Rightarrow x_i \neq x_j$, which by (3.32) implies that the network is nonblocking.

3.4.3.2 The Reverse Omega Network
The reverse omega network is obtained from the left half of the Clos network. An example of reverse omega network is shown in Fig. 3.36. Using the numbering scheme as shown in the figure, it is easy to see that for a reverse omega network to be externally nonblocking, we must have

$$x_i = x_j \quad \Rightarrow \quad y_i \neq y_j, \tag{3.35}$$

and it is internally nonblocking if and only if the active inputs satisfy

$$a_i = a_j \quad \Rightarrow \quad x_i \neq x_j. \tag{3.36}$$

The nonblocking conditions for the reverse omega network are like the reverse of those for the omega network and are given by the following theorem.

Theorem 3.9 (Nonblocking Conditions for Reverse Omega Networks). The reverse omega network is internally nonblocking if the output destinations $d_0, ..., d_{n-1}$ ($d_j > d_i$ if $j > i$) and their corresponding inputs $s_0, ..., s_{n-1}$ satisfy the following:

1. *Concentrated outputs*: Any output between two active outputs is also active. That is, if d_i and d_j are active and $d_i \leq w \leq d_j$, then w is active.
2. *Monotonic inputs*: $s_0 < s_1 < \cdots < s_{n-1}$ or $s_0 > s_1 > \cdots > s_{n-1}$.

Proof. The proof is similar to the one in Theorem 3.8. Denote two active inputs s_i and s_j by $a_i b_i$ and $a_j b_j$ and their corresponding output destinations d_i and d_j by $y_i x_i$ and $y_j x_j$, respectively. Without loss of generality, assume that $s_i < s_j$. If the inputs are monotonic and outputs are concentrated, we have

$$a_j b_j - a_i b_i \geq y_j x_j - y_i x_i.$$

If $a_i = a_j$, the two inputs are on the same input module and we have

$$q - 1 \geq b_j - b_i = a_j b_j - a_i b_i \geq y_j x_j - y_i x_i.$$

However, $x_i = x_j$ implies

$$y_j x_j - y_i x_i = (y_j - y_i) \cdot q \geq q,$$

which is impossible. Therefore, the reverse omega network is internally nonblocking with ordered and concentrated output according to condition (3.36). □

Similar to the omega network, the set of active outputs on the reverse omega network needs only be concentrated in a modulo fashion. This can be seen by considering the reverse omega network as the mirror image of the omega network.

Both the omega and the reverse omega network exhibit simple self-routing schemes. This feature is extremely important when the networks are used to construct fast packet switches.

For the omega network, if the outputs are numbered by the 2-tuple numbering scheme as shown in Fig. 3.35, then the routing tag for a call destined to $x_i y_i$ is given by (x_i, y_i). In other words, the call takes the $(x_i + 1)$th output in the first-stage switch and the $(y_i + 1)$th output in the second-stage switch. Likewise for the reverse omega network, if the outputs are numbered by the 2-tuple numbering scheme as shown in Fig. 3.36, the routing tag for a call destined to $y_i x_i$ is given by (x_i, y_i). If the nonblocking conditions are satisfied, the connection requests can route through the networks in a rapid and nonblocking manner.

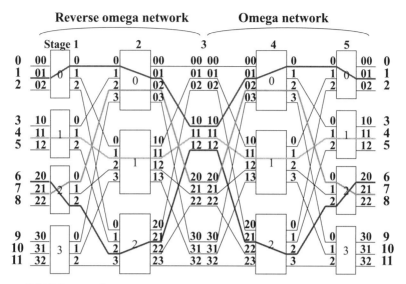

FIGURE 3.37 Combining a reverse omega network and an omega network.

3.4.3.3 Cascade Combination of Half-Clos Networks

Consider the network formed by combining a reverse omega network and an omega network. An example is shown in Fig. 3.37. The network has five stages but the central stage (stage 3) is dummy and no switching is performed in that stage. Input packets are first routed through the reverse omega network and then fed into the omega network for routing to route to their final destinations. This network is nonblocking if the inputs and their corresponding outputs are monotonic. This is because given monotonic inputs, we can always route the packets to a set of concentrated ports in stage 3 so as to avoid blocking in the reverse omega network. Afterward, when the packets enter the omega network, no blocking will occur because the inputs are concentrated and their outputs are monotonic. Any sets of concentrated ports in stage 3 can be used, including cyclically concentrated sets. It is observed that due to the shuffle stage at the output of the reverse omega network, routing the packets to a set of concentrated ports in stage 3 is equivalent to routing them to consecutive modules in stage 2. This can be done with the help of the rank of each connection request.

Since stage 3 is dummy, we can eliminate it and stage 2 and stage 4 are combined into a single stage. The resulting network is a three-stage Clos network. A nonblocking route assignment can therefore be obtained by assigning the connection requests having consecutive ranks to consecutive middle-stage modules in a modulo fashion. This is the basis of the rank-based assignment algorithm.

In general, any multistage interconnection network can be combined with its reverse network to form a general Clos-type network. The scenario is similar to combining the omega with the reverse omega network. In an MIN, there is a unique path for each input–output pair. If the MIN is $N \times N$ and is composed of 2×2 switch elements, the number of alternative paths for each input–output pair in the combined network will be equal to $N/2$. These alternative paths are the product of the paths in the MINs.

3.4.3.4 Recursiveness Properties Both the omega and the reverse omega network can be recursively constructed and share the same order-preserving property. The following is a proof of the order-preserving property of the omega network. This proof can be similarly applied to the reverse omega network.

Theorem 3.10 (Order-Preserving Property of Omega Network). If the connection requests to an omega network satisfy the concentrated inputs and monotonic outputs conditions, then the set of connection requests to each switch module in the network also satisfy the concentrated inputs and monotonic outputs conditions. That is, the order of their ranks is preserved.

Proof. Without loss of generality assume that the inputs to an omega network are concentrated and their corresponding outputs are monotonically increasing. The proof is divided into two parts.

(1) *First-stage modules:* We first show that the inputs to a switch module in the first stage are concentrated and their corresponding outputs on that module are monotonic. Consider two active ports b_i and b_j on a first-stage module M, with $b_i < b_j$. They are numbered by the 2-tuple addresses Mb_i and Mb_j. Suppose there exists a port b_k, where $b_i < b_k < b_j$, which is inactive. Due to the shuffle stage, the corresponding input addresses of these ports are $b_i M$, $b_k M$, and $b_j M$, respectively (see Fig. 3.38(a)). We have

$$b_k M - b_i M = (b_k - b_i) \cdot q \geq q \quad \text{and}$$
$$b_j M - b_k M = (b_j - b_k) \cdot q \geq q.$$

Therefore, $b_i M < b_k M < b_j M$. In other words, between two active inputs, there is an inactive port and this contradicts with the concentrated inputs assumption. Therefore, the active inputs on M are concentrated.

Suppose the destinations of inputs $b_i M$ and $b_j M$ ($b_i M < b_j M$) are $x_i y_i$ and $x_j y_j$, respectively. On module M, the two connection requests will take outputs x_i and x_j, respectively, due to the self-routing property. We want to show that given $b_i < b_j$ we have $x_i < x_j$. In other words, the active outputs on M are monotonically increasing.

With the concentrated inputs and monotonic outputs assumptions, we have

$$b_i M < b_j M \Rightarrow x_i y_i < x_j y_j. \tag{3.37}$$

88 FUNDAMENTAL PRINCIPLES OF PACKET SWITCH DESIGN

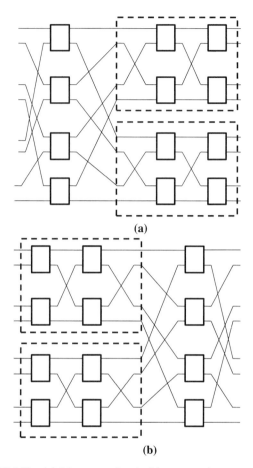

FIGURE 3.38 (a) A banyan network; (b) a reverse banyan network.

This implies that $x_i \leq x_j$. But since the omega network is nonblocking and $a_i = a_j$, by (3.32) we have $x_i \neq x_j$. Therefore, we have $x_i < x_j$ and the outputs on M are monotonically increasing.

(2) *Second-stage modules:* Consider a second-stage switch module M' in an omega network. Since the active outputs of the omega network are monotonically increasing, the active outputs on M are obviously monotonically increasing.

Denote the active outputs on M' by $x_1 y_1, x_2 y_2, \ldots, x_k y_k$ ($x_1 y_1 < x_2 y_2 < \cdots < x_k y_k$) and suppose that they originate from inputs $b_1 a_1, b_2 a_2, \ldots, b_k a_k$, respectively (see Fig. 3.38(b)). Obviously, $b_1 a_1 < b_2 a_2 < \cdots < b_k a_k$ and they are concentrated. In other words, their least significant digit a_i's are consecutive (modulo q) because a_i's are of base q. Let $m_1, m_2, \ldots, m_k$ be the active inputs on M corresponding to $x_1 y_1, x_2 y_2, \ldots, x_k y_k$. If we imagine routing from the outputs to the inputs instead, the network becomes a reverse omega network

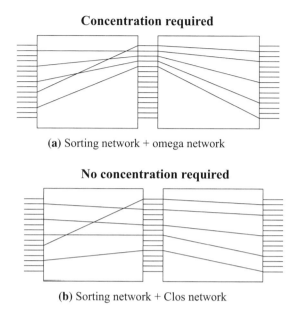

FIGURE 3.39 Construction of nonblocking self-routing switches.

and it is easy to see that the $m_1 = a_1, m_2 = a_2, ..., m_k = a_k$. Therefore, the set of active inputs on M is concentrated (modulo q). □

3.4.4 Sort-Clos Principle

A nonblocking self-routing point-to-point switch can be constructed with an omega network or a Clos network. As illustrated in Fig. 3.39(a), we can cascade a sorting network and an omega network to form a nonblocking self-routing switch. The sorting network sorts the input packets based on their output destinations and concentrates them so that the set of input packets to the omega network is concentrated and their corresponding output destinations are monotonic. The omega network then routes the packets to their destinations in a nonblocking manner. This is the basic sort-banyan principle. In a similar manner, a nonblocking self-routing switch can be constructed by cascading a sorting network and a Clos network, as shown in Fig. 3.39(b). The sorting network, however, is only required to provide a set of monotonic connection requests to the Clos network. *No concentration is required*. Thus, the contention resolution consists only of two phases. By using the rank-based assignment algorithm, a nonblocking route assignment can be easily found and the packets can be self-routed to their destinations. We denote this as the *sort-Clos principle*. It is the generalization of the sort-banyan principle. With this switching principle, the construction of nonblocking self-routing switch is *no longer limited to* 2×2 *switch elements*.

90 FUNDAMENTAL PRINCIPLES OF PACKET SWITCH DESIGN

PROBLEMS

3.1 Consider an 8×8 reverse shuffle-exchange network.
  (a) Show how the bits in the output address are used to route a packet from input 001 to output 110. Show how a packet is routed from input 110 to output 010.
  (b) If this network is used as a loss system, do you think its loss probability would be higher or lower than a shuffle-exchange network?

3.2 For the class of Banyan networks, suppose a central controller knowing the destinations of all the input packets is used to perform the routing and setting the states of the switch elements. Can the loss probability be improved with respect to that in the self-routing Banyan networks? If your answer is no, explain why. If your answer is yes, give an example in which a centralized routing decision leads to fewer packets being dropped.

3.3 Consider an 8×8 Banyan network. Suppose with probability 0.75 a packet is destined for outputs 000, 001, 010, or 011, and with probability 0.25 it is destined for the other four outputs. Within each group of outputs, the packet is equally likely to be destined for any of the four outputs. Is the loss probability higher in this case than when a packet is equally likely to be destined for any of the eight outputs? Modify the analysis in the text to answer this question.

3.4 In Section 3.2.3, it is mentioned that each of the $N^{N/2}$ states of the Banyan network corresponds to a unique input–output mapping by the unique-path property. Explain why. In Section 3.2.2, it is mentioned that the uniform-traffic assumption also means that a packet is equally likely to be destined for either of the two outgoing links of a switch element. Explain why.

3.5 Show how to use the reverse banyan (reverse shuffle-exchange) network as a concentrator. That is, we want to concentrate packets from N inputs to M outputs where $M < N$. It does not matter which output a packet goes to as long as the packet gets to an output. As long as there are no more than M incoming packets, none of them will be dropped or buffered at the input.

3.6 Use the order-preserving property to show that a bitonic sorter that works for zero–one sequence also works for sequences of arbitrary numbers.

3.7 A circular bitonic sequence is a sequence that can be made bitonic by shifting the sequence in a circular fashion. For example, the sequence $\langle 3, 1, 1, 2, 3, 4, 5, 4 \rangle$ is circular bitonic because after two circular left shift it becomes the bitonic sequence $\langle 1, 2, 3, 4, 4, 5, 3, 1 \rangle$. The bitonic sequences are a subset of circular bitonic sequences.
  (a) Show that a bitonic sorter also sorts circular bitonic sequences of arbitrary numbers correctly. (*Hint*: Use a proof similar to that of the zero–one principle, but first argue that if the numbers are limited to 0's and 1's, there is no difference between circular bitonic and bitonic sequences.)

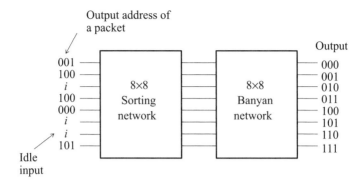

FIGURE 3.40 A set of input packets for a sort-banyan network.

(b) Theorem 3.4 concerns the decomposition of a zero–one bitonic sequence into two zero–one bitonic sequences. For a bitonic sequence of arbitrary numbers, the decomposition does not necessarily yield two bitonic sequences; one of the sequences could be circular bitonic but not bitonic. Give an example to illustrate this.

3.8 Prove the bitonic sorting algorithm without relying on the zero–one principle.

3.9 The proof of the odd–even merging algorithm in the text assumes that n is even. Repeat the proof for odd n.

3.10 For n a power of 2, compute the number of 2×2 comparators in an overall sorting network based on the odd–even merging algorithm.

3.11 Show that the odd-even merging algorithm can merge two sorted input sequences of different sizes. That is, the input sequences are $\langle a_1, \ldots, a_s \rangle$ and $\langle b_1, \ldots, b_t \rangle$ where $s \ne t$.

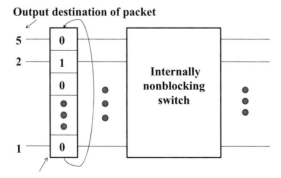

FIGURE 3.41 Ring contention-resolution scheme.

3.12 Consider an $N \times N$ banyan network with $n = \log_2 N$ stages, where n is even. Suppose that there are N input packets, all destined for different outputs. Show that in the worst case, there may be $\sqrt{N}$ packets wanting to cross the same link inside the network in order to reach their outputs. That is, the same link is in $\sqrt{N}$ input–output paths of the packets.

3.13 Consider a Batcher–banyan network that uses the three-phase contention-resolution scheme.
  (a) How many bits of delays (in terms of N) are needed before the completion of the first two phases, assuming the acknowledgments take only one-bit delay?
  (b) Suppose that the information payload of the packets is 53-byte long. Express the overhead in terms of the ratio of the contention-resolution delay to the amount of time needed to route one batch of packets to their destinations. For what N does the ratio become 50%?

3.14 Consider an 8×8 Batcher–banyan switch with input packets as depicted in Fig. 3.40.
  (a) Draw the 8×8 Batcher bitonic sorting network and show how the paths taken by the input packets.
  (b) Which packets will win contention in the three-phase switching scheme?
  (c) Show the paths taken by the winning packets in the banyan network.

3.15 When the sorting network is used for contention resolution, an unlucky packet may end up at a sorting-network output lower than other packets destined for the same switch output repeatedly in successive time slots. For fairness, we can add several priority bits to the header so that the priority level of a packet will be raised in the next time slot each time a packet loses contention. Explain how these bits should be used in the sorting network so that the packet that has lost contention the most number of times will be the winner among all packets destined for the same switch output in each time slot.

3.16 We can use a ring contention–resolution scheme for a sort-banyan network. The ring contention-resolution scheme is very much like the medium access control (MAC) of the token-ring local-area network, except that there are N tokens, one for each output. The tokens circulate around the inputs, as illustrated in Fig. 3.41. Each token has only one bit that is set to 0 initially. The token for output i is positioned at input i at the beginning. If input i has a packet that is destined for output i, it sets the token to 1, capturing the right to send to output i. The tokens then undergo a circular shift so that the token for output i is now positioned at input $i + 1 \pmod{N}$. If input $i + 1 \pmod{N}$ has a packet destined for output i, it first must make sure that the token is not already set to 1. If yes, it has lost the contention to another input. Otherwise, it sets token to 1 and captures the right to send a packet to output i. After N steps and each token has the chance to visit each token once, the contention-resolution

process is completed. The winning inputs can then send their packets to the outputs.
   (a) How does this scheme compare with the three-phase scheme in terms of contention-resolution overhead?
   (b) Describe how the basic scheme can be modified to obviate the need for the banyan network by letting the tokens circulate around the inputs twice.

4

SWITCH PERFORMANCE ANALYSIS AND DESIGN IMPROVEMENTS

This chapter first focuses on the performance analysis of some simple switch designs. From the analysis, we shall see that head-of-line (HOL) blocking as well as output contention among packets may severely limit the switch throughput. This chapter will then present several fundamental switch design principles to alleviate these problems and achieve performance improvement. Each switch design principle will be illustrated using specific implementations. Note, however, that many implementations are possible based on each switch design concept, and one should not confuse implementations with concepts.

4.1 PERFORMANCE OF SIMPLE SWITCH DESIGNS

This section analyzes the delay and throughput performance of some simple switch designs with an internally nonblocking structure. Since the packets are fixed-length, the arrivals can be aligned (as discussed at the beginning of the preceding chapter) with a maximum alignment delay of one-packet duration. With the alignment, we may assume in our analysis that time is slotted and that the packets arrive to the inputs at the beginning of each time slot.

For simplicity, we shall also assume a uniform-traffic distribution. Packet arrivals at the inputs are described by a simple Bernoulli process: in any time slot, the probability that a packet will arrive on a particular input is ρ_o. The parameter ρ_o is also referred to as the offered load. Each packet is destined for any given output with equal probability $1/N$ and the output destinations of successive packets are independent.

We shall engage much of the analysis in the domain of generating functions. The generating function of any random variable R, $E[z^R]$, will be denoted by

---

Principles of Broadband Switching and Networking, by Tony T. Lee and Soung C. Liew
Copyright © 2010 John Wiley & Sons, Inc.

$R(z)$. If conditional probabilities are used in taking the expectation, we shall write $R(z|\text{conditions})$.

4.1.1 Throughput of an Internally Nonblocking Loss System

Let us first derive the maximum throughput of a loss system. There are no queues in the system, and packets that have lost contention in a time slot are simply dropped.

Consider a particular output i. In any time slot, the probability that an arbitrary output has a packet destined for it is ρ_o/N. Thus, the probability that none of the N inputs has a packet destined for it is $(1 - \rho_o/N)^N$. This value approaches $e^{-\rho_o}$ very quickly as N increases. The throughput is therefore

$$\rho = 1 - e^{-\rho_o} \tag{4.1}$$

for large N. Since the function increases with ρ_o, the maximum throughput, or the maximum carried load, is obtained when the offered load $\rho_o = 1$, and it is

$$\rho^* = 1 - e^{-1} \approx 0.632. \tag{4.2}$$

Instead of dropping packets that have lost contention, we may buffer them at the inputs so that they can attempt to access their outputs at a later time slot. This further reduces the maximum throughput to 0.586 because of HOL blocking, as shown in the next subsection.

4.1.2 Throughput of an Input-Buffered Switch

The analysis of a waiting system is more involved. Suppose that the input queues are first-in-first-out. Then only the HOL packets may contend for output access. The scenario is depicted in Fig. 4.1. In addition to the N input queues, we may imagine for

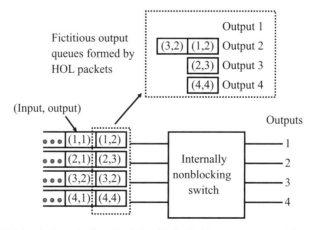

FIGURE 4.1 An input-buffered switch with the fictitious queues used for analysis.

PERFORMANCE OF SIMPLE SWITCH DESIGNS   97

analytical purposes that there are also N fictitious queues, one for each output, whose elements consist of the HOL packets destined for the N outputs. Thus, a packet joins a fictitious queue only after it has proceeded to the front of its input queue.

Let us consider the fictitious queue associated with a particular output i. Let C_m^i be the number of packets in it at the start of time slot m. Let A_m^i be the number of packets entering the fictitious queue at the start of time slot m. The system evolves according to the equation

$$C_m^i = \max(0, C_{m-1}^i - 1) + A_m^i$$
$$= B_{m-1}^i + A_m^i, \tag{4.3}$$

where $B_{m-1}^i = \max(0, C_{m-1}^i - 1)$ is the number of packets remaining in the fictitious queue at the end of time slot $m - 1$. To interpret the equation, note that if the queue was not empty at the beginning of time slot $m - 1$, then one packet would be cleared by the end of it, hence the term $C_{m-1}^i - 1$. On the other hand, if the queue was empty ($C_{m-1}^i = 0$), no packet would be cleared, and the queue remained empty at the end of the time slot, hence the term $\max(0, C_{m-1}^i - 1)$. To this we add the number of newly arriving packets at the beginning of time slot m, A_m^i, to obtain C_m^i.

To study the system dynamic, we need to know A_m^i. For the calculation of the maximum allowable throughput, let us assume that all the input queues are saturated so that as soon as a HOL packet is cleared, its position is taken by a subsequent packet in the next time slot. In this way, the system is loaded at its maximum capacity. At the beginning of time slot m, the inputs can be divided into two groups:

1. Those that won contention in the previous time slot.
2. Those that lost contention in the previous time slot.

Let F_{m-1} be the number of inputs that won contention in time slot $m - 1$. This is also the number of new HOL packets in time slot m under the saturation condition. The total number of backlogged packets at the HOL at the end of time slot $m - 1$ is $\sum_{i=1}^{N} B_{m-1}^i$. By conservation,

$$F_{m-1} + \sum_{i=1}^{N} B_{m-1}^i = N.$$

The probability that a new HOL packet is destined for output i is $1/N$. Thus, considering all the F_{m-1} new HOL packets, the probability that k of them are destined for output i is

$$\Pr\{A_m^i = k \mid F_{m-1}\} = \binom{F_{m-1}}{k} \left(\frac{1}{N}\right)^k \left(1 - \frac{1}{N}\right)^{F_{m-1}-k}. \tag{4.4}$$

The above packet arrival process is difficult to deal with analytically because F_{m-1} (and therefore A_m^i) depends on B_{m-1}^i. It is known in queueing theory that whenever the arrival process depends on the system state, the analysis becomes complicated. Fortunately, the concern goes away when N is very large so that we can at least get an analytical result for that case.

To see this, let us first look at the effect of B_{m-1}^i on F_{m-1}. Recall that F_{m-1} is the number of cleared packets in time slot $m-1$. Certainly, given B_{m-1}^i, the range of values that F_{m-1} can adopt is $1, \ldots, N - B_{m-1}^i$, since we know that at least B_{m-1}^i HOL packets addressed for output i did not get cleared. When N is small (e.g., $N = 2$), the effect of B_{m-1}^i on F_{m-1} is rather significant. However, when N is large, since B_{m-1}^i is the backlogged packets of one output among many outputs, B_{m-1}^i is typically much smaller than N and $N - B_{m-1}^i \approx N$. The dependence between B_{m-1}^i and F_{m-1} is negligible for large N. In the limit $N \to \infty$, B_{m-1}^i and A_m^i become independent. Therefore, for simplicity, we shall assume that N is very large in the following analysis.

Intuitively, the binomial arrival process in (4.4) approaches a Poisson process as $N \to \infty$. However, F_{m-1} changes from time slot to time slot. As a result, the binomial arrival process as described in (4.4) is a modulated binomial process in which the number of packet sources F_{m-1} changes with time. We now argue that to the extent that $N \to \infty$, A_m^i approaches an "unmodulated" Poisson process with rate ρ, where ρ is the maximum load yet to be determined.

To avoid cluttered notation, we shall drop the superscript i in A_m^i, B_m^i, and C_m^i in the following with the implicit understanding that they all refer to the particular output i. The generating function of A_m given F_{m-1} is

$$A_m(z \mid F_{m-1}) \stackrel{\text{def}}{=} E[z^{A_m} \mid F_{m-1}] = \sum_{k=0}^{F_{m-1}} z^k \Pr\{A_m = k \mid F_{m-1}\}$$

$$= \sum_{k=0}^{F_{m-1}} z^k \binom{F_{m-1}}{k} \left(\frac{1}{N}\right)^k \left(1 - \frac{1}{N}\right)^{F_{m-1}-k}$$

$$= \left[1 + \frac{(z-1)}{N}\right]^{F_{m-1}}. \tag{4.5}$$

Now, let

$$G_{m-1} = \frac{F_{m-1}}{N}. \tag{4.6}$$

The knowledge of G_{m-1} is equivalent to the knowledge of F_{m-1}, so that $A_m(z \mid F_{m-1}) = A_m(z \mid G_{m-1})$. Since the value of G_{m-1} is finite (between 0 and 1), we may take the following limit for large N:

$$A_m(z \mid G_{m-1}) = \left[1 + \frac{(z-1)}{N}\right]^{NG_{m-1}}$$

$$\to e^{(z-1)G_{m-1}}$$

$$= 1 + (z-1)G_{m-1} + \frac{(z-1)}{2!}G_{m-1}^2 + \cdots \quad (4.7)$$

We want to uncondition $A_m(z \mid G_{m-1})$ on G_{m-1}. Define

$$X_m^j \stackrel{\text{def}}{=} \begin{cases} 1, & \text{if the HOL packet on input } j \text{ is cleared in time slot } m, \\ 0, & \text{otherwise.} \end{cases} \quad (4.8)$$

Thus, by definition,

$$F_m = \sum_{j=1}^{N} X_m^j.$$

Suppose that all inputs are treated equally in the contention-resolution process. Then, by symmetricity, for all j, $E[X_m^j] = \Pr\{X_m^j = 1\} = \rho$ in the steady state as $m \to \infty$, where ρ is the maximum carried load. In general,

$$E[(X_m^j)^k] = 0^k \Pr\{X_m^j = 0\} + 1^k \Pr\{X_m^j = 1\} = \rho$$

for all $k \geq 1$. Thus, we can write

$$E[G_{m-1}] = E\left[\frac{F_{m-1}}{N}\right] = \frac{1}{N}\sum_j E[X_{m-1}^j] = \rho$$

and

$$E[G_{m-1}^2] = \frac{1}{N^2}\sum_j \sum_k E\left[X_{m-1}^j X_{m-1}^k\right]$$

$$= \frac{1}{N^2}\left(\sum_j E\left[\left(X_{m-1}^j\right)^2\right] + \sum_j \sum_{k \neq j} E\left[X_{m-1}^j X_{m-1}^k\right]\right).$$

When N is large, the dependence between the clearance of packets at inputs j and k can be neglected, and X_{m-1}^j and X_{m-1}^k are independent as $N \rangle \infty$. We have

$$E[G_{m-1}^2] \approx \frac{1}{N^2}\left(\sum_j E\left[(X_{m-1}^j)^2\right] + \sum_j \sum_{k \neq j} E\left[X_{m-1}^j\right] E\left[X_{m-1}^k\right]\right)$$

$$= \frac{1}{N^2}[N\rho + N(N-1)\rho^2]$$

$$\to \rho^2.$$

In general, we can show that

$$E[G_{m-1}^k] \to \rho^k. \tag{4.9}$$

Thus,

$$A_m(z) = E_{G_{m-1}}[A_m^j(z \mid G_{m-1})]$$

$$= 1 + (z-1)E[G_{m-1}] + \frac{(z-1)^2}{2!}E[G_{m-1}^2] + \cdots$$

$$\to 1 + (z-1)\rho + \frac{(z-1)^2}{2!}\rho^2 + \cdots$$

$$= e^{(z-1)\rho}. \tag{4.10}$$

Note that in the steady state, the arrival process A_m is independent of m and therefore we can replace $A_m(z)$ with $A(z)$. It can be easily verified that $A(z)$ is the generating function of a Poisson process:

$$\Pr\{A = k\} = \frac{\rho^k}{k!}e^{-\rho}. \tag{4.11}$$

In the steady state, the subscript m in B_m and C_m can also be dropped. The generating function of $B = \max(0, C-1)$ is

$$B(z) = \sum_{k=0}^{\infty} z^k \Pr\{B = k\}$$

$$= \Pr\{B = 0\} + z \Pr\{B = 1\} + z^2 \Pr\{B = 2\} + \cdots$$

$$= \Pr\{C = 0\} + \Pr\{C = 1\} + z \Pr\{C = 2\} + z^2 \Pr\{C = 3\} + \cdots$$

$$= \Pr\{C = 0\} + z^{-1}(z \Pr\{C = 1\} + z^2 \Pr\{C = 2\} + z^3 \Pr\{C = 3\} + \cdots$$

$$= (1 - z^{-1}) \Pr\{C = 0\} + z^{-1}(\Pr\{C = 0\} + z \Pr\{C = 1\} + z^2 \Pr\{C = 2\} + \cdots$$

$$= (1 - z^{-1})(1 - \rho) + z^{-1}C(z), \tag{4.12}$$

PERFORMANCE OF SIMPLE SWITCH DESIGNS

where we have made use of the fact that $1 - \Pr\{C = 0\} = \rho$ (the throughput is the probability that a packet is cleared, and a packet will be cleared so long as $C > 0$).

We have already shown that A becomes independent of B as $N \to \infty$. The generating function of the sum of two independent random variables is the product of the individual generating functions. Thus, taking the generating functions on both sides of (4.3) yields

$$C(z) = [(1 - z^{-1})(1 - \rho) + z^{-1}C(z)]A(z). \quad (4.13)$$

Rearranging the above, we obtain

$$C(z)(z - A(z)) = (z - 1)A(z)(1 - \rho). \quad (4.14)$$

Differentiating with respect to z twice, substituting $z = 1$, and noting that $C(1) = A(1) = 1$, we have

$$2C'(1)(1 - A'(1)) - A''(1) = 2A'(1)(1 - \rho). \quad (4.15)$$

Now, C is the number of packets at the fictitious queue of output i. The sum of the packets at all the fictitious queues is N, since these are the HOL packets from the N inputs. Thus, on average each fictitious queue has one packet, and therefore $E[C] = C'(1) = 1$. Substituting this into the above, we obtain

$$2(1 - A'(1)) - A''(1) = 2A'(1)(1 - \rho). \quad (4.16)$$

It is easy to derive from $A(z) = e^{(z-1)\rho}$ that $A'(1) = \rho$ and $A''(1) = \rho^2$. Substituting into the above and rearranging, we get a quadratic equation for ρ

$$\rho^2 - 4\rho + 2 = 0. \quad (4.17)$$

The roots of the equation are $2 \pm \sqrt{2}$, the smaller of which is the maximum throughput

$$\rho^* = 2 - \sqrt{2} \approx 0.586. \quad (4.18)$$

This is the maximum achievable throughput when there are infinite buffers in the input queues. It is time to sit back and ponder the physical meaning of this parameter. As we have shown earlier, if packets losing contention are dropped, the maximum throughput is 0.632, which is higher than the above. Intuitively, the lower throughput here is attributable to the memory effect that heavy contention tends to carry over to the next time slot. For instance, if in the current time slot, there are many packets destined for a particular output, in the next time slot there will be only *one fewer* packet destined for this output, and the contention remains intense. With the loss

TABLE 4.1 Maximum Throughput for Input-Buffered Switches of Different Sizes

N	ρ^*
2	0.75
3	0.68
4	0.66
5	0.64
6	0.63
7	0.62
8	0.62
∞	0.59

system, all these packets would have been dropped and the next time slot is a whole new ball game.

The carried load ρ is $\rho_o(1 - P_{drop})$, where ρ_o is the offered load and P_{drop} is the probability that a packet is dropped. When $\rho_o < \rho^*$, P_{drop} can be made arbitrarily small with sufficiently large buffer size, making it possible to carry all the offered load (i.e., $\rho \approx \rho_o$). In contrast, when $\rho_o \geq \rho^*$, no matter how large the buffer size is, the carried load cannot be larger than ρ^*. This is because the throughput limitation is due to switch design, not the buffer size.

The analysis presented applies only for large N. When N is small, we can use a Markov chain analysis to find the maximum throughput. One of the exercises asks you to do that for a 2×2 switch. Unfortunately, the number of states in the Markov chain grows exponentially with N and the analysis becomes very complicated. Thus, the analysis is simple when N is small or very large, but not somewhere in between. That the performance analysis can be made simpler by letting certain system parameter approach infinity is also observed in many other systems. One example is the class of multiple-access communication networks, which is often analyzed assuming the number of users is infinite. Fortunately for the input-buffered switch, the maximum throughput approaches 0.586 very quickly as N increases, as shown in Table 4.1. The analysis of many cases with finite N is therefore unnecessary.

It should be noticed that our analysis did not assume any particular contention-resolution discipline. It does not matter whether the packets at the fictitious queue are served in a first-in-first-out (FIFO) fashion, random fashion, or any other manner. As long as the queue is a *work-conserving* system in which a packet is cleared in each time slot when the queue is not empty, the same maximum throughput will be observed. The delay experienced by a packet, on the other hand, does depend on the contention-resolution discipline. What this means is that the delay will go to infinity as the offered load approaches the maximum throughput, but for offered load below the maximum throughput different contention-resolution disciplines yield different delays.

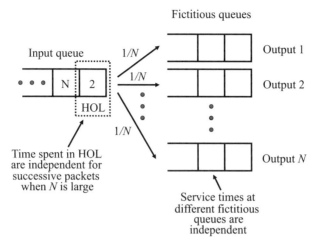

FIGURE 4.2 Queueing scenario for the delay analysis of the input-buffered switch.

4.1.3 Delay of an Input-Buffered Switch

We shall again assume the simple input traffic in which arriving packets are independent and equally likely to be destined for any of the N outputs. There are several methods for the derivation of the delay statistics. We present a method that focuses on the busy periods; Problems 4.9 and 4.10 in the problem set show you an alternative derivation for the mean delay. The derivation here is divided into two parts. First of all, we derive the waiting time of a packet in the input queue in terms of the delay at the fictitious queues (i.e., contention delay experienced at the HOL). We then find the delay statistics of the fictitious queues assuming a FIFO contention-resolution discipline.

With reference to Fig. 4.2, let us focus on a particular input queue. Whenever a packet gets to the head of line, it will enter one of the N fictitious queues. For large N it is unlikely that successive HOL packets will enter the same fictitious queue, since each packet is equally likely to be destined for any of the N fictitious queue. Thus, we may assume that the *service times* (the delays at the HOL) of different packets are independent.

The input alternates between busy period and idle period over time. Figure 4.3 is a graph depicting a typical busy period. The y-axis is the amount of work remaining (or number of packets in the input queue) $U(t)$. Each busy period is in turn divided into service intervals $X_0, X_1, \ldots, X_i, \ldots$. The packets that are cleared in interval X_i are those that arrive during interval X_{i-1}. Interval X_0 is started off by one or more packets that arrive during the previous idle period.

For consistency, let us assume for the purpose of analysis that packets arrive only at the beginning of the time slots ($t = 1, 2, 3, \ldots$), but service starts and completes slightly after the time-slot boundaries ($t = 1^+, 2^+, 3^+, \ldots$). Thus, X_0 begins in the same time slot as the packets arrive, and those packets that arrive just as X_{i-1} ends

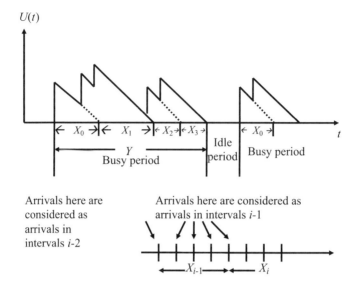

FIGURE 4.3 The busy periods and interpretations for delay analysis of an input queue.

and X_i starts will be considered as having arrived in X_{i-1} and therefore will be served in X_i (see Fig. 4.3). Define

$$Y \stackrel{\text{def}}{=} \text{length of busy period} = X_0 + X_1 + \cdots,$$

$$N_i \stackrel{\text{def}}{=} \text{number of arrivals during interval } X_i,$$

$$S \stackrel{\text{def}}{=} \text{service time at the HOL}.$$

Suppose that $N_{i-1} = n$ packets arrive during interval $i-1$ with $X_{i-1} = x$. Then X_i is the sum of the service times of the n packets. The service times of the packets are independent and identically distributed. Thus, the generating function of their sums is the product of their generating functions. We can write

$$X_i(z \mid X_{i-1} = x, N_{i-1} = n) = S^n(z).$$

Unconditioning on N_{i-1}, we have

$$X_i(z \mid X_{i-1} = x) = \sum_{n=0}^{\infty} S^n(z) \Pr\{N_{i-1} = n \mid X_{i-1} = x\}.$$

Notice that we allow the possibility of $n = 0$ and $X_i = 0$. We note that the above summation is similar to the computation of the conditional generating function of

N_{i-1} except that z has been replaced by $S(z)$. Therefore,

$$X_i(z \mid X_{i-1} = x) = N_{i-1}[S(z) \mid X_{i-1} = x].$$

Unconditioning on X_{i-1} gives

$$X_i(z) = \sum_{x=0}^{\infty} N_{i-1}[S(z) \mid X_{i-1} = x] \Pr\{X_{i-1} = x\}.$$

Let

$$K \stackrel{\text{def}}{=} \text{number of arrivals in an arbitrary time slot}.$$

We shall assume that arrivals at different time slots are independent and identically distributed. The total number of arrivals over x time slots therefore has the generating function $K^x(z)$. Noting that $N_{i-1}(z \mid X_{i-1} = x) = K^x(z)$, we can write

$$X_i(z) = \sum_{x=0}^{\infty} K^x[S(z)] \Pr\{X_{i-1} = x\} = X_{i-1}\{K[S(z)]\}. \tag{4.19}$$

The physical meaning of the above relationship can be unveiled by expanding the generating functions in terms of the power of z. Intuitively, since X_i is the sum of the service durations of those packets arriving in interval $i - 1$, we can expect X_{i-1} to correlate positively with X_i. Each time slot in interval $i - 1$ produces a random of K packets to be served in interval i, and each of these packets has a random service time of S. Equation (4.19) is simply a succinct way of summarizing this dependency. We shall come back to use the above relationship later. Let us now focus on one particular packet that arrives in service interval i. Define

$W \stackrel{\text{def}}{=}$ waiting time of the packet before joining HOL,

$R_i \stackrel{\text{def}}{=}$ the residual time of interval i upon the arrival of the packet,

$M_i \stackrel{\text{def}}{=}$ the number of arrivals in the same interval but in time slots prior to the one in which the packet of focus arrives,

$L \stackrel{\text{def}}{=}$ the number of arrivals in the same time slot as the packet of focus that will be served before it.

In queueing theory, the waiting time customarily refers to the time from the arrival to the start of service, and the delay is the waiting time plus the service time. Thus, defining the "start of service" as the time the packet proceeds to the HOL, the delay here is the waiting time plus time spent at the HOL. The definitions of the other

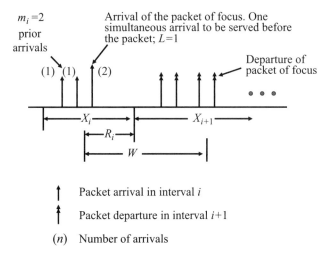

FIGURE 4.4 Illustration of the meanings of the random variables used in the delay analysis of an input queue.

random variables are illustrated in Fig. 4.4. Note also that for a Bernoulli arrival process at most one packet can arrive in any given time slot, and so $L = 0$. However, such is not the case for the fictitious queue. In order that we can borrow the results here for the analysis of the fictitious queue later, the general dependency of W on L is retained intentionally in the following.

Suppose $M_i = m$, $L = l$, and $R_i = r$ are given. Then the waiting time is the sum of the residual time of interval i and service times of the $m + l$ packets. We have

$$W(z \mid i, X_i = x, R_i = r, M_i = m, L = l) = z^r S^{m+l}(z).$$

We assume the arrivals on different time slots are independent and therefore L and M_i are independent. Unconditioning on M_i and L, we have

$$W(z \mid i, X_i = x, R_i = r) = z^r \sum_{l=0}^{\infty} \sum_{m=0}^{\infty} S^{m+l}(z) \Pr\{M_i = m \mid i, X_i = x, R_i = r\}$$
$$\times \Pr\{L = l \mid i, X_i = x, R_i = r\}$$
$$= z^r L[S(z) \mid i, X_i = x, R_i = r]$$
$$\times M_i[S(z) \mid i, X_i = x, R_i = r]$$
$$= z^r L[S(z)] \, M_i[S(z) \mid i, X_i = x, R_i = r], \quad (4.20)$$

where we note in the last line that the L relates to the arrivals in one particular time slot and therefore is independent of the lengths of X_i and R_i. To uncondition on X_i

PERFORMANCE OF SIMPLE SWITCH DESIGNS   107

and R_i, we first find

$$\Pr\{X_i = x, R_i = r \mid i\} = \Pr\{R_i = r \mid i, X_i = x\} \Pr\{X_i = x \mid i\}$$

$$= \frac{1}{x} \times \frac{x}{E[X_i]} \Pr\{X_i = x\}. \quad (4.21)$$

In the above, given that the packet arrives in a service interval of length x, it is equally likely for the residual time R_i to be 0, 1, 2, ..., or $x - 1$. Also, it is more likely for the packet to fall into a larger interval than a smaller one. Therefore, the interval size observed by the packet is not identically distributed as X_i. To see this, let us consider a very long stretch of time with a large number of busy periods. Let N be the number of busy periods we observe. For large N, the number of busy periods with $X_i = x$ is $N \Pr\{X_i = x\}$. The expected number of arrivals in these intervals is $\rho x N \Pr\{X_i = x\}$. The expected number of arrivals in intervals i of all N busy periods, regardless of X_i, is $\sum_{x=0}^{\infty} \rho x N \Pr\{X_i = x\}$. Thus, the probability that a packet entering in an ith interval will find the interval to be of length x is

$$\Pr\{X_i = x \mid i\} = \lim_{N \to \infty} \frac{N \Pr\{X_i = x\} x \rho}{\sum_{x=0}^{\infty} N \Pr\{X_i = x\} x \rho} = \frac{x \Pr\{X_i = x\}}{E[X_i]}.$$

We now use (4.21) to uncondition (4.20)

$$W(z \mid i) = \sum_{x=1}^{\infty} \sum_{r=0}^{x-1} z^r L[S(z)] M_i[S(z) \mid i, X_i = x, R_i = r] \times \frac{\Pr\{X_i = x\}}{E[X_i]}$$

$$= \frac{L[S(z)]}{E[X_i]} \sum_{x=1}^{\infty} \Pr\{X_i = x\} z^{x-1} \sum_{r=0}^{x-1} \frac{M_i[S(z) \mid i, X_i = x, R_i = r]}{z^{x-1-r}}. \quad (4.22)$$

Recall that K is the random variable for the number of arrivals in one time slot. Given $X_i = x$ and $R = r$, there are $x - r - 1$ earlier time slots in which the M_i packets can arrive. Therefore, $M_i[z \mid i, X_i = x, R_i = r] = K^{x-r-1}(z)$. Substituting into (4.22) and simplifying, we have

$$W(z \mid i) = \frac{L[S(z)]}{E[X_i]} \sum_{x=1}^{\infty} \frac{\Pr\{X_i = x\}\{K^x[S(z)] - z^x\}}{K[S(z)] - z}$$

$$= \frac{L[S(z)]}{E[X_i]} \frac{(X_i\{K[S(z)]\} - X_i(z))}{K[S(z)] - z}$$

$$= \frac{L[S(z)]}{E[X_i]} \frac{(X_{i+1}(z) - X_i(z))}{K[S(z)] - z}, \quad (4.23)$$

where in the last line we have made use of the relationship (4.19). Given that the packet enters in a busy period, the probability that it falls into interval i is $E[X_i]/E[Y]$.

Unconditioning on i, we have

$$W(z \mid \text{enter in busy period}) = \sum_{i=0}^{\infty} W(z \mid i) \frac{E[X_i]}{E[Y]}$$

$$= \frac{L[S(z)]}{E[Y]\{K[S(z)] - z\}} \sum_{i=0}^{\infty} [X_{i+1}(z) - X_i(z)]$$

$$= \frac{L[S(z)]}{E[Y]\{K[S(z)] - z\}} [1 - X_0(z)], \qquad (4.24)$$

where we have made use of the fact that $\lim_{i \to \infty} X_i(z) = 1$ (since with probability one a busy period eventually will terminate and therefore $\lim_{i \to \infty} \Pr\{X_i = 0\} = 1$). For the input queue, X_0 is the service time of one packet because at most one packet can arrive in a time slot. If multiple packets can arrive simultaneously, such as the case with the fictitious queue, things are more complicated. To analyze the general situation, let N_{-1} be the number of simultaneous arrivals that start off a new busy period. Given that $N_{-1} = n$, the subsequent X_0 consists of the service times of n packets. Thus,

$$X_0(z \mid N_{-1} = n) = S^n(z).$$

Now N_{-1} is the number of arrivals in a time slot given that there is at least one arrival. Therefore,

$$\Pr\{N_{-1} = n\} = \begin{cases} 0, & \text{if } n = 0, \\ \dfrac{\Pr\{K = n\}}{1 - \Pr\{K = 0\}}, & \text{if } n \geq 1. \end{cases}$$

This gives

$$X_0(z) = \sum_{n=1}^{\infty} X_0(z \mid N_{-1} = n) \frac{\Pr\{K = n\}}{1 - \Pr\{K = 0\}}$$

$$= \sum_{n=1}^{\infty} S^n(z) \frac{\Pr\{K = n\}}{1 - \Pr\{K = 0\}}$$

$$= \frac{K[S(z)] - \Pr\{K = 0\}}{1 - \Pr\{K = 0\}}. \qquad (4.25)$$

Substituting the above into (4.24), we get

$$W(z \mid \text{enter in busy period}) = \frac{L[S(z)]}{E[Y]\{K[S(z)] - z\}} \left\{ \frac{1 - K[S(z)]}{1 - \Pr\{K = 0\}} \right\}. \qquad (4.26)$$

There are several ways to find $E[Y]$. One of the exercises asks you to derive it based on a direct probabilistic argument. If our derivation $W(z \mid \text{enter in busy period})$ is correct up to this point, we can use a normalization method. Specifically, any generating function substituted with $z = 1$ is equal to 1. The normalization method should always be used with care because the mistakes in the derivation of $W(z \mid \text{enter in busy period})$, if any, will not be uncovered; whereas if we use both the direct and normalization methods, we can verify that the two results are the same. Knowing that this checking is done in the exercise, we now normalize $W(z \mid \text{enter in busy period})$, giving

$$E[Y] = \frac{S'(1)K'(1)}{[1 - S'(1)K'(1)](1 - \Pr\{K = 0\})} = \frac{\overline{S}\rho}{(1 - \overline{S}\rho)(1 - \Pr\{K = 0\})}. \quad (4.27)$$

Substituting the above into (4.24) gives

$$W(z \mid \text{enter in busy period}) = \frac{1 - \overline{S}\rho}{\overline{S}\rho} \left\{ \frac{L[S(z)]\{1 - K[S(z)]\}}{K[S(z)] - z} \right\}.$$

Over a long stretch of time t, ρt packets arrive, and each of them contributes an average of $\overline{S}$ to the busy periods. Thus, the fraction of time the system is in busy periods is $\lim_{t \to \infty} \rho t \overline{S}/t = \rho \overline{S}$. Therefore, with probability $\rho \overline{S}$ a packet will enter during a busy period and with probability $1 - \rho \overline{S}$ it will enter during an idle period, in which case the generating function of the waiting time is $L[S(z)]$. We have

$$W(z) = (1 - \rho \overline{S})L[S(z)] + \rho \overline{S} W(z \mid \text{enter in busy period})$$

$$= (1 - \rho \overline{S})L[S(z)] \left\{ \frac{1 - z}{K[S(z)] - z} \right\}. \quad (4.28)$$

For the input queue with Bernoulli arrival process,

$$L(z) = 1, \quad (4.29)$$

$$K(z) = 1 + (z - 1)\rho. \quad (4.30)$$

Substituting into (4.28), we get

$$W(z) = (1 - \rho \overline{S}) \frac{1 - z}{(1 - z) + [S(z) - 1]\rho}. \quad (4.31)$$

The mean waiting time $\overline{W}$ can be obtained by differentiating (4.31). This yields

$$\overline{W} = \frac{\rho \overline{S(S - 1)}}{2(1 - \rho \overline{S})}. \quad (4.32)$$

Equations (4.28) and (4.31) are only valid when the input queue is FIFO (the fictitious queue does not have to be). Equation (4.32), on the other hand, applies even when the

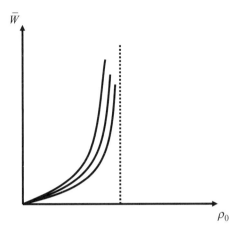

FIGURE 4.5 Different contention-resolution policies have different waiting time versus load relationships, but a common maximum load at which waiting time goes to infinity.

service discipline is not FIFO; it can be shown by focusing on the remaining work $U(t)$ (see Fig. 4.3) that all work-conserving systems have the same average waiting time.

The service discipline of the fictitious queue (i.e., the contention-resolution policy), however, does affect $\overline{W}$ through the second moment of the service time $\overline{S(S-1)}$. Figure 4.5 shows qualitatively the relationship between waiting times of different service disciplines and load. As shown, all service disciplines have the same maximum load at which waiting time goes to infinity (recall that our saturation throughput calculation did not assume any particular service discipline). However, their waiting times at load below the maximum load differ.

Let us focus on a FIFO fictitious queue. We want to make use of (4.28) for the analysis here, too. In the fictitious queue, each packet takes exactly one slot to be cleared. Thus, the service time S in (4.28) should be replaced as

$$S(z) \to z \quad \text{and} \quad \overline{S} \to 1.$$

Also, S in the input queue is the delay in the fictitious queue. The waiting time in the fictitious queue is $S-1$, Thus, replacing the waiting time in (4.28) with the waiting time in the fictitious queue means

$$W(z) \to \frac{S(z)}{z}.$$

Equation (4.28) now takes the form

$$S(z) = (1-\rho)\frac{z(1-z)L(z)}{K(z)-z}. \tag{4.33}$$

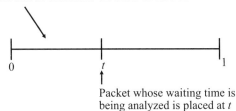

FIGURE 4.6 A unit line for determining the order of service among simultaneously arriving packets.

We argued previously in the analysis of maximum throughput that the arrival process is Poisson when N is very large and the system is saturated. A similar argument can be used to show that the arrival process is still Poisson even when the system is not saturated. Thus,

$$K(z) = e^{(z-1)\rho}. \tag{4.34}$$

Several packets can arrive simultaneously when the arrival is Poisson. Thus, L is not always zero. Suppose simultaneous packets are served in a random order. We can imagine a continuous line of unit length (see Fig. 4.6) onto which the arrivals are placed at random. The arrivals placed before the packet of focus will be served first. Given that the packet whose waiting time is being analyzed is placed at t, the number of packets placed before is Poisson with rate ρt. Thus,

$$L(z \mid t) = e^{(z-1)\rho t}. \tag{4.35}$$

The above is a well-known fact for people who are familiar with the Poisson process, but one of the exercises shows you how to derive it yourself. Given the random nature, t is equally likely to be placed anywhere between 0 and 1. Therefore,

$$L(z) = \int_0^1 e^{(z-1)\rho t} dt = \frac{e^{(z-1)\rho} - 1}{(z-1)\rho}. \tag{4.36}$$

By differentiation, we can obtain $\overline{S} = S'(1)$ and $\overline{S(S-1)} = S''(1)$ from (4.33). After going through the tedious computation, we get

$$\overline{S} = 1 + \frac{\rho}{2(1-\rho)} \tag{4.37}$$

and

$$\overline{S(S-1)} = \frac{\rho(\rho^2 - 4\rho + 6)}{6(1-\rho)^2}. \tag{4.38}$$

Substituting the above into (4.32) and solving, we obtain

$$\overline{W} = \frac{\rho^2(\rho^2 - 4\rho + 6)}{6(1-\rho)(\rho^2 - 4\rho + 2)}. \qquad (4.39)$$

The average delay experienced by a packet from arrival to departure is

$$\overline{D} = \overline{W} + \overline{S} = \frac{\rho^2(\rho^2 - 4\rho + 6)}{6(1-\rho)(\rho^2 - 4\rho + 2)} + \frac{2-\rho}{2(1-\rho)}. \qquad (4.40)$$

As an example, for offered load $\rho_o = \rho = 0.5$, $\overline{S} = 1.5$ and $\overline{W} \approx 1.41$. The average delay $\overline{D} \approx 3$. That is, a packet experiences an average delay of three packet durations in the input queue. We see that the delay is rather small if the offered load is slightly below the maximum throughput. The variance of delay is also an important parameter, especially for real-time communication services. The variance can be obtained from the higher derivatives of $W(z)$ and $S(z)$.

4.1.4 Delay of an Output-Buffered Switch

As mentioned earlier in the chapter, the throughput of a switch can be improved by increasing the group size, which is the number of packets that can access a common output in a given time slot. The throughput as a function of the group size will be studied in the next chapter.

Let us now examine the ideal case in which all N packets (where N is the number of input ports) can access the same output in the same time slot if they so desire. All packets are switched to their destined outputs immediately upon their arrivals and none of them are dropped at the inputs. Therefore, there is no throughput limitation like that in the input-buffered switch. This ideal switch also obviates the need for input buffers. In addition, instead of the fictitious output queues discussed previously, we have real output queues here. These output queues are needed because the output link may not be able to transmit all the simultaneously arriving packets immediately.

The study of the output queues is similar to that of the the fictitious queue. In fact, the justification for the Poisson arrival process is more straightforward for the output queues because its packet sources are the inputs, whose number does not change with time; the packet sources of the fictitious queue are those inputs whose HOL packets were cleared in the previous time slot, and their number may change from time slot to time slot.

Given that the arrival process to each input is Bernoulli with rate ρ_o, the arrival process to an output is binomial: $\Pr\{A = k\} = \binom{N}{k}\left(\frac{\rho_o}{N}\right)^k\left(1 - \frac{\rho_o}{N}\right)^{N-k}$, which approaches the Poisson distribution with rate ρ_o as $N \to \infty$. We can therefore immediately use the result obtained for the fictitious queue, replacing ρ with ρ_o. Note that for the fictitious queue, the carried load ρ may not be the same as the offered load ρ_o because of input queueing; but for the output queue, since all packets are immediately switched to the output with no loss, the offered load is the same as the carried load so long as the output buffer does not overflow.

Let us consider a particular output and assume that it has infinite buffering space. From (4.14), the number of backlogged packets at the beginning of each time slot is given by

$$C(z) = \frac{(z-1)A(z)(1-\rho_o)}{(z-A(z))}, \qquad (4.41)$$

where $A(z) = e^{(z-1)\rho_o}$. Unlike for the fictitious queue in which $\overline{C} \leq 1$ because of the constraint imposed by the input queues, the value of $\overline{C}$ for the output queue depends on ρ_o, and it can be obtained from (4.41) by differentiation:

$$\overline{C} = C'(1) = \rho_o + \frac{\rho_o^2}{2(1-\rho_o)}. \qquad (4.42)$$

The average delay in the output queue is given by the Little's law:

$$\overline{S} = \frac{\overline{C}}{\rho_o} = 1 + \frac{\rho_o}{2(1-\rho_o)}. \qquad (4.43)$$

The delay can also be obtained directly by considering the waiting time. Replacing ρ with ρ_o in (4.33), (4.34), and (4.36), we get

$$S(z) = \frac{z(1-\rho_o)(1-e^{(z-1)\rho_o})}{\rho_o(e^{(z-1)\rho_o} - z)}. \qquad (4.44)$$

From (4.44), we can derive (4.43) and other higher moments of S.

4.2 DESIGN IMPROVEMENTS FOR INPUT QUEUEING SWITCHES

The preceding section showed that input-buffered switch with single FIFO queues, on the contrary, is good for its simplicity but is constrained by its low throughput of 58.6%. It is the result of head-of-line (HOL) blocking at an input queue, forbidding packets other than the HOL one to be transmitted when more than one HOL packet from different inputs contends at the same output (see Fig. 3.28). For throughput improvement, we need to remove the FIFO constraint on the input queues to relieve the HOL blocking.

4.2.1 Look-Ahead Contention Resolution

An earlier scheme for improving the performance of an input-buffered switches is look-ahead selection of packets in FIFO input queues. In the look-ahead scheme, the first w packets of each input will be examined during the contention-resolution

TABLE 4.2 Maximum Throughput of a 32 × 32 Input-Buffered Switch for Various Contention Window Sizes

w	ρ^*
1	0.59
2	0.70
3	0.76
4	0.80
5	0.83
6	0.85
7	0.87
8	0.88
∞	1 (by analytical argument)

phase. The contention-resolution procedure is divided into w cycles. The first cycle is like the original scheme and at the end of it a set of winning packets out of the HOL packets are determined. In the second cycle, the scheme looks at the second set of packets of the inputs that have lost contention in the first cycle to see if they are destined for any outputs that are as yet unclaimed. A second set of winning packets is then determined. In this way, each cycle looks one packet deeper into the queue than the previous cycle. After w cycles, the w sets of winning packets are then sent to their outputs. Each input still sends at most one packet, and each output still receives at most one packet, in a time slot. Certainly, as w increases, we expect the maximum throughput to increase accordingly.

There is no known exact analytical method for deriving the maximum throughput for $w > 1$. Table 4.2 lists the simulation results for various values of w for a 32 × 32 switch. In general, it can be easily shown (see Problem 4.12) that as $w \to \infty, \rho^* \to 1$, assuming the input queues have infinite amounts of buffers (note that this is another example that the analysis is simple when a relevant system parameter is either small or very large; but it becomes difficult when the parameter is somewhere in between). Table 4.2 shows, however, that ρ^* approaches 1 very slowly.

It should be pointed out the throughput improvement does not come without cost. If the three-phase sort-banyan network switch is used, the first two phases must be performed w times, resulting in higher contention-resolution overhead. Let ϵ be the overhead associated with one round of contention resolution measured in time slot. The actual throughput is then $\rho^*/(1 + w\epsilon)$. As w is increased, a point may be reached at which the throughput improvement is more than offset by the contention-resolution overhead, bringing about a lower actual throughput.

A second factor that may decrease the throughput improvement is the input traffic characteristics. We have assumed that successive packets on an input are independently and equally likely to be destined for any output. In many situations, successive packets may not be independent and are destined for only a small subset of the outputs. The input look-ahead scheme will yield smaller throughput improvement in

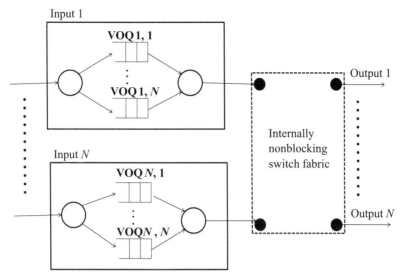

FIGURE 4.7 Virtual output queueing scheme.

such situations because looking deeper into the queue may often find packets with the same destination.

4.2.2 Parallel Iterative Matching

In order to get rid of HOL blocking in an input-buffered switch, we can modify the above buffer organization in such a way that every input port maintains N logical queues, named virtual output queues (VOQs), instead of keeping a single FIFO queue. This is illustrated in Fig. 4.7. In this setup, each of these N logical queues is dedicated for one of the N output ports so that packets with different destinations can take part in the contention resolution simultaneously, thus entirely eliminating HOL blocking at input ports. This technique is known as virtual output queueing because the switch behaves as if it is an output queueing switch. In order to further increase the switch throughput, a scheduler is needed for resolving input and output conflicts with the existing connection requests. The arbitration of the scheduler is then passed to the switching fabrics, and connection among input and outputs will be set up accordingly.

The question that arises now is how we can schedule the transmission of the queued packets so as to achieve a high throughput. In actuality, choosing a set of packets for transmission in an input-buffered switch can be formulated as a matching problem in a bipartite graph as shown in Fig. 4.8, where inputs and outputs of the switch can be considered as the two sets of nodes for matching. It is obvious that if we could always maximize the number of matches for the matching requests given in the graph, then we can achieve an optimum throughput performance for the switch. However, owing to the fact that finding a maximum matching is computationally complex, an

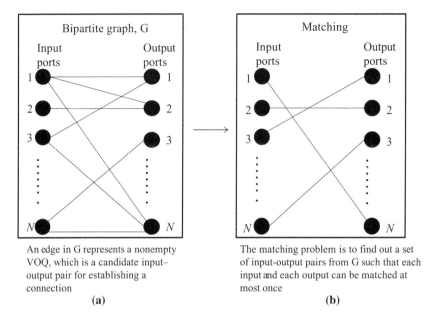

FIGURE 4.8 Relationship between switch scheduling and bipartite graph matching.

algorithm, called parallel iterative matching (PIM), is proposed to find a maximal matching for the graph. Note that a maximal matching is a matching in which no unmatched input has a queued packet destined for an unmatched output (i.e., no more matches can be made without rearranging the existing matching), while a maximum matching is one with the maximum number of matching pairs of inputs and outputs.

PIM uses parallelism, randomness, and iteration to find a maximal matching between the inputs that have buffered packets for transmission and the outputs that have queued packets (at the inputs) destined for them. Based on this maximal matching, we can determine which inputs transmit packets over the nonblocking switch to which outputs in the next time slot. Specifically, this matching algorithm iterates the following three steps until a maximal matching is found or a fixed number of iterations are performed.

Request:
Each unmatched input sends a request to every output for which it has a queued packet.

Grant:
If an unmatched output receives any requests, it grants to one by randomly selecting a request uniformly over all requests.

Accept:
If an input receives grants, it accepts one by selecting an output among those that granted to this input.

DESIGN IMPROVEMENTS FOR INPUT QUEUEING SWITCHES 117

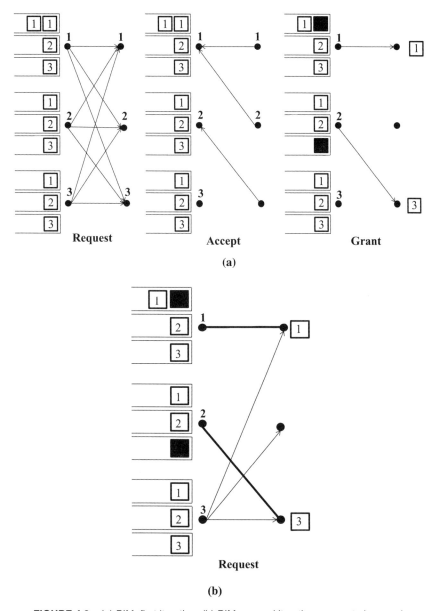

FIGURE 4.9 (a) PIM: first iteration; (b) PIM: second iteration, request phase only.

Figure 4.9(a) illustrates how PIM works in one iteration. First, in the request phase, every unmatched input port sends a request to every output port whenever there is backlogged packet for it. As depicted in Fig. 4.9, input port 1 sends requests to output ports 1, 2, and 3 simultaneously. Then, in the grant phase, if an unmatched output receives multiple requests, it grants one randomly. For example, output 2 grants input 1 in the grant phase below. Finally, in the accept phase, an unmatched input will accept one grant randomly on receiving multiple grants. In this case, input 1 accepts the grant from output 1 and input 2 accepts that from output 3. After the first iteration, input 3 remains unmatched, and therefore, it will send requests to output ports 1, 2, and 3 in the request phase of the second iteration (see Fig. 4.9(b)). Since outputs 1 and 3 have already been matched, they will neglect those requests and postpone scheduling of the corresponding packets to the next time slot. Only output 2 will consider the request sent by input 3 in this iteration. Thus, two iterations are required for PIM to find a maximal matching in this example.

In the following, we analyze the maximum throughput of an input-buffered switch with respect to the number of iterations of PIM invoked. Let us assume that the switch is very large in size and all the logical queues are saturated so that as soon as a HOL packet is cleared, its position is taken by a subsequent packet in the next time slot. In this way, the switch is loaded at its maximum capacity and each output port receives a request from every unmatched input port during the request phase.

Consider that in the first iteration, the output port receive a request from each of N input ports and will select one of them randomly to grant. The probability that a particular input port does not receive any grant is

$$\Pr\{\text{no grant received}\} = \left(1 - \frac{1}{N}\right)^N. \tag{4.45}$$

Thus, the probability that an input port will receive at least one grant for matching in the first iteration is given by

$$\hat{\rho} = 1 - \Pr\{\text{no grant received}\} = 1 - \left(1 - \frac{1}{N}\right)^N. \tag{4.46}$$

When N is large,

$$\hat{\rho} \approx 1 - e^{-1} = 0.632. \tag{4.47}$$

This is the maximum throughput of an input-buffered switch with invoking only one iteration of PIM.

After the first iteration, there are totally $\hat{N} = (1 - \hat{\rho})N$ input–output pairs not yet to be matched. We can use the same approach to find the number of pairs that could be matched up in the second iteration. That is,

$$\left[1 - \left(1 - \frac{1}{\hat{N}}\right)^{\hat{N}}\right] \cdot \hat{N} = \hat{\rho} \cdot \hat{N} = \hat{\rho} \cdot (1 - \hat{\rho})N. \tag{4.48}$$

TABLE 4.3 Maximum Throughput for Input-Buffered Switches Invoking Different Number of Iterations of PIM

Number of iterations	1	2	3	4	5
Maximum throughput	0.632	0.865	0.950	0.981	0.993

In other words, the proportion of pairs that can be matched up, or the maximum throughput that can be achieved in the first two iterations of PIM, is

$$\text{max. throughput}_2 = \hat{\rho} + \hat{\rho}(1 - \hat{\rho}), \tag{4.49}$$

where $\hat{\rho} = 0.632$.

Similarly, for $K \geq 2$ iterations, it is easy to show that the maximum throughput is

$$\text{max. throughput}_K = \hat{\rho} \sum_{k=1}^{K} (1 - \hat{\rho})^{k-1}, \tag{4.50}$$

when N is sufficiently large. The analytical results of the maximum throughput are shown in Table 4.3 for different number of iterations of PIM invoked.

4.3 DESIGN IMPROVEMENTS BASED ON OUTPUT CAPACITY EXPANSION

Output capacity expansion refers to a class of design strategies that alleviate the contention problem by allowing more than one packet to reach the same output in the same time slot. Therefore, the throughput of a switch can be much increased.

4.3.1 Speedup Principle

Perhaps the most straightforward output capacity expansion strategy is to speed up the operation of the switch so that each switch cycle is equal to a fraction of a time slot. For example, by operating an internally nonblocking switch at twice the external link rate, each switch cycle is equal to half a time slot. In the first half of a time slot, a contention-resolution scheme elects the winning packets for output access. In the second half, the losing packets, together with the new HOL packets, attempt to access their outputs. It is easily seen that the maximum throughput is now 2×0.586, and since the offered load by definition cannot be more than 1, the input queues will not saturate.

For such a switch, a packet incurs delays at an input queue and an output queue (see Fig. 4.10). The output queue is needed because two packets may arrive at an output in a time slot, during which only one packet can be transmitted on the output link. The input queue is still required because there remains the chance that more than two HOL input packets are destined for the same output.

120 SWITCH PERFORMANCE ANALYSIS AND DESIGN IMPROVEMENTS

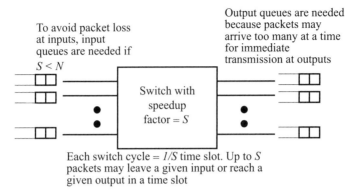

FIGURE 4.10 The speedup principle.

The delay may still be too large even though the throughput limitation has been removed. To reduce the delay further, we can speed up the switch operation by more than two times. However, there is a practical limit on the speedup factor. For example, assuming a 53-byte packet size and an input or output link rate of 155 Mbps, each time slot is about 2.7 μs, which is already straining the switch design even without speedup.

A method to achieve the effect of speedup without actually speeding up switch operation is to use multiple switches. Figure 4.11(a) shows a duplex switch that achieves a speedup factor of 2 by directing packets to two switches alternately. There are two switch cycles to a time slot, one for each switch. The switch cycle in the second switch lags behind that in the first switch by an amount of time needed for the first switch to resolve packet contention. After the contention-resolution process, the winning packets then access their outputs via the first switch, and the losing packets and the new HOL packets contend with each other to access their outputs via the second switch. In this way, the two switches take turns switching the packets.

Figure 4.11(b) depicts an alternative that uses *packet slicing*. For speedup of two, each packet is cut into half, with the complete output address attached to each half. The two halves are sent simultaneously into two switches that work at the same speed as the external link. Thanks to its half size, each packet takes approximately half a time slot to traverse the switch.

For comparison with other switches, we may express the complexity of a speedup switch by the product of the speedup factor and the complexity of a similar switch without speedup. Suppose that the complexity measure is the crosspoint count and that the underlying switch is Batcher–banyan. Then the total crosspoint count is $\frac{N}{4} \log_2 N (\log_2 N + 1) + \frac{N}{2} \log_2 N$. For a speedup factor of S, the order of complexity for a speedup switch is $SN \log_2^2 N$ for large N. Certainly, switch performance improves with larger S, but the complexity also becomes higher. From an engineering viewpoint, the problem is to find the minimum complexity such that certain performance criteria, such as delay and throughput, can be satisfied. We will find this theme recurring again and again in this chapter.

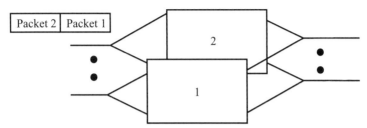
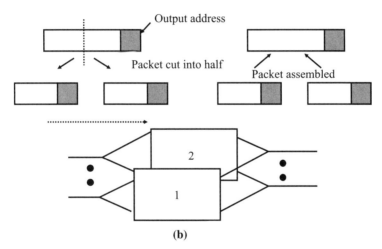

FIGURE 4.11 Methods for achieving speedup effect without speeding up switch operation: (a) using multiple switches; (b) using packet-slicing concept.

4.3.2 Channel-Grouping Principle

The channel-grouping principle is another output capacity expansion method. In the speedup method, it is possible to send out S packets from an input in a time slot (e.g., when there are S packets in a particular input queue and no packet in all other input queues). Channel grouping refers to a method in which an input may send out at most one packet, but multiple packets may be received by an output if these packets are from different inputs. The number of packets that can be received in a time slot is referred to as the *group size*.

Generally, given that the group size R is equal to the speedup factor S, the switch that employs channel grouping is less complex than the one that employs speedup. On

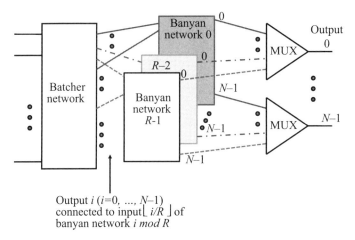

FIGURE 4.12 A Batcher–R-banyan network that implements the channel-grouping principle.

the other hand, the speedup switch has better performance because of its capability to switch multiple packets of an input within the same time slot.

4.3.2.1 Channel Grouping Using Batcher–banyan Networks

A Batcher–R-banyan network that utilizes the channel-grouping concept is shown in Fig. 4.12. As shown, the structure consists of a Batcher network connected to R parallel banyan networks. The outputs of the Batcher network are connected to the inputs of the R banyan network in an interleaving fashion as shown in the figure: the first output of the Batcher network is connected to the first input of the first banyan network, the second output to the first input of the second banyan network, and so on. In general, the output i ($i = 0, 1, 2, \ldots, N - 1$) of the Batcher network is connected to input $\lfloor i/R \rfloor$ (the integer part of i/R) of banyan network $i \bmod R$.

If at most R input packets are addressed to the same destination (output of the overall switch), then the Batcher network and the interleaving connections between it and the banyan networks ensure that the input packets to each banyan network are sorted, concentrated, and have nonconflicting output addresses. The banyan networks are therefore nonblocking. Outputs i ($i = 0, \ldots, N - 1$) of different banyan networks correspond to the same logical address and they are collected and fed into a multiplexer for transmission on the output port.

To ensure that no more than R packets are addressed to the same output, the underlying contention-resolution scheme must be modified. Consider for example the three-phase scheme in which the Batcher network is used to resolve contention. At the end of the first phase, output i of the Batcher network examines output $i - R$ (as opposed to output $i - 1$ without channel grouping) for possible conflict. If the destination addresses of these two packets are the same, then at least $R + 1$ packets have the same destination address, thanks to the sorting performed by the Batcher network. That is, packets at outputs $i - R$ to i all have the same destination addresses.

Output i's packet is then declared a loser and queued at its input at the Batcher network, waiting for the next switch cycle to reattempt output access. On the other hand, if the destination addresses of the packets at outputs i and $i - R$ are different, the packet at output i is declared a winner and will be switched in the third phase. In this way, no more than R packets for the same destination will be declared the winner, thus ensuring the nonblocking operation in the third phase.

Instead of the R banyan networks, we may also use the expansion banyan network structure shown in Fig. 4.13. In this configuration, each output of the Batcher network is connected to a 1-to-R expander. The outputs of the N expanders are in turn connected to R truncated banyan networks. Compared to the preceding structure, this structure has the advantage that the outputs belonging to the same group are adjacent to each other. Recall that the packets from each output group must be multiplexed and buffered in an output queue because the output link may not be able to transmit all arriving packets immediately. If each truncated banyan network is implemented in a chip, then the multiplexer and the output queue associated with each group can also be implemented in the same chip.

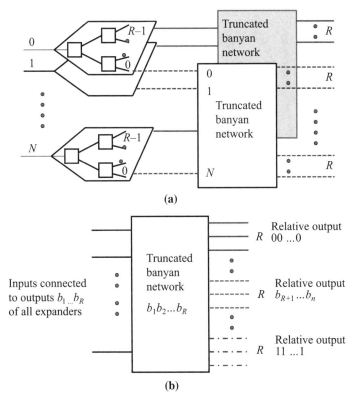

FIGURE 4.13 (a) The expansion banyan network; (b) labeling of the truncated banyan network and its output groups.

The expander has the structure of a binary tree. Routing in the expander is similar to that in an ordinary banyan network. Assuming R is a power of 2, a 1-to-R expander has $r = \log_2 R$ stages and it uses the r most significant bits of the output address for routing. A truncated banyan network is a banyan network with the last r stages removed and it uses the remaining $s = \log_2 N/R$ bits for routing. Recall that as a packet travels through a banyan network, it sees a smaller and smaller subnetwork, and the outputs it can reach are restricted to a smaller and smaller subset. If it is not necessary to reach any particular output of an output group (i.e., any output within a group will do), then there is no need to travel farther through the banyan network beyond the stages needed to distinguish between the different groups, hence the idea behind truncation.

To see why the expander truncated banyan network is nonblocking, first note that no conflict occurs in the expanders since at most one packet travels through each of them. Packets that exit from expander outputs $b_1 b_2 \cdots b_r$ (labeled in a binary fashion) are for destinations $b_1 b_2 \cdots b_r 00 \cdots 0, b_1 b_2 \cdots b_r 00 \cdots 1, \ldots$, or $b_1 b_2 \cdots b_r 11 \cdots 1$. These packets are routed to a particular truncated banyan network labeled $b_1 b_2 \cdots b_r$, as shown in Fig. 4.13(b). The input packets to any particular truncated banyan network are monotone and concentrated, thanks to the preceding Batcher network. The proof of the following theorem completes our argument.

Theorem 4.1. The R-truncated-banyan network is nonblocking if the active inputs $x_1, \ldots, x_n$ ($x_j > x_i$ if $j > i$) and their corresponding output addresses $y_1, \ldots, y_m$ satisfy the following:

1. *Limited output conflict*: At most R packets are destined for any given output.
2. *Monotonic outputs*: $y_1 \leq y_2 \leq \cdots \leq y_m$ or $y_1 \geq y_2 \geq \cdots \geq y_m$.
3. *Concentrated inputs*: Any input between two active inputs is also active. That is, $x_i \leq w \leq x_j$ implies input w is active.

Proof. The output address of a packet has only $s = n - r$ remaining bits for routing. The *relative* output address of an output group is therefore labeled $b_{r+1} \cdots b_n$, where $n = \log_2 N$. The complete address of the output group, $b_1 \cdots b_r b_{r+1} \cdots b_n$, is obtained by concatenating the truncated banyan network label $b_1 b_2 \cdots b_r$ with the relative output address $b_{r+1} \cdots b_n$. To show that conflict does not occur in any of the s stages of the truncated banyan network, let us assume the contrary: two packets at a node in stage k, $1 \leq k \leq s$, want to access the same outgoing link. Let the input and relative output addresses of the packet 1 be $x = a_n \cdots a_1$ and $y = b_{r+1} \cdots b_n$, respectively, and those of packet 2 be $x' = a'_n \cdots a'_1$ and $y' = b'_{r+1} \cdots b'_n$. As in the proof of Theorem 3.1, their conflict in stage k implies

$$a_{n-k} \cdots a_1 = a'_{n-k} \cdots a'_1 \qquad (4.51)$$

and

$$b_{r+1} \cdots b_{r+k} = b'_{r+1} \cdots b'_{r+k}. \qquad (4.52)$$

Since the packets are concentrated, the total number of packets between inputs x and x', inclusively, is

$$|x' - x| + 1.$$

By condition 1, at most R packets can be destined for any particular output address. By condition 2, the largest and the smallest addresses of the $|x' - x| + 1$ packets must be y and y', respectively, or y' and y, respectively. Hence, the number of distinct output addresses among the $|x' - x| + 1$ packets is

$$|y - y'| + 1 \geq (|x' - x| + 1)/R. \tag{4.53}$$

But according to (4.51) and (4.52),

$$\begin{aligned} |x' - x| &= |a'_n \cdots a'_1 - a_n \cdots a_1| \\ &= |a'_n \cdots a'_{n-k+1} 0 \cdots 0 - a_n \cdots a_{n-k+1} 0 \cdots 0| \\ &\geq 2^{n-k} \end{aligned} \tag{4.54}$$

and

$$\begin{aligned} |y' - y| &= |b'_{r+1} \cdots b'_n - b_{r+1} \cdots b_n| \\ &= |b'_{r+k+1} \cdots b'_n - b_{r+k+1} \cdots b_n| \\ &\leq 2^{n-r-k} - 1 = \frac{2^{n-k}}{R} - 1 \end{aligned} \tag{4.55}$$

Combining (4.54) and (4.55), we have

$$|y' - y| \leq |x' - x|/R - 1. \tag{4.56}$$

But (4.53) and (4.56) contradict each other and therefore the packet conflict could not have occurred given that the three conditions are satisfied. □

To illustrate that channel-grouped switches are less complex than speedup switches when $R = S$, let us compare the number of crosspoints needed. For the Batcher–truncated banyan network, the N expanders have $N(R - 1)$ crosspoints, the R truncated banyan networks have $\frac{RN}{2} \log_2 \frac{N}{R}$ crosspoints, and the Batcher network has $\frac{N}{4} \log_2 N(\log_2 N + 1)$ crosspoints. If $R = S < N$, which is the case for large N since R and S need only to be fixed at a constant value independent of N to achieve satisfactory performance (see the next subsection), then the complexity of the Batcher network dominates. Since only one Batcher network is needed here and S Batcher networks are needed in the speedup scheme, we conclude that the channel-grouped switch is less complex than the speedup switch when the underlying switch structures are Batcher–banyan.

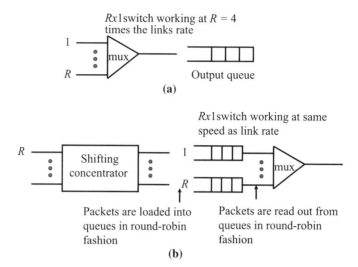

FIGURE 4.14 (a) Multiplexer and output queue at an output of a channel-grouped switch. To accommodate simultaneous packet arrivals, the multiplexer must work R times faster than link rate. (b) An implementation of a logical FIFO queue such that the multiplexer only have to work at same speed as link rate.

Both speedup and channel-grouping schemes require a multiplexer and an output queue in each output. Figure 4.14(a) shows an implementation that employs an $R \times 1$ switch that runs at R times the link speed. In each time slot, the multiplexer scans across all the inputs and switches the packets, if any, one by one to the output queue. The access speed of the memory that implements the queue must also be fast enough to keep up with the scanning.

Figure 4.14(b) is an alternative scheme that achieves the same function without employing speedup. It consists of an $R \times R$ shifting concentrator (described in Section 3.2.4) followed by R queues and a multiplexer. For explanation, let us pretend that the structure does not have the shifting concentrator for the time being. The multiplexer works at the same speed as the output link. It scans the queues in a round-robin fashion and in each time slot and selects a packet from one of the queues for transmission on the output link. In R time slots, each queue has at least one chance to send a packet. It is possible for successive packets of a communication session to arrive at different output queues due to the interference of packets from other inputs: the number of other packets contending with the session's packets, and their inputs, may change from time slot to time slot. If by chance a packet of the session reaches a queue that is shorter than the one reached by an earlier packet, it will be transmitted first, causing the packets to be received out of order at the destination.

The purpose of the shifting concentrator is to maintain the sequence order of the packets. It concentrates and loads the arriving packets in a round-robin fashion into the queues. In the very first time slot of the switch operation, the shifting concentrator loads the arriving packets into queues 1, 2, and so on. Similarly, packets will be read out from the queues by the multiplexer starting from queue 1 in a round-robin manner. Suppose that queue i is the last queue to which a packet is loaded and that $K \leq R$ packets arrive in the next time slot. Then the shifting concentrator sends the newly arriving packets to queues $i+1, \ldots, i+K$ (mod R). In the subsequent time slot, packets will be loaded into the queues starting from queue $i+K+1$ (mod R). Meanwhile the multiplexer reads packets out of the queues in a round-robin fashion, and if it finds the queue to be read empty in a time slot, it simply waits until the shifting concentrator loads a packet into the queue—in this situation, the other queues are also empty, and the queue is the one to be loaded next. It can be easily seen that the R queues are logically equivalent to a FIFO queue, because it is impossible for a packet of a session to depart from the switch before an earlier packet does. As for the shifting concentrator, it has already been shown in Section 3.2.4 that it can be implemented using a reverse banyan network.

4.3.2.2 Performance of Switches Based on Channel Grouping

Let us consider a waiting system in which if more than R packets are destined for the same output, the excess packets are buffered at the inputs. The special case $R = 1$ has been considered in the preceding chapter. The basic steps are the same in the general case. Therefore, we will go through the derivations rather briefly.

We made the same assumption that each packet is equally likely to be destined for any of the N outputs and packet destinations are independent. To derive the maximum throughput, as before, consider the fictitious queue that consists of HOL packets destined for a particular output. Let C_m be the number of packets in the queue and A_m be the number of arriving packets at the start of time slot m. The system evolves according to the equation

$$C_m = \max(0, C_{m-1} - R) + A_m. \tag{4.57}$$

As before, we can argue that for large N, A_m is Poisson and independent of m in the steady state. For the steady-state analysis, the subscript m is dropped. Its generating function is

$$A(z) = e^{(z-1)\rho}, \tag{4.58}$$

where ρ is the throughput of the system. Using the same steps, we can show that

$$C(z) = \frac{A(z) \left[\sum_{k=0}^{R-1} (z^R - z^k) \Pr\{C = k\} \right]}{z^R - A(z)}. \tag{4.59}$$

We have expressed $C(z)$ in terms of $\Pr\{C = 0\}, \Pr\{C = 1\}, \ldots, \Pr\{C = R-1\}$. These probabilities cannot be found so easily, and they can only be derived indirectly numerically.

The term $\sum_{k=0}^{R-1}(z^R - z^k) \Pr\{C = k\}$ in the numerator of $C(z)$ is a polynomial of degree R in z. We can therefore alternatively express it as $K(z-1)(z-z_1)\ldots$

$(z - z_{R-1})$, where $1, z_1, \ldots, z_{R-1}$ are roots of the polynomial and $K = \sum_{k=0}^{R-1} \Pr\{C = k\}$. Substitute this form into (4.59). Differentiating $C(z)$ with respect to z and taking the limit $z \to 1$ we obtain (see Problem 4.21 for details)

$$\overline{C} = C'(1) = \rho + \frac{\rho^2 - R(R-1)}{2(R-\rho)} + \sum_{k=1}^{R-1} \frac{1}{1 - z_k}, \qquad (4.60)$$

where 1 and $z_k, k = 1, \ldots, R-1$, are the R zeros of the numerator of $\sum_{k=0}^{R-1}(z^R - z^k)\Pr\{C = k\}$. But we do not know $\Pr\{C = 1\}, \ldots, \Pr\{C = R-1\}$ and therefore z_k cannot be found by solving for the zeros in a direct manner! Fortunately, there is an indirect method to derive z_k without the knowledge of the probabilities. In fact, it is the other way round: the probabilities can be found from z_k after the indirect derivation.

The first step of this indirect method makes use of a result in complex analysis called the Rouche's theorem to show that the denominator of $C(z)$, $z^R - A(z)$, contains exactly R zeros with magnitude less than or equal to 1. The use of Rouche's theorem is beyond the scope of this book, and we will simply accept this fact here. It can be easily verified that $z = 1$ is a zero of $z^R - A(z)$ by substitution. Denote the other $R - 1$ zeros by $z'_1, z'_2, \ldots, z'_{R-1}$.

The second step makes use of the fact that any generating function, including $C(z)$, must be analytical (does not go to infinity) for $|z| \le 1$.[1] Therefore, the R zeros of the denominator with magnitude not more than one, $1, z'_1, \ldots, z'_{R-1}$, must also be the R zeros $1, z_1, \ldots, z_{R-1}$ in the numerator—otherwise substituting $z = z'_k$ in $C(z)$ will make $C(z)$ go to infinity because the denominator becomes zero while the numerator is nonzero; but this is impossible because we have just argued that substituting any z with $|z| \le 1$ into $C(z)$ should not make it go to infinity. Thus, $z_k, k = 1, \ldots, R-1$, can be found numerically by solving for the roots of the denominator, and this is given by solving the following $(R-1)$ complex equations:

$$z\left(\cos\frac{2k\pi}{R} + i\sin\frac{2k\pi}{R}\right) - A^{\frac{1}{R}}(z) = 0, \qquad k = 1, \ldots, R-1. \qquad (4.61)$$

As before, we can argue that $C'(1) = 1$ when the input queues saturate. Substituting this into (4.60), we get

$$\rho + \frac{\rho^2 - R(R-1)}{2(R-\rho)} + \sum_{k=1}^{R-1} \frac{1}{1 - z_k(\rho)} = 1. \qquad (4.62)$$

---

[1] To see this intuitively, consider the fact that a generating function is a weighted sum of the associated probabilities with z^k being the weights. When $z = 1$, the probabilities sum to 1 by definition. When $|z| < 1$, the weighted sum can only be smaller.

TABLE 4.4 Maximum Throughput for Input-Buffered Switches with Group Size R

R	ρ^*
1	0.586
2	0.885
3	0.975
4	0.996
8	1.000

Note that we have explicitly put in the dependence of z_k found from (4.61) on ρ through the notation $z_k(\rho)$. Equation (4.62) relates the throughput ρ and the group size R. To find the saturation throughput ρ numerically, we start with an initial guess of ρ and find the corresponding roots $z_k(\rho)$ using (4.61). The roots are then substituted into (4.62) to get a quadratic equation of ρ, from which we obtain a new ρ. This new ρ is then substituted into (4.61) for a new set of roots $z_k(\rho)$. This process is iterated until the required accuracy of ρ is obtained. The resulting throughput is the saturation or maximum achievable throughput ρ^*. Table 4.4 lists the values of ρ^* for various R and shows that it indeed approaches to 1 very quickly with R.

The derivation of the waiting time can largely be proceeded as before. In particular, the general form of (4.31) still applies:

$$W(z) = (1 - \rho\overline{S})\frac{1-z}{(1-z) + [S(z) - 1]\rho}, \quad (4.63)$$

where $W(z)$ and $S(z)$ are, respectively, the generating functions of the waiting time and the service time at the HOL. The mean waiting time is

$$\overline{W} = \frac{\rho\overline{S(S-1)}}{2(1-\rho\overline{S})}. \quad (4.64)$$

These equations are expressed in terms of S, the service time at HOL. But as for $C(z)$, there is no known way of finding $S(z)$ or $\Pr\{S\}$ in closed form. To find $S(z)$ numerically, let the number of packets at the end of a time slot in the steady state be B. Then,

$$B(z) = \frac{C(z)}{A(z)} = \frac{\left[\sum_{k=0}^{R-1}(z^R - z^k)\Pr\{C = k\}\right]}{z^R - A(z)}. \quad (4.65)$$

As for $C(z)$, we can argue that for $B(z)$ the R roots of the numerator are the same as the R roots of the denominator that are less than or equal to 1. For a given $\rho < \rho^*$, therefore, we can use (4.61) to find the roots $z_k(\rho)$, $k = 1, \ldots, R-1$, and then substitute them into

$$B(z) = \frac{(z-1)\prod_{k=1}^{R-1}(z-z_k(\rho))}{z^R - A(z)} \left(\sum_{k=0}^{R-1} \Pr\{C = k\} \right), \quad (4.66)$$

where $\sum_{k=1}^{R-1} \Pr\{C = k\}$ can be found from $B(1) = 1$.

To find $S(z)$, let us consider a FIFO fictitious queue. The tricky part here is that in each time slot, there may be up to R packets being cleared. As in the $R = 1$ case, let B be the number of other packets already in the queue found by an arriving packet, and let L be the number of other packets arriving together with the packet that are scheduled to be served first. Define $S' = B + L + 1$, hence

$$S'(z) = B(z)L(z)z, \quad (4.67)$$

where $L(z)$ is given by (4.36) using the same argument in that section:

$$L(z) = \frac{e^{(z-1)\rho} - 1}{(z-1)\rho}. \quad (4.68)$$

Note that $S = S'$ if $R = 1$. In general,

$$S = \begin{cases} 1, & \text{if } S' \leq R, \\ 2, & \text{if } R < S' \leq 2R, \\ \vdots \\ i, & \text{if } (i-1)R < S' \leq iR, \\ \vdots \end{cases} \quad (4.69)$$

By expanding $S'(z)$ in (4.67) as a power series of z, we can find $\Pr\{S' = k\}$, from which we can then get

$$\Pr\{S = i\} = \sum_{k=(i-1)R+1}^{iR} \Pr\{S' = k\} \quad \text{for } i = 1, 2, \ldots \quad (4.70)$$

In particular, for a given $\rho < \rho^*$, we can approximate $S(z) = \sum_{i=1}^{\infty} \Pr\{S = i\}z^i$ with a truncated series. From this approximate $S(z)$, we can then estimate $W(z)$ and $\overline{W}$ from (4.63) and (4.64). To decide how the series should be truncated for the derivation of $\overline{W}$, note that it depends on $\overline{S(S-1)} = S''(1)$, Therefore, we can write a computer program to test when $i(i-1)\Pr\{S = i\}$ becomes too small to be an important term in $S''(1)$.

As mentioned before, when $R > 1$, delays are also incurred by packets at the output queues. When R is small, it is difficult to analyze the delay at an output queue because the packet arrivals at successive time slots are strongly correlated. In general, however, one would engineer R to be large enough so that the switch does not have any significant throughput limitation (say, $R > 4$ from Table 4.4). In this situation, the delay at the input queue is usually much smaller than the delay at the output, and we can make an approximation that all packets are immediately switched to their outputs. Output delay can then be analyzed using the approach in Section 4.1.4.

4.3.3 Knockout Principle

The performance analysis above has assumed channel-grouped switches operated as waiting systems. As R becomes large, the switch throughput approaches one, and we might ask whether in fact almost all packets are immediately switched to their outputs upon arrival. Furthermore, if the probability of queueing at the inputs is small enough, perhaps we can do away with the input queues and simply drop the very few packets that have lost contention. Such switches operate as loss systems, and the underlying principle is called the knockout principle.

The knockout principle states that the likelihood of a packet being dropped due to contention can be made arbitrarily small with sufficiently large R, and furthermore, for a given loss probability requirement, there exists an R that is independent of the switch size N such that the actual loss probability is not larger than the requirement.

To demonstrate the knockout principle, let us first calculate the loss probability as a function of R. Assume that the destinations of packets on different inputs are independent and that they are equally likely to be destined for any of the N outputs, then the probability of k packets destined for a particular output in a time slot is binomial:

$$P_k = \binom{N}{k} \left(\frac{\rho_o}{N}\right)^k \left(1 - \frac{\rho_o}{N}\right)^{N-k}, \qquad k = 0, 1, \ldots, N, \tag{4.71}$$

where ρ_o is the offered load or the probability that an input has a packet. The probability of a packet being dropped at an output is the expected number of packets dropped divided by the offered load:

$$P_{\text{loss}} = \frac{1}{\rho_o} \sum_{k=R+1}^{N} (k-R) \binom{N}{k} \left(\frac{\rho_o}{N}\right)^k \left(1 - \frac{\rho_o}{N}\right)^{N-k}. \tag{4.72}$$

One can show that P_{loss} increases with N. The proof is messy and does not yield much insight, and it will be omitted here. The reader can verify this fact by trying out several values of ρ_o and R. Returning to (4.72), taking the limit as $N \to \infty$, we have

$$P_{\text{loss}} = \frac{1}{\rho_o} \sum_{k=R+1}^{\infty} (k-R) \frac{\rho_o^k}{k!} e^{-\rho_o}$$

$$= \frac{1}{\rho_o} \left[\sum_{k=0}^{\infty} (k-R) \frac{\rho_o^k}{k!} e^{-\rho_o} - \sum_{k=0}^{R} (k-R) \frac{\rho_o^k}{k!} e^{-\rho_o} \right]$$

$$= \frac{1}{\rho_o}\left[(\rho_o - R) - \sum_{k=0}^{R} k\frac{\rho_o^k}{k!}e^{-\rho_o} + R\sum_{k=0}^{R}\frac{\rho_o^k}{k!}e^{-\rho_o}\right]$$

$$= \frac{1}{\rho_o}\left[(\rho_o - R) - \rho_o\sum_{k=0}^{R-1}\frac{\rho_o^k}{k!}e^{-\rho_o} + R\sum_{k=0}^{R}\frac{\rho_o^k}{k!}e^{-\rho_o}\right]$$

$$= \frac{1}{\rho_o}\left[(\rho_o - R) - (\rho_o - R)\sum_{k=0}^{R}\frac{\rho_o^k}{k!}e^{-\rho_o} + \rho_o\frac{\rho_o^R}{R!}e^{-\rho_o}\right]$$

$$= \left(1 - \frac{R}{\rho_o}\right)\left(1 - \sum_{k=0}^{R}\frac{\rho_o^k}{k!}e^{-\rho_o}\right) + \frac{\rho_o^R}{R!}e^{-\rho_o}.$$

(4.73)

It can be easily shown from the above expression that P_{loss} decreases as R increases. That is, P_{loss} decreases monotonically as $R \to N$, and when $R = N$, $P_{\text{loss}} = 0$ (it is physically obvious that if up to N packets can access the same output simultaneously, the loss probability must be zero). This means that for any given loss probability requirement, we can always find a sufficiently large R to achieve a loss probability that is not larger than the requirement. To illustrate that a finite R would do for all N, consider a loss probability requirement of 10^{-6}. We can compute from (4.73) that $R = 8$ is enough. Furthermore, this R applies for all N, since (4.73) is an upper bound on the loss probability for finite N.

The channel-grouped Batcher–banyan networks described in the previous subsection can adopt the knockout principle and operate as a loss system. As in the waiting system, a contention-resolution scheme (e.g., the three-phase scheme) is used to select up to R packets for each output for switching. The excess packets, instead of being queued, are dropped at the inputs.

An alternative to separating the contention-resolution phase and the switching phase is to deal with the contention on-the-fly while the packets are being routed. This eliminates the overhead associated with a dedicated contention-resolution phase. Figure 4.15 shows a Batcher–banyan switch design in which a reverse banyan concentrator is inserted between the Batcher and the banyan networks. In this network, packets are launched in their entireties into the Batcher network at the beginning of each time slot (as opposed to only the headers in the three-phase scheme). As discussed before, winning and losing packets are determined at the outputs of the Batcher network by address comparison. The losing packets are dropped. This means that the remaining winning packets may not be concentrated anymore. Internal conflict may arise if they are launched into the banyan networks directly. To avoid such conflict, a reverse banyan network is a used as a concentrator to remove the gaps between the packets.

The outputs of the reverse banyan concentrator are connected in an interleaving fashion to R $N \times N$ banyan networks (as in the connection from the Batcher network to the banyan networks in Fig. 4.12): output i ($i = 0, \ldots, N - 1$) of the concentrator is connected to input $\lfloor i/R \rfloor$ of banyan network i mod R. It is easily seen that no conflict occurs in any of the $N \times N$ banyan network because the packets are concentrated and

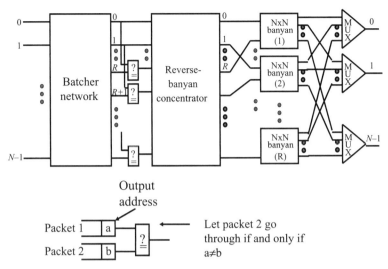

FIGURE 4.15 A Batcher–banyan knockout switch.

have distinct monotonic addresses. Outputs i ($i = 0, \ldots, N-1$) of different banyan networks correspond to the same output address and they are collected and fed into a multiplexer with a logical FIFO (see Fig. 4.14(b)) to maintain packet sequence integrity.

The Batcher–banyan knockout switch is a good example of how the basic building blocks that we have learned so far can be used to construct a sophisticated packet switch. A number of parallel distributed algorithms—sorting in the Batcher network, concentration in the reverse banyan network, routing in the banyan networks, and multiplexing in the logical FIFO queues at the outputs—are executed simultaneously. Every essential step seems to have been parallelized except for one thing. In the discussion of the reverse banyan concentrator in Section 3.2.4, we have ignored a critical problem that surfaces when N is large, namely, how does a packet at an input find out the number of arriving packets at other inputs so that its output address at the shifting concentrator can be determined accordingly? After all, the concentrator output assigned to a packet corresponds to the number of active packets above it: from top to bottom, the first active packet is assigned output 0, second, output 1, third, output 2, and so on. When the number of inputs N is small, a central controller that polls the inputs and returns the assigned output addresses at the beginning of each time slot can be incorporated (see Fig. 4.16(a)). But this takes up an amount of time proportional to the number of inputs. When the number of inputs N is large, this approach is not viable. Hence, a parallel architecture that performs such polling and address assignment in a distributed fashion is needed.

For each input, the task is to compute the number (or the running sum) of packets above it. Figure 4.16(b) shows an 8×8 running-adder address generator for such computation. An $N \times N$ running-adder address generator comprises $\log_2 N$ stages of N adders each. At the inputs, a $\log_2 N$-bit header is attached to each packet for the

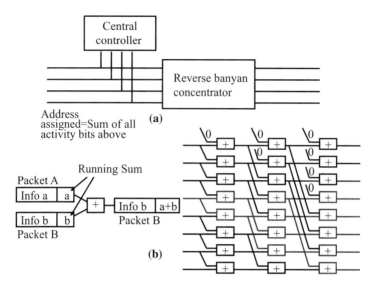

FIGURE 4.16 (a) A central controller for computing the assignments of packets to concentrator outputs; (b) a running-adder address generator that computes the assignments in a parallel and distributed manner.

storage of the running sum. An adder is a 2×1 device that takes two input packets, adds the running sum of the upper input to that of the lower input, and outputs the lower packet with the new running-sum header. At stage k ($k = 1, \ldots, \log_2 N$), link i ($i = 0, \ldots, N - 1$) is connected to two adders: the lower input of adder i and the upper input of adder $i + 2^{k-1}$.

Initially, the running-sum headers of active input packets to the address generator are set to one. By construction, some of the inputs to the adders at various stages may not be connected to any previous-stage outputs (see Fig. 4.16(b)), and the running sums of these inputs are set to zero. To understand the structure, let us trace through the path taken by a packet on input i and examine the steps for computing its running sum. Let

$$X_j = \begin{cases} 1, & \text{if input } j, j = 0, \ldots, N - 1, \text{ has an active packet,} \\ 0, & \text{if input } j, j = 0, \ldots, N - 1, \text{ is inactive,} \\ 0, & \text{if } j < 0. \end{cases}$$

At the output of stage 1, the running sum of the packet at input i is $S_i[1] = X_i + X_{i-1}$; at the output of stage 2, the running sum is $S_i[2] = S_i[1] + S_{i-2}[1] = X_i + X_{i-1} + X_{i-2} + X_{i-3}$; and in general, at the output of stage $k \leq \log_2 N$, $S_i[k] = \sum_{j=0}^{2^k - 1} X_{i-j}$, so that the running sum of the packet at the output of the address generator is

$$S_i[\log_2 N] = X_i + X_{i-1} + \cdots + X_{i-N-1} = X_i + X_{i-1} + \cdots + X_0 \qquad (4.74)$$

Since the running sum $S_i[\cdot]$ includes packet i, one is subtracted from $S_i[\log_2 N]$ at the output of the address generator to obtain the concentrator-output assignment of the packet.

Let us consider the order of complexity of the overall Batcher–banyan knockout switch. The complexity of the address generator is of order $N \log_2 N$ because it has $N \log_2 N$ adders. Considering its crosspoint count, the reverse banyan concentrator also has a complexity of order $N \log_2 N$. The crosspoint count of the R $N \log_2 N$ banyan networks is $\frac{RN}{2} \log_2 N$. Since R can be fixed at a constant independent of N to satisfy any loss probability requirement, the order of complexity is $N \log_2 N$ for large N. By the same token, the N logical FIFO queues are of complexity order N. For large N, therefore, the order of complexity of the Batcher network $N \log_2^2 N$ dominates. Thus, overall, this is still a $N \log_2^2 N$ network.

Figure 4.17 shows yet another realization of the knockout concept. This is the design proposed by the original proponents of the knockout principle. It uses N buses to broadcast all input packets to all outputs. At each output, filters are used to select the packets addressed to it. Contention resolution is performed by a "knockout" concentrator at the output that selects R packets to feed into a logical FIFO queue. The excess packets are dropped (knocked out) in the knockout concentrator. An advantage of this structure is that multicast capability (sending copies of a packet to several outputs simultaneously) is inbuilt: the same packet can be selected by filters at different outputs concurrently. A disadvantage is that the implementation complexity is higher than that of the Batcher–banyan design.

The knockout concentrator is implemented as shown in Fig. 4.18. It is divided into R sections, each corresponding to a tournament that selects one winning packet. The N inputs of the concentrator are connected to the first section on the left. The losing packets at each section are fed into the next section so that after R sections, R or fewer

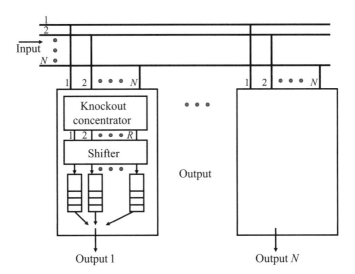

FIGURE 4.17 A knockout switch based on broadcast buses and knockout concentrators.

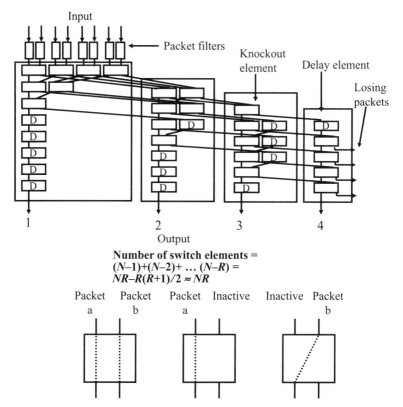

FIGURE 4.18 An 8×4 knockout concentrator and operation of its component 2×2 switch elements.

packets (if there are fewer than R input packets) are selected as winners. As shown in the figure, each section is a binary-tree interconnection of 2×2 switch elements. The delay elements are used to synchronize packet arrivals at the switch elements and at the concentrator outputs.

At each switch element, if there is only one packet at the inputs, it is declared the winner and sent out on the left output. If there are two packets at the inputs, the switch element is set to the bar state to make the left packet winner and right packet loser. The switch element can also be designed so that it alternates between favoring the left and right inputs in successive switch cycles to achieve a certain degree of fairness.

The number of switch elements in the kth section is $N - k$. The total number of switch elements is $(N - 1) + (N - 2) + \cdots + (N - R) \approx NR$ for $N \gg R$. Therefore, considering all outputs, the total number of switch elements in all N knockout concentrators is $N^2 R$. Hence, the complexity of this design is of order N^2.

That the knockout concept can be implemented using very different switch architectures points out that it is important to separate principles from realizations so that we have a clear idea of what the realizations aim to achieve. The same switching principle can be implemented in different ways, each with its own advantages and disadvantages. When comparing switch architectures, better insight can be obtained if we have a clear idea of whether the differences are due to different implementations of the same switching principle or different underlying switching principles (e.g., waiting system versus loss system).

4.3.4 Replication Principle

Switches that employ the channel-grouped and knockout principles are internally nonblocking. In fact, they are more than internally nonblocking in that multiple packets can be routed to the same output destination simultaneously. Many other switching networks, such as the Banyan networks, are internally blocking. It turns out that we can tackle the problems of internal and external conflicts concurrently in these networks using two techniques that are similar to channel grouping. The first technique employs the replication principle and is discussed in this subsection. The second technique is based on the dilation principle and will be discussed in the next subsection.

To explain the replication principle, let us apply it to the class of Banyan networks. Figure 4.19 depicts a parallel Banyan network. The basic idea is to have multiple copies of Banyan networks so that a packet has several alternative routes, one in each Banyan network, to its destination. There are two ways to operate the parallel network.

The first alternative is based on random routing. Suppose that there are K parallel networks. An incoming packet is routed randomly to one of the K networks. The load to each Banyan network is reduced by a factor of K, giving rise to a correspondingly lower P_{loss}. Using (3.5), with the load set to ρ_0/K, we find that

$$P_{\text{loss}} = \frac{n\rho_0}{n\rho_0 + 4K}. \qquad (4.75)$$

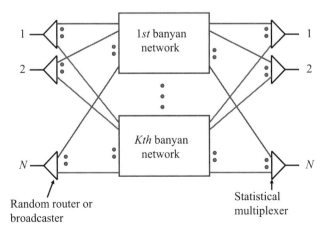

FIGURE 4.19 A parallel banyan network.

Suppose we fix the loss probability requirement and ask what is the value of K that is needed to satisfy that requirement. From (4.75), we have

$$K = \frac{(P_{\text{loss}}^{-1} - 1)n\lambda}{4} \approx \frac{\lambda}{4} \frac{\log N}{P_{\text{loss}}} \tag{4.76}$$

for small P_{loss}. Each Banyan network has $\frac{N}{2} \log N$ switch elements. In terms of N, therefore, the parallel Banyan network has a complexity order of $N \log^2 N$.

Instead of random routing, we can use broadcast routing. An incoming packet is broadcast to all the K parallel Banyan networks. Since multiple copies of the packet may be received at the output, filters are needed to remove redundant packets. With broadcast routing, the load to each Banyan network is still ρ_o, but a packet is lost only if all its replicas fail to reach its destination output. For this strategy to work properly, we must adopt a random contention-resolution scheme in each of the parallel Banyan networks so that when two packets attempt to access the same output of a 2×2 switch element, the winning packet will be chosen at random. Otherwise, with all the Banyan networks using the same fixed contention-resolution scheme (e.g., always choose the packet from the upper input port), packets that are dropped in one Banyan network will also be dropped in other Banyan networks.

Even with the random contention-resolution scheme, the event of a packet being dropped in one Banyan network is not independent of the events of its replicas being dropped in the other Banyan networks, because all Banyan networks have the same set of input packets. Given a packet is dropped in one Banyan network, for instance, it is more likely that the packet will be dropped in another Banyan network because the knowledge that it is dropped in the first network implies that there is at least one other packet contending with it. For simplicity, if we further make the assumption that the contention-resolution processes in different Banyan networks are independent, then

$$P_{\text{loss}} = \left(\frac{n\rho_o}{n\rho_o + 4} \right)^K. \tag{4.77}$$

From this, we get

$$K = \frac{\log P_{\text{loss}}}{\log \left(1 - \frac{4}{n\rho_o + 4}\right)} \approx \frac{(4 + \rho_o \log N)}{4} (-\ln P_{\text{loss}}). \tag{4.78}$$

The number of parallel Banyan networks needed for broadcast routing is less than the number needed for random routing. In either case, however, the order of complexity is not less than that of the Batcher–banyan network.

4.3.5 Dilation Principle

We shall focus on dilated Banyan networks operated as loss systems. The basic idea of the dilated Banyan network is to expand the internal link bandwidth in order

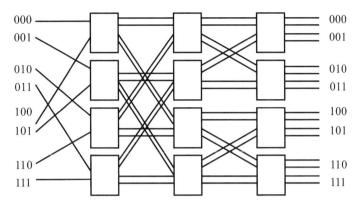

FIGURE 4.20 An 8 × 8 banyan network with dilation degree 2.

to reduce the likelihood of a packet being dropped. Figure 4.20 shows an 8 × 8 dilated Banyan network with a *dilation degree* of 2. The overall interconnection structure of a dilated-Banyan network is the same as that in the Banyan network, except that connected switch elements are linked by a multiplicity of d channels in the d-dilated-Banyan network. Thus, the regular Banyan network can be considered as a special case of the d-dilated-Banyan network with $d = 1$. The switch elements are themselves $2d \times 2d$ switches with two outgoing addresses. Each outgoing address has d associated outgoing links. Consequently, up to a maximum of d packets can be forwarded to the same outgoing address in any given time slot.

In a loss system, if more than d packets are destined for the same outgoing address of a switch element, then d packets would be forwarded and the remaining packets dropped from the system. Thus, by making d sufficiently large, we can achieve arbitrarily small packet loss probability. The drawback, of course, is that the switch becomes complex as d increases.

To see how large d should be for a given loss probability requirement P_{loss}, let us first consider an exact calculation. Assume that input packets to the overall network are independent and equally likely to be destined for any of the N outputs. Let $R_m(j)$ be the probability that j packets are forwarded to an outgoing address of a switch element at stage m, where $0 \le j \le d$. Only d packets are forwarded when more than d packets are destined for the output address. The probability of i packets entering a switch element at stage $m + 1$ is

$$S_{m+1}(i) = \sum_{k=0}^{i} R_m(k) R_m(i - k), \qquad 0 \le i \le 2d, \tag{4.79}$$

since $R_m(k)R_m(i - k)$ is the probability that there are k and $i - k$ packets on the upper and lower input-channel groups, respectively. The probability that j of these packets are destined for a particular output address is $\binom{i}{j}2^{-i}$. Thus,

$$R_{m+1}(j) = \begin{cases} \sum_{i=j}^{2d} S_{m+1}(i)\binom{i}{j}2^{-i}, & \text{if } j < d, \\ \sum_{i=d}^{2d} S_{m+1}(i) \sum_{k=d}^{i} \binom{i}{k}2^{-i}, & \text{if } j = d. \end{cases} \quad (4.80)$$

With the initial condition

$$R_0(j) = \begin{cases} 1 - \rho_o, & \text{if } j = 0, \\ \rho_o, & \text{if } j = 1, \\ 0, & \text{if } j > 1, \end{cases}$$

where ρ_o is the offered load, $R_n(j)$ can be found recursively. The packet loss probability for the overall switch is simply

$$P_{\text{loss}} = 1 - \frac{\sum_{j=0}^{d} j R_n(j)}{\rho_o}, \quad (4.81)$$

where $n = \log_2 N$.

The above is not a closed-form result and it does not relate P_{loss} to n, d, and ρ_o explicitly. The approximate analysis below will be used to study how dilation degree d is related to n for given P_{loss} and ρ_o.

There are two groups of d incoming links for each switch element. Strictly speaking, the links belonging to the same group are not independent in the sense that finding a packet on one link is correlated with finding packets on other links. For analytical tractability, however, we will make the simplifying assumption that the input ports are independent. Let P_m denote the probability that there is a packet at an input channel of a switch element at stage $m + 1$. With the independence assumption, the probability of finding a packet at an arbitrary outgoing link of this switch element (or an incoming link of a switch element in the $m + 2$ stage) is

$$\begin{aligned} P_{m+1} &= \frac{1}{d}\left\{\sum_{k=1}^{d} k \binom{2d}{k}\left(\frac{P_m}{2}\right)^k \left(1 - \frac{P_m}{2}\right)^{2d-k}\right. \\ &\quad \left. + d\sum_{k=d+1}^{2d} \binom{2d}{k}\left(\frac{P_m}{2}\right)^k \left(1 - \frac{P_m}{2}\right)^{2d-k}\right\} \\ &= P_m - \frac{1}{d}\sum_{k=d+1}^{2d} (k-d)\binom{2d}{k}\left(\frac{P_m}{2}\right)^k \left(1 - \frac{P_m}{2}\right)^{2d-k}. \quad (4.82) \end{aligned}$$

DESIGN IMPROVEMENTS BASED ON OUTPUT CAPACITY EXPANSION

The complexity of the dilated Banyan network can be estimated from the packet loss probability for a switch element at stage m given by

$$Q_m = 1 - \frac{P_m}{P_{m-1}}$$

$$= \frac{1}{P_{m-1}d} \sum_{k=d+1}^{2d} (k-d) \binom{2d}{k} \left(\frac{P_{m-1}}{2}\right)^k \left(1 - \frac{P_{m-1}}{2}\right)^{2d-k}, \quad m > \log_2 d.$$

(4.83)

We limit the application of the above formula to $m > \log_2 d$ because it can be seen from the structure of the switch that $Q_m = 0$ for $m \leq \log d$: there can be no more than $d/2$ packets at each input group of switch elements at stages 1 through $\log d$; that is, even under full loading conditions, when all input ports of the overall network have a packet, it takes at least $\log d + 1$ stages before contentions between packets may occur.

It is easy to see from (4.82) that P_m decreases monotonically with m. Intuitively, the loss probability Q_m also decreases monotonically with m:

$$Q_1 \geq \cdots \geq Q_n.$$

(4.84)

The above monotonic sequence is easy to interpret: As we progress through the stages of the network and as packets are dropped due to contention, fewer and fewer packets are left. This means the contention level is lighter in the later stages and therefore Q_m decreases as m increases. Problem 4.24 derives (4.84) formally.

In the following, all logarithms are taken to the base two unless otherwise noted: $\log = \log_2$. The loss probability P_{loss} can be bounded from above as follows:

$$P_{\text{loss}} = \frac{P_0 - P_n}{P_0}$$

$$= \frac{(P_0 - P_1) + (P_1 - P_2) + \cdots + (P_{n-1} - P_n)}{P_0}$$

$$= Q_1 + Q_2 \frac{P_1}{P_0} + \cdots + Q_{\log d} \frac{P_{\log d - 1}}{P_0} + Q_{\log d + 1} \frac{P_{\log d}}{P_0} + \cdots + Q_n \frac{P_{n-1}}{P_0}$$

$$\leq (n - \log d) Q_{\log d + 1},$$

(4.85)

where the last line is obtained by observing that $Q_1 = Q_2 = \cdots = Q_{\log_d} = 0$ and that Q_m and P_m decrease monotonically with m. This is basically a union bound (i.e., the overall loss probability is upper-bounded by the sum of loss probabilities at different stages), which is likely to be very good when P_{loss} (and therefore Q_m) is small. To find a bound for $Q_{\log d+1}$, substituting $P_{\log d} = \rho_0/d$ into (4.83) gives

$$Q_{\log d+1} = \frac{1}{\rho_0} \sum_{j=d+1}^{2d} (j-d) \binom{2d}{j} \left(\frac{\rho_0}{2d}\right)^j \left(1 - \frac{\rho_0}{2d}\right)^{2d-j}$$

$$= \frac{(\rho_0/2d)^d}{\rho_0} \sum_{i=1}^{d} i \binom{2d}{i+d} \left(\frac{\rho_0}{2d}\right)^i \left(1 - \frac{\rho_0}{2d}\right)^{d-i}, \qquad (4.86)$$

where we have made the index change $i = j - d$. Now,

$$\binom{2d}{i+d} = \binom{2d}{d}\binom{d}{i} \times \frac{d!\,i!}{(i+d)!} = \binom{2d}{d}\binom{d}{i} \times \frac{i}{i+d}\frac{i-1}{i-1+d} \times \cdots$$

$$\times \frac{1}{1+d} \leq \binom{2d}{d}\binom{d}{i} 2^{-i} \qquad \text{for } i \leq d.$$

Substituting the above into (4.86) and simplifying by Stirling's formula

$$n! = n^n e^{-n} \sqrt{2\pi n}(1 + \epsilon(n)), \qquad (4.87)$$

where $\epsilon(n) > 0$ is a decreasing function of n, we get

$$Q_{\log d+1} \leq \frac{1}{4} \left(\frac{\rho_0}{d}\right)^d \left(1 - \frac{\rho_0}{4d}\right)^{d-1} \frac{2^d}{\sqrt{\pi d}}.$$

Thus,

$$P_{\text{loss}} \leq (n - \log d)\frac{1}{4}\left(\frac{\rho_0}{d}\right)^d \left(1 - \frac{\rho_0}{4d}\right)^{d-1} \frac{2^d}{\sqrt{\pi d}}. \qquad (4.88)$$

To get a lower bound for P_{loss}, continuing from the first line of (4.85), we have

$$P_{\text{loss}} \geq \frac{P_n}{P_0}(Q_{\log d+1} + Q_{\log d+2} + \cdots + Q_n)$$

$$\geq (1 - P_{\text{loss}})(n - \log d)Q_n. \qquad (4.89)$$

From (4.83) and with the index change $i = k - d$, we obtain

$$Q_n = \frac{(P_{n-1}/2)^d}{P_{n-1}d} \sum_{i=1}^{d} i\binom{2d}{i+d}\left(\frac{P_{n-1}}{2}\right)^i \left(1 - \frac{P_{n-1}}{2}\right)^{d-i}. \qquad (4.90)$$

Now,

$$\binom{2d}{i+d} = \binom{d}{i} \times \frac{2d \cdots (d+1)}{(d+i) \cdots (i+1)} \geq \binom{d}{i} 2^{d-i}.$$

Substituting into (4.90) and simplifying

$$Q_n \geq \frac{[(\rho_o/d)(1-P_{\text{loss}})]^d}{4} \left(1 - \frac{\rho_o}{4d}\right)^{d-1}, \qquad (4.91)$$

where we have made use of the facts that $P_{n-1} \geq P_n = (\rho_o/d)(1 - P_{\text{loss}})$ and $P_{n-1} \leq \rho_o/d$. Substitution of (4.91) into (4.89) gives

$$\frac{P_{\text{loss}}}{(1-P_{\text{loss}})^{d+1}} \geq \frac{1}{4} \left(\frac{\rho_o}{d}\right)^d \left(1 - \frac{\rho_o}{4d}\right)^{d-1} (n - \log d). \qquad (4.92)$$

If we are only interested in very small P_{loss}, then

$$\frac{P_{\text{loss}}}{(1-P_{\text{loss}})^{d+1}} \approx P_{\text{loss}}(1 + (d+1)P_{\text{loss}}) \approx P_{\text{loss}}.$$

Making this approximation, we obtain

$$P_{\text{loss}} \geq \frac{1}{4} \left(\frac{\rho_o}{d}\right)^d \left(1 - \frac{\rho_o}{4d}\right)^{d-1} (n - \log d). \qquad (4.93)$$

Combining (4.88) and the above, we have

$$\frac{(n - \log d)}{4} \left(\frac{\rho_o}{d}\right)^d \left(1 - \frac{\rho_o}{4d}\right)^{d-1} \leq P_{\text{loss}} \leq \frac{(n - \log d)}{4} \left(\frac{\rho_o}{d}\right)^d \left(1 - \frac{\rho_o}{4d}\right)^{d-1} \frac{2^d}{\sqrt{\pi d}}. \qquad (4.94)$$

The above can be used to estimate P_{loss} for given d and n. We can turn the problem around and find the required dilation degree d for given loss probability requirement P_{loss} and switch size $N = 2^n$. It can be shown (see Problem 4.25) that

$$d \log d = \log \log N - \log P_{\text{loss}} + O(d), \qquad (4.95)$$

where $O(d)$ is a term that increases not more than linearly with d: that is, a constant C can be found such that $O(d) \leq Cd$.

Each $2d \times 2d$ switch element in the dilated Banyan network have only two outgoing addresses, and its implementation can be simpler than a regular $2d \times 2d$ switch element with $2d$ outgoing addresses. Figure 4.21 shows how the $2d \times 2d$ switch element can be implemented. An incoming packet is switched by a 1×2 element to its desired output group. A $2d \times d$ concentrator is then used to concentrate the packets

144 SWITCH PERFORMANCE ANALYSIS AND DESIGN IMPROVEMENTS

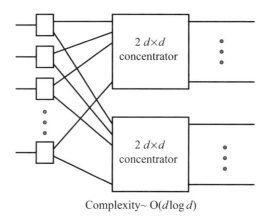

FIGURE 4.21 An implementation of $2d \times 2d$ switch element with order of complexity $d \log d$.

destined for the same group to d outgoing links. Excess packets are dropped if more than d packets arrive for the same output group.

We have learned earlier in this chapter that the concentrator can be realized using a running-adder address generator and a reverse Banyan network. The order of complexity is $d \log d$. The order of complexity of each switch element is therefore $d \log d$. Since there are altogether $\frac{N}{2} \log N$ such switch elements, the overall dilated network is of order $\frac{N}{2} \log N \times d \log d$. For large N, $d \log d$ is of order $\log \log N$, in terms of N. The dilated network is therefore of order $N \log N \log \log N$. This order of complexity is less than those of the parallel Banyan network and Batcher–banyan network. However, since order-of-complexity measure is meaningful only for large N, it cannot be argued that this network is simpler to implement in practice, especially if one considers that for small d, it may not be worthwhile to use the running-adder-and-reverse banyan approach to build the $2d \times 2d$ switch.

PROBLEMS

4.1 Consider a 2×2 switch operating as a waiting system. Each input has a queue. If the two packets at the heads of the queues are destined for the same output, one of them is chosen at random to be routed and the other will remain in the queue. Consider the saturation situation in which there are always packets in the queues. Draw a four-state Markov chain in which state (i, j) means that the packet at the head of queue 1 is destined for output i and the packet at the head of queue 2 is destined for output j. Each transition corresponds to the passing of a time slot. Derive the maximum switch throughput.

4.2 Explain intuitively why input-buffered switches of smaller dimensions have higher maximum throughput than those of larger dimensions when the packets on each input are equally likely to be destined for any of the outputs.

4.3 Consider two $N \times N$ internally nonblocking packet switches where N is very large. One is operated as a waiting system with very large input buffers and the other is operated as a loss system without input buffers. For the former, packets are dropped when the input buffers overflow, and for the latter, packets that lost contention are dropped.

(a) When the load is 0.5, which system has a lower packet loss probability? How about when the load is 1?

(b) Roughly, what is the "transition point" for the load at which the relative loss performance of the systems is reversed?

4.4 Consider an 8×8 input-buffered switch with nonuniform traffic distributions: Packets from inputs 1 to 4 are equally likely to be destined for outputs 1 to 4 but none of them is destined for outputs 5 to 8. Similarly, packets from inputs 5 to 8 are equally likely to be destined for outputs 5 to 8 but none of them is destined for outputs 1 to 4. Do you expect the maximum throughput to be higher or lower than the case in which packets from all inputs are equally likely to be destined for all outputs. Explain.

4.5 Explain qualitatively why a switch operated as a loss system generally has a higher maximum throughput than the same switch operated as a waiting system. For example, for an internally nonblocking switch, the maximum throughput is 0.632 for the former and 0.586 for the latter. Should we then operate all switches as loss systems?

4.6 For a large input-buffered switch, we derived in the text that the average backlog in a fictitious queue is

$$C'(1) = \frac{2\rho(1-\rho) + \rho^2}{2(1-\rho)},$$

where ρ is the throughput.

(a) Explain why $C'(1)$ is between 0 and 1 in general.

(b) Show that the maximum throughput is indeed achieved when the input queue is saturated (i.e., when $C'(1) = 1$).

(c) Plot qualitatively the throughput ρ as a function of the offered load ρ_o for ρ_o between 0 and 1. That is, ρ_o is the probability that a packet arrives at an input queue in a given time slot.

4.7 The text uses the normalization method to derive the expected length of a busy period equation (4.27). We now wish to derive it directly. Consider a long stretch of time from 0 to t.

(a) Argue that the total amount of time the system is in busy periods is approximately $\rho t \overline{S}$, and therefore, the total amount of idle time is approximately $t - \rho t \overline{S}$.

(b) Show that the average duration of idle periods is $\frac{1}{1-\Pr\{K=0\}}$, and therefore the total number of idle periods is approximately $(t - \rho t \overline{S})(1 - \Pr\{K = 0\})$.

(c) From (a) and (b) show that $E[Y] = \frac{\overline{S}\rho}{(1-\overline{S}\rho)(1-\Pr\{K=0\})}$.

4.8 Prove the argument that gives rise to Eq. (4.35):
  (a) Consider a Poisson process with rate ρ. Show that given that at least one packet arrives, the probability that there are N other arrivals is still Poisson with rate ρ (i.e., the arrivals of other packets are not affected by the arrival of the packet of focus).
  (b) Given that N is Poisson with rate ρ, show that if the N arrivals are randomly placed on a unit line, the number of arrivals placed before point t is Poisson with rate ρt. (*Hint*: Find the conditional probability that L packets are placed before t given that N packets arrive. Then uncondition on N.)

4.9 The text uses the busy-period method to arrive at the waiting time at an input queue of an input-buffered switch (Eqs. (4.31) and (4.32)). This problem considers the number of packets in the input queue upon the arrival of a new packet. The mean waiting time $\overline{W}$ in (4.32) can then be obtained by Little's law.
  (a) Argue that the number of packets H found by an arriving packet (not including the arriving packet) has the same probability distribution as the number of packets D left behind by a departing packet (not including the departing packet).
  (b) Show that $D_{m+1} = \max(0, D_m - 1) + A_{m+1}$, where D_m is the number of packets left behind by the mth packet that departs from the system and A_m is the number packets that arrive during its service time.
  (c) Derive $D(z)$ from the above and obtain $\overline{D}$ from it.
  (d) Argue that the number of packets found by an arriving packet has the same probability distribution as the number of packets found by an arbitrary observer at arbitrary time. Use Little's law to obtain the average waiting time from $\overline{D}$ (i.e., derive (4.32)).

4.10 Let us consider an alternative derivation of the first and second moments of the service time of an input queue (Eqs. (4.37) and (4.38)). Let B be the number of packets found by an arriving packet to the fictitious queue and L be the number of simultaneously arriving packets scheduled to be served before the arriving packet.
  (a) Show that $S = B + L + 1$.
  (b) Express $\overline{S}$ and $\overline{S(S-1)}$ in terms of the moments of B and L.
  (c) Find $B(z)$ in terms of $A(z) = e^{(z-1)\rho}$, the moment-generating function of the arrival process. (*Hint*: See Eqs. (4.12) and (4.13).)
  (d) Find the moments of B and L, and express $\overline{S}$ and $\overline{S(S-1)}$ in terms of only ρ.

4.11 True or false? The look-ahead contention-resolution scheme in the input-buffered switch cannot maintain the sequence order of the cells of a point-to-point VC because the packets in the input queues may not be served in a first-in-first-out fashion.

4.12 For the input-buffered switch operated with a look-ahead scheme with window size w (see the last few paragraphs of Section 4.1.2):
(a) Argue that if the input queues have infinite amounts of buffer, as $w \to \infty$, the maximum throughput approaches 1.
(b) Show that the average delay of the input-window scheme can never be smaller than that in the ideal output-buffered switch of Section 4.1.4. (*Hint*: Consider the packet backlog and use Little's law.)

4.13 Consider the following analytical derivation of the maximum throughput of an input-buffered switch with a look-ahead window size of $w = 2$. Under saturation, since we are considering the first two packets from each input queue, the average number of packets in the fictitious queue of a particular output is $C'(1) = 2$. Substituting this into the formula that relates $C'(1)$ to the throughput from the analysis of the fictitious queue, $C'(1) = \frac{(2\rho - \rho^2)}{2(1-\rho)}$, and solving for ρ, we get $\rho = 3 - \sqrt{(5)} = 0.76$, which is different from the simulation result of 0.70. Give an example showing how the above argument fails to reflect the real situation in the look-ahead scheme.

4.14 Consider a three-phase Batcher–banyan switch with a look-ahead contention-resolution scheme. With reference to the maximum throughputs given in Table 4.2 (these throughputs do not take into account the contention-resolution overhead), for what value of w is the throughput maximized for a 32 × 32 switch if the contention-resolution overhead is taken into consideration. Assume a packet size of 53 bytes plus 6 bits (5 bits for output address and 1 bit for activity bit), and assume that checking of conflict at the outputs of the sorting network takes one-bit time and the backpropagating acknowledgments take one-bit time.

4.15 Consider an improved input look-ahead scheme. For an input queue, if the first packet loses contention in the first round, the second packet will be examined in the second round. However, if its output address is the same as that of the first packet, we proceed immediately to the third packet (since the output has been taken by a packet at another input during the first-round contention), and so on until a packet with a different output address is found. Thus, we can be sure that the packets presented by an input for contention in the w rounds have distinct addresses.
(a) Describe qualitatively why you expect to see improvement over the look-ahead scheme described in the text. Do you expect much improvement if successive packets are independent and the switch size N is large? What if the output addresses of successive packets are strongly correlated and likely to be the same?

(b) What is wrong with the following analysis? Assume N is very large and all packets are independent. Let $\rho^* = 0.586$ be the maximum throughput when $w = 1$. When $w = 2$, after the first round, a set of outputs have been assigned to a set of inputs, and the fraction of remaining inputs and outputs is approximately $1 - \rho^*$. It is as if we have a smaller $N(1 - \rho^*) \times N(1 - \rho^*)$ switch in the second round. Therefore, the additional throughput obtained with the second-round contention is $(1 - \rho^*)\rho^*$. Arguing this way, the total throughput in general is $\rho^* + (1 - \rho^*)\rho^* + \cdots + (1 - \rho^*)^{w-1}\rho^*$.

4.16 Which switch has the better delay performance when the input offered load is 0.2?

1. The generic nonblocking input-buffered (with no speedup, no look-ahead, etc.) where queueing delay is incurred only at the inputs.

2. The generic nonblocking output-buffered switch in which all incoming packets are immediately forwarded to their desired outputs and queued there.

4.17 Consider the packet-slicing switch design (see Fig. 4.11(b)). Suppose that the switch is 256×256 and the packet size is 53 bytes not including the output-address header. If a speedup factor of 8 is to be achieved, what is the ratio of switch cycle to time slot?

4.18 Consider an 8×8 channel-grouped Batcher–banyan switch with $R = 2$. Suppose that the three-phase contention-resolution scheme is adopted. Draw diagrams showing how the following set of packets will be routed through the switch: $(0, 1), (2, 4), (4, 2), (5, 1), (6, 2), (7, 1)$, where (i, j) means a packet at input i destined for output j. Do not show the inside of the Batcher and banyan networks. Only show the positions of the packets at the inputs and outputs of the networks.

4.19 Show how to build an $N \times M$ expansion Batcher–banyan switch, where $M = NK > N$, K an integer, using an $N \times N$ Batcher network and an expansion banyan structure similar to that of Fig. 4.13. There is no channel grouping and each of the M outputs corresponds to a distinct address. Show how the routing is done and argue that the switch is internally nonblocking.

4.20 Consider an $N \times MR$ switch operating as a channel-grouped waiting system. There are M output addresses, and a group of R output ports belong to the same address. Rank these switches according to their maximum throughputs per input: (i) $N = 16, M = 16, R = 2$; (ii) $N = 16, M = 16, R = 1$; (iii) $N = 16, M = 8, R = 2$; (iv) $N = 8, M = 8, R = 2$.

4.21 This problem shows how to derive Eq. (4.60) from Eq. (4.59) in the text. From Eq. (4.59), we have $C(z) = N(z)/D(z)$, where $D(z) = z^R A^{-1}(z) - 1$ and $N(z) = \sum_{k=0}^{R-1}(z^R - z^k)\Pr\{C = k\}$. The text argues that $N(z)$ can be written alternatively as $N(z) = K(z - 1)(z - z_1(\rho)) \cdots (z - z_{R-1}(\rho))$, where $z_k(\rho)$ is the $(R - 1)$ complex roots of $z^R - A(z)$ with modulus less than 1.

(a) Show that $D(1) = N(1) = 0$.
(b) By differentiating the left and right sides of $D(z)C(z) = N(z)$ once, show that $D'(1) = N'(1)$. From this, show that $K = (R - \rho)/\prod_{i=1}^{R-1}(1 - z_i(\rho))$.
(c) By differentiating $D(z)C(z) = N(z)$, show that $C'(1) = (N''(1) - D''(1))/2D'(1)$.
(d) Find $N''(1)$, $D''(1)$, and $D'(1)$ to substitute into the above to derive Eq. (4.60).

4.22 Find the mean delay at an input queue of a channel-grouped waiting system in which $R = 2$ and $\rho = 0.8$.

4.23 For an 8×8 Batcher–banyan knockout switch with $R = 2$ (as the one depicted in Fig. 4.15), draw diagrams showing how the following set of packets will be routed through the switch: (0, 1), (2, 4), (4, 2), (5, 1), (6, 2), (7, 1), where (i, j) means a packet at input i destined for output j. Do not show the inside of the Batcher and banyan networks, but show how the running-adder address generator, including its internal operations, and the reverse banyan network are used to concentrate the packets.

4.24 We want to derive (4.84). First, rewrite (4.82) in the form $P_{m+1} = P_m - \frac{1}{d}f(P_m)$, where $f(x) = \sum_{k=d+1}^{2d}(k-d)\binom{2d}{k}(\frac{x}{2})^k(1-\frac{x}{2})^{2d-k}$.
(a) Show that $f(x)$ increases monotonically with x for $0 \le x \le 1$.
(b) Show that $Q_{m+1} - Q_m = (P_m^2 - P_{m+1}P_{m-1})/P_{m-1}P_m \le 0$.

4.25 Derive (4.95) as follows:
(a) Taking logarithms on both sides of (4.88), show that

$$\log n - \log P_{\text{loss}} \ge d \log d + f(d),$$

where

$$f(d) = 2 - d \log \rho_o - d - \frac{1}{2}\log \pi + \frac{1}{2}\log d.$$

(b) Taking logarithms on both sides of (4.93), show that

$$\log P_{\text{loss}} - (d+1)\log(1 - P_{\text{loss}}) \ge \log n + \log\left(1 - \frac{\log d}{n}\right) - 2$$
$$+ d\log \rho_o - d \log d + (d-1)\log\left(1 - \frac{\rho_o}{4d}\right) + d.$$

From

$$(d-1)\log\left(1 - \frac{\rho_o}{4d}\right) \ge (d-1)\log\left(1 - \frac{\rho_o}{4}\right)$$

since $d \geq 1$, and

$$\log\left(1 - \frac{\log d}{n}\right) \geq \log\left(1 - \frac{\log d}{2^{d \log d + f(d)}}\right)$$

by substitution from the result of part (a), show that

$$\log n - \log P_{\text{loss}} \leq d \log d + g(d),$$

where

$$g(d) = -\log\left(1 - \frac{\log d}{2^{d \log d + f(d)}}\right) + 2 - d \log \rho_0 -$$
$$(d-1) \log\left(1 - \frac{\rho_0}{4}\right) - d - (d+1) \log(1 - P_{\text{loss}}).$$

(c) From parts (a) and (b), conclude that

$$d \log d = \log \log N - \log P_{\text{loss}} + O(d).$$

4.26 Which of the following switches operated as a loss system has the highest throughput (or lowest loss probability)?
1. Knockout switch with knockout parameter = 4.
2. Dilated banyan network with dilation degree = 4.
3. Parallel banyan network with four banyan networks in parallel.

4.27 Which input-buffered switch has the higher throughput?
1. Speedup switch with speedup factor = 3.
2. Channel-grouped switch with group size = 3.

5

ADVANCED SWITCH DESIGN PRINCIPLES

The preceding chapter has introduced various basic switch design principles to solve packet contention problem. In this chapters, we shall first present some advance switch design principles to further alleviate this problem. We shall see that despite these advanced switch designs, switch dimensions continue to be severely limited by both technological and physical constraints. We close this chapter by providing some key principles to construct large switches out of modest-size switches, without sacrificing overall switch performance.

5.1 SWITCH DESIGN PRINCIPLES BASED ON DEFLECTION ROUTING

One way to solve the packet contention problem without having to buffer the losing packets is to use deflection routing. The basic idea is to route (deflect) the losing packets to "wrong" outgoing links rather than drop them. Redundancy is built into the switch design so that deflected packets can be routed in later switching stages in a way that corrects the earlier mistakes.

5.1.1 Tandem-Banyan Network

The tandem-Banyan switching network consists of K Banyan networks connected in series (see Fig. 5.1). Except for the last Banyan network, each output of a Banyan network is connected to both an input of the subsequent Banyan network and a concentrator (statistical multiplexer). With this setup, a packet would be routed to the concentrator if it reaches the correct output, and to the subsequent Banyan network otherwise. Thus, each packet can have up to K attempts to reach its destined output.

Principles of Broadband Switching and Networking, by Tony T. Lee and Soung C. Liew
Copyright © 2010 John Wiley & Sons, Inc.

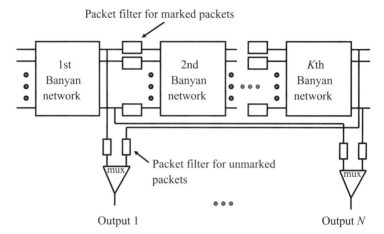

FIGURE 5.1 The tandem-Banyan network.

Deflection routing is employed within each Banyan network: whenever there is a conflict at a 2×2 switch element, one packet would be routed correctly while the other would be marked and routed in the wrong direction. To optimize the number of correctly routed packets, the marked packet would have a lower priority than an unmarked one for the rest of its journey within the Banyan network. In other words, because of the unique-path property, a deflected packet will reach the wrong output no matter how it is routed in the later stages of the Banyan network, and we might as well try to route the unmarked packets correctly when they contend with the marked packets.

If a packet remains unmarked when it reaches the output of the Banyan network, it has reached the correct destination. Therefore, it is removed and forwarded to the concentrator associated with the output destination. On the other hand, a marked packet will be unmarked and forwarded to the next Banyan network, and a new attempt to route the packet to its desired output is initiated. A packet is considered lost if it still fails to reach the desired output after passing through all the K Banyan networks.

Let D be the delay suffered by a packet as it travels through a Banyan network. Certainly, a packet that reaches its correct destination at a later Banyan network experiences a larger delay than one that does so at an earlier Banyan network. Thus, correctly routed packets to an output address may not reach its concentrator simultaneously. To compensate for the delay differences, one can insert delay elements with varying delays at different places: for the links that connect the N outputs of Banyan network i to the N concentrators, one can introduce a delay of $(K - i)D$. In this way, correctly routed packets from all Banyan networks experience a delay of KD and arrive at the inputs of the concentrators simultaneously.

To study the functional dependence of K on N for a given P_{loss}, let $L_k = \rho_k/\rho_o$ be the probability that a packet still fails to reach its destination after traveling through

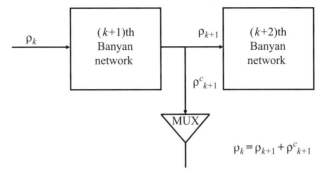

FIGURE 5.2 Relationship between the offered load ρ_k, carried load ρ^c_{k+1}, and rejected load ρ_{k+1} of the $(k+1)$ Banyan network in a tandem-Banyan network.

k Banyan networks, where ρ_0 is the initial offered load and ρ_k is the offered load to the input of $(k+1)^{\text{th}}$ Banyan network, and let ρ^c_k be the carried load on each output of the kth Banyan network (see Fig. 5.2). It follows that

$$\rho_{k+1} = \rho_k - \rho^c_{k+1} \approx \rho_k - \frac{4\rho_k}{n\rho_k + 4} = \frac{n\rho_k^2}{n\rho_k + 4}. \quad (5.1)$$

We then immediately have the recursive formula for the loss probability of a tandem-Banyan switch with $k+1$ Banyan networks:

$$L_{k+1} = (\rho_{k+1}/\rho_0) = \frac{n\rho_0(\rho_k/\rho_0)^2}{n\rho_0(\rho_k/\rho_0) + 4} = \frac{aL_k^2}{aL_k + 4}, \quad (5.2)$$

where $a = \rho_0 n = \rho_0 \log N$. From the above,

$$L_{k+1} - L_k = \frac{-4L_k}{aL_k + 4}. \quad (5.3)$$

Using the Taylor series approximation technique introduced earlier, we can transform the above difference equation into the following differential equation:

$$\frac{dL_k}{dk} = \frac{-4L_k}{aL_k + 4}.$$

Integration and matching of the boundary conditions, $L_0 = 1$ and $L_K = P_{\text{loss}}$, gives

$$K = \frac{\rho_0 \log N}{4}(1 - P_{\text{loss}}) - \ln P_{\text{loss}}. \quad (5.4)$$

In the analysis, we have made the assumption that packets at successive Banyan networks are uncorrelated with each other. Whether two packets collide in a Banyan network depends on whether there are common links between their paths, which in turn are determined by their inputs and outputs by the unique-path property. Even if the input packets to the first Banyan network are uncorrelated, they become correlated in the succeeding Banyan networks because of contention and path overlaps in the preceding Banyan networks. Consequently, packet output destinations are not independent anymore after the first Banyan network. Recall that there are many possible network structures within the class of Banyan networks (e.g., see Fig. 3.4). The Banyan network in the tandem-Banyan network can be any of them provided the associated routing algorithm is also used. It turns out that the loss probability of the overall tandem network depends rather strongly on which Banyan network is used, thanks to the fact that different Banyan networks generate different packet correlations at subsequent stages of the tandem network. The study of this dependency is complicated and will not be treated here. We point out, however, that the noncorrelation assumption in our analysis is one that yields an optimistic performance estimate.

In any case, the analysis does reveal the asymptotic complexity order of the switch. From (5.4), the order of complexity of the tandem-Banyan network for a fixed loss probability is $N(\log N)^2$, the same as that of the Batcher–banyan switch! That is, even under the optimistic assumption, the complexity of the tandem-Banyan network is still of order $N(\log N)^2$.

5.1.2 Shuffle-Exchange Network

If a packet is deflected in a Banyan network of the tandem-Banyan switch, say at the ith stage, it must go through another $n - i = \log_2 N - i$ stages before it exits the Banyan network and an attempt to route it in the next Banyan network can be started. The journey of a packet to its destination can be represented by a finite-state machine (see Fig. 5.3) wherein the state, called the *distance*, is the remaining number of stages that the packet must travel before reaching its desired output. Certainly, at the entrance of the tandem-Banyan switch, all packets have a distance of n. As shown in Fig. 5.3, each time a packet is routed correctly, the distance is decremented by 1. However, each time a packet in state j, $j \leq n$, is deflected, its distance increases by $n - 1$ to $n + j - 1$. This increase in distance is called the *deflection distance*.

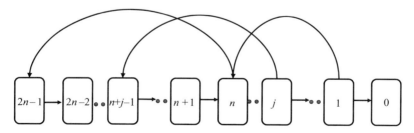

FIGURE 5.3 The state-transition diagram of the tandem-banyan network.

SWITCH DESIGN PRINCIPLES BASED ON DEFLECTION ROUTING   155

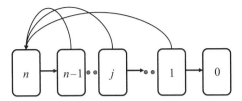

FIGURE 5.4 The state-transition diagram of the shuffle-exchange network.

The number of Banyan networks required to meet a P_{loss} requirement, hence the switch complexity, is closely related to the number of steps required for a packet to reach its destinations in the finite-state machine model. The number of steps is a random variable depending on the number of deflections and the deflection penalties. With K Banyan networks in the tandem structure, a packet must reach its destination within Kn steps, or else it will be dropped. Thus, minimizing deflection distance has the effect of decreasing the packet loss probability. By the same token, for a fixed loss probability requirement, minimizing deflection distance reduces the number of Banyan networks required, hence the switch complexity.

A question one might raise is whether attempts can be made immediately upon deflections to route packets to their destinations rather than having to waste $j - 1$ more steps in the current Banyan network. In other words, can one realize a finite-state machine model depicted in Fig. 4.18 wherein each deflection sets the distance of the packet back to n rather than $n + j - 1$? This requires the stages in the Banyan network to be homogeneous and leads us to the shuffle-exchange network introduced in Chapter 3 (see Fig. 3.4).

We mentioned in Chapter 3 that the regular shuffle-exchange network with n stages is isomorphic to the banyan network without a formal proof. To show that the same routing mechanism applies and to introduce some key concepts and notation, let us first consider the structure of a single stage, shown in Fig. 5.5.

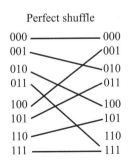
Packet on link $x_3 x_2 x_1 \rightarrow x_2 x_1 x_3$

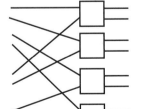
Packet on link $x_3 x_2 x_1 \rightarrow x_2 x_1 d_3$
d_3 = *routing bit*

FIGURE 5.5 Algebraic operations of shuffle and exchange.

156 ADVANCED SWITCH DESIGN PRINCIPLES

The term "shuffle" refers to the way adjacent switch stages are interconnected: it is analogous to the shuffling of a deck of cards. Suppose we take the top and bottom halves of a deck of cards and merge them by taking cards one by one alternately from the two halves from top to bottom. We are said to have performed a "perfect shuffle." Imagine a deck of N cards riding on the outgoing links of a switch stage, one on each link. These cards "travel" to the incoming links of the next stage through the interconnection. It can be easily seen from Fig. 5.5 that the interconnection performs a perfect shuffle on the cards.

What is interesting from the switching viewpoint is that perfect shuffle achieves a cyclic left shift of the link label. Let us label the links from top to bottom in a binary fashion. Thus, the top link is $00\cdots 0$ and the bottom link is $11\cdots 1$. It can be shown that (or verified from the example in Fig. 5.5) outgoing link $x_n x_{n-1} \cdots x_1$ of a stage is connected to incoming link $x_{n-1} \cdots x_1 x_n$ of the next stage. In other words, a packet on link $x_n x_{n-1} \cdots x_1$ occupies link $x_{n-1} \cdots x_1 x_n$ after the shuffle.

The term "exchange" refers to the operation performed by 2×2 switch elements. Let us label the 2×2 switch elements in any stage with $(n-1)$-bit binary numbers. The two incoming, as well as the two outgoing, links connected to switch element $x_{n-1} \cdots x_1$ are labeled $x_{n-1} \cdots x_1 0$ and $x_{n-1} \cdots x_1 1$. The 2×2 switch element forwards an input packet to output link $x_{n-1} \cdots x_1 0$ if the routing bit is 0 and to link $x_{n-1} \cdots x_1 1$ if the routing bit is 1. In other words, the switch element *exchanges* the least significant bit of the link label with the routing bit. Thus, if the incoming link occupied by a packet is $x_{n-1} \cdots x_1 x_n$ at a switching stage and d_n is the routing bit, then the packet will occupy link $x_{n-1} \cdots x_1 d_n$ at the output of the stage (see Fig. 5.5).

The algebraic operations of the shuffling and switching stages provide us with a way to understand why the overall routing mechanism works. Let the source and destination addresses of a packet be $S = s_n \cdots s_1$ and $D = d_n \cdots d_1$, respectively. The destination address will be used for routing starting from the most significant bit to the least significant bit. Initially, the packet occupies link $s_n \cdots s_1$ at the entrance to the shuffle-exchange network. After the first shuffle, its link label is $s_{n-1} \cdots s_1 s_n$ at the input to the first-stage switching node. Bit d_n is used to switch this packet to outgoing link $s_{n-1} \cdots s_1 d_n$. We see that s_n has been replaced by d_n. By another shuffle and exchange, the packet then occupies link $s_{n-2} \cdots s_1 d_n d_{n-1}$ at the output of the next stage. Repeating this process, we see that the output links of the successive switching stages traversed by a packet are

$$S = s_n \cdots s_1 \rightarrow s_{n-1} \cdots s_1 d_n \rightarrow \cdots \quad (5.5)$$

$$\rightarrow s_{n-i} \cdots s_1 d_n \cdots d_{n-i+1} \rightarrow \cdots \quad (5.6)$$

$$\rightarrow d_n \cdots d_1 = D. \quad (5.7)$$

In other words, the sequence of links is embedded in the binary string $s_n \cdots s_1 d_n \cdots d_1$ and can be revealed by an n-bit window moving one bit per stage from left to right. After the last step, the packet reaches link $d_n \cdots d_1$, which is the desired destination. In this way, bit by bit, the packet is routed to its targeted output. Figure 5.6 gives an example of how a packet is routed in an 8×8 regular shuffle-exchange network.

SWITCH DESIGN PRINCIPLES BASED ON DEFLECTION ROUTING 157

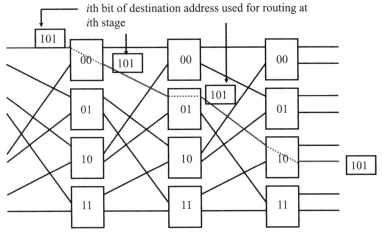

Sliding window routing: packet with source–destination label $s_n...s_{n-i}...s_1 d_n...d_{n-i+1}...d_1$ occupies link $s_{n-i}...s_1 d_n...d_{n-i+1}...d_1$ after stage i

FIGURE 5.6 Routing in shuffle-exchange network.

The above describes the route traversed by a packet when there is no contention from other packets. When deflection routing is used to tackle the contention problem, instead of a regular shuffle-exchange network with n stages, we have an elongated shuffle-exchange network with $L \geq n$ stages, as depicted in Fig. 5.7. As discussed

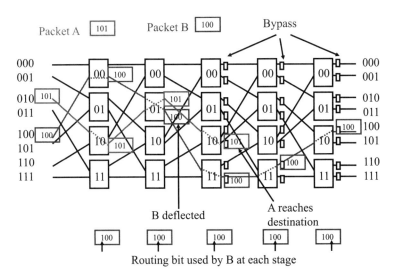

FIGURE 5.7 A shuffle-exchange network with $n = 3$ and $L = 5$.

158 ADVANCED SWITCH DESIGN PRINCIPLES

before, the output address of a packet is represented in binary form for routing, and the most significant bit is used in the first stage, the second most significant bit in the second stage, and so on. When all the bits are used, the packet has reached its desired destination, and it will be removed from the elongated shuffle-exchange network and forwarded to the concentrator of the associated output so that it will bypass the remaining stages. The mechanism for doing so will be referred to as the *bypass* mechanism. When a deflection occurs, routing is started from the most significant bit again. An example is given in Fig. 5.7. It is possible for a packet to reach its destination after stages $n, n+1, \cdots,$ or L, depending on the number of deflections and the deflection distances. Therefore, bypass mechanisms must be installed after each of these stages.

Figure 5.4 shows the state-transition diagram associated with the elongated shuffle-exchange network. Each deflection sets the distance of a packet back to n. Because of the reduced penalty associated with deflections, we would expect the complexity of this switch to be less than that of the tandem-Banyan network. Although this is true, it can be shown that the asymptotic complexity *order* of this network is still not better than that of the tandem-Banyan network: $N(\log_2 N)^2$. The reason is that the deflection penalty is still too large: a packet that is already very close to its destination may still be deflected back to state n. One might wonder if it is possible to construct a network in which the deflection penalty is a constant that is independent of n. In fact, one can show that the network complexity will then be of order $N \log N$. It turns out that the dual shuffle-exchange network to be described in Section 5.1.5 can indeed achieve this. The network can be understood more easily after we have discussed the feedback version of the shuffle-exchange network and its bidirectional variant.

5.1.3 Feedback Shuffle-Exchange Network

Let us consider a feedback shuffle-exchange network that has only one switching stage. The outgoing links, instead of being forwarded to a next stage, are fed back to the same stage after a shuffle. Figure 5.8 gives an example of such a network with four nodes.

Feedback and feedforward networks are also called the undirected and directed networks, respectively, in the computer science literature. Both are used to allow the processors and memories in a parallel computer architecture to communicate with each other. Each input or output link of the directed network is connected to a processor or memory. Its nodes are just simple switching elements. Each node in the undirected network, on the other hand, is associated with a processor. Nodes rather than input and output links are the sources and destinations of packets.

Suppose that there are N nodes in the feedback shuffle-exchange network whose addresses are labeled with a binary number $x_n \cdots x_1$. Each node has two incoming links and two outgoing links. The links can be labeled with an $(n+1)$-bit binary number $x_n \cdots x_1 x_0$. The outputs of the nodes undergo a perfect shuffle before they are fed back to the inputs. Thus, a packet that is switched to output d_n of node $x_n \cdots x_1$ will reach node $x_{n-1} \cdots x_1 d_n$ after the shuffle. Using the destination address $D = d_n \cdots d_1$

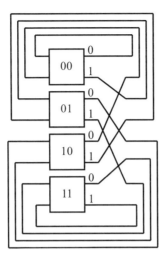

FIGURE 5.8 A four-node feedback shuffle network.

as the routing tag, a packet will reach its destination node after n undeflected steps. From this, we see that the operation of the feedback network is similar to that in the feedforward network.

Notice, however, that whereas the design issue in the feedforward network is the number of stages L required to meet a loss probability requirement, the situation here is slightly different in that a packet can be routed indefinitely until it reaches its destination node: packet loss probability due to contention is zero! Moreover, the number of stage is fixed at one. Is there some magic going on here? The answer is we have not gained something for nothing. It turns out that the performance issue in the feedback network is throughput rather than packet loss. Whereas the maximum throughput per input or output of the feedforward network is close to one when the loss probability is small, which can be achieved by having a sufficient number of stages, the fact that there is only one stage in the feedback network severely limits the network throughput when N is large. We shall derive this throughput limitation analytically later in this subsection.

Recall that with the multistage feedforward shuffle-exchange network, a node forwards a packet to the next stage as soon as it has received enough bits of the header for it to make the switching decision. It is not necessary to wait for the arrival of the entire packet.[1] With the feedback version of the network, because of the feedback loop, the information bits of the previous packets are still being transmitted at the outputs when their headers arrive at the inputs of the nodes. Therefore, the arriving packets must wait until all the bits of the previous packets have been transmitted before they can be forwarded to the outputs. As shown in Fig. 5.9, a one-packet buffer is located at each input link of a node to buffer the arriving packet while it waits for

---

[1]This is sometimes called virtual cut-through routing.

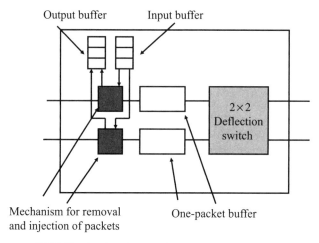

FIGURE 5.9 A node in the feedback shuffle network.

the next time slot to be transmitted. In addition, there is a mechanism at the inputs for the removal of packets destined for the node and the injection of packets originating from the node. There is also an input queue for the storage of packets originating from the node. A packet in the queue must wait for an empty time slot on the input links (i.e., when one of the inputs does not carry an active packet) before it can enter the network. In the following, we assume a greedy policy in which as soon as an empty slot is available in one of the input links, the packet at the head of the input queue will be injected into the network.

As in the feedforward shuffle-exchange network, whenever a packet is deflected, routing of the packet starts anew, beginning with the most significant bit of the destination address again. To implement self-routing, a possibility is to have two fields in the header of a packet: a routing tag consisting of the destination address $d_n \cdots d_1$ and a $\log_2 n$-bit pointer indicating the current routing bit d_i to be used. Originally, the pointer points to bit d_n. Whenever the packet is routed correctly, the pointer value is decremented, and whenever the packet is deflected, the pointer is reset to point to d_n again.

To analyze the delay and throughput of the system, let us make the simplifying assumption that the packets arriving at the two inputs of a node are independent and that they are equally likely to be destined for either of the two outputs. We assume a simple contention-resolution policy in which the winning packet is chosen at random. Assume further that the deflection probability q is independent of the distance of the packet—this is a reasonable simplifying assumption when there are a large number of packets entering and leaving the system in each time slot. Let T_i denote the expected additional number of steps taken by a packet in state i before reaching its destination node. We have

$$T_i = 1 + pT_{i-1} + qT_n, \qquad 1 \le i \le n, \tag{5.8}$$
$$T_0 = 0,$$

where $p = (1 - q)$ is the probability of not being deflected. Equation (5.8) is a linear difference equation in i. It is a well-established result that the general solution of a linear difference equation is a constant times the homogeneous solution plus a particular solution. The homogeneous solution to $T_i - pT_{i-1} = 0$ is $T_i = p^i$ and a particular solution to $T_i - pT_{i-1} = 1 + qT_n$ is $T_i = (1 + qT_n)/(1 - p) = (1 + qT_n)/q$. Therefore, we have

$$T_i = cp^i + \frac{1 + qT_n}{q}, \qquad (5.9)$$

where c is a constant to be found by matching the boundary condition. The boundary condition $T_0 = 0$ yields $c = -(1 + qT_n)/q$. Thus,

$$T_i = \frac{(1 + T_n)(1 - p^i)}{q}, \qquad 0 \le i \le n. \qquad (5.10)$$

Substituting $i = n$ in the above, we get

$$T_n = \frac{1 - p^n}{p^n q}. \qquad (5.11)$$

Substituting this into (5.9), we have

$$T_i = \frac{1 - p^i}{p^n q}, \qquad 0 \le i \le n. \qquad (5.12)$$

Let ρ be the probability of finding a packet at an input link at the beginning of a time slot; this is the link loading. Then, with probability $\rho/2$ an input packet will encounter a packet at the other input of the same node that desires the same output. Thus, with the random contention-resolution scheme, the packet will be deflected with probability $q = \rho/4$. The total throughput of the network Λ is given by Little's law:

$$\Lambda = \frac{\text{Expected number of packets in the whole network}}{\text{Expected packet delay}}$$

$$= \frac{2N\rho}{T_n} = \frac{8Nq^2 p^n}{1 - p^n}$$

$$= \frac{8N(\rho/4)^2 (1 - \rho/4)^n}{1 - (1 - \rho/4)^n}. \qquad (5.13)$$

It can be shown from the above equation that for $n \le 4$, Λ is maximum when $\rho = 1$. This is achieved when the input queues at the nodes are saturated so that as soon as a packet reaches its destination, a new packet is injected into the network. In this case, the maximum throughput of the system $\Lambda_{\max}$ is equal to its saturation throughput

162 ADVANCED SWITCH DESIGN PRINCIPLES

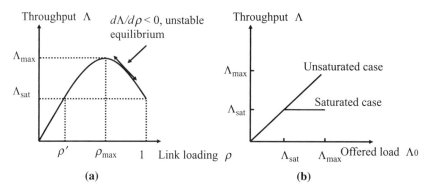

FIGURE 5.10 (a) Throughput as a function of link loading; (b) throughput as a function of offered load; for feedback shuffle-exchange network with $n \geq 5$.

Λ_{sat}. As long as the offered load to the system is lower than Λ_{max}, the input queues will not saturate.

Things become more complicated when $n \geq 5$ because the maximum throughput is not obtained when the input queues are saturated (see Problem 5.4). In other words, we can find a link loading $\rho^* < 1$ such that $\Lambda(\rho = \rho^*) > \Lambda(\rho = 1) = \Lambda_{sat}$. As shown in Problem 5.4, $d\Lambda/d\rho$ becomes negative when ρ becomes too large. Thus, beyond a certain offered load, the carried load of the system actually decreases. Intuitively, when there are too many packets in the system, packet deflections due to contention become very likely, and it may take so many steps for packets to reach their destinations that the throughput decreases. The reason why this does not happen for smaller n is that the distance to destination is bounded by n, and the penalty associated with deflections is not so high when n is small.

The fact that $\Lambda_{max} > \Lambda_{sat}$ also reveals that the system can run into unstable congestion behavior. To explain this, let ρ_{max} be the value of ρ when Λ_{max} is achieved. Each $\Lambda \in [\Lambda_{sat}, \Lambda_{max})$ can be achieved with two ρ values (see Fig. 5.10(a)). Let ρ' be the lower ρ value at which Λ_{sat} is achieved. Supposed that the offered load to each and every node is Λ_o/N so that the total offered load to the network is Λ_o. We consider three regions of operation: $\Lambda_o > \Lambda_{max}$, $\Lambda_{sat} \leq \Lambda_o \leq \Lambda_{max}$, and $\Lambda_o < \Lambda_{sat}$.

When $\Lambda_o > \Lambda_{max}$, the input queues saturate because the input rate of packets is greater than the sustainable output rate, the throughput. Because the queues are saturated, as soon as a packet reaches its destination and leaves the system, another packet will enter the system (since we assume the greedy policy that a packet from a nonempty queue will be injected into the system as soon as an empty slot is available). Therefore, the achieved throughput $\Lambda = \Lambda_{sat}$, corresponding to the full link-loading situation when $\rho = 1$.

When $\Lambda_o < \Lambda_{sat}$, the queues will not saturate, and throughput equals offered load: $\Lambda = \Lambda_o$. When $\Lambda_{sat} \leq \Lambda_o \leq \Lambda_{max}$, the system can operate in either the stable region $\rho \in [\rho', \rho_{max}]$—a value of ρ between ρ' and ρ_{max}—or the unstable region

$\rho \in [\rho_{\max}, 1]$. In the stable region, throughput equals offered load: $\Lambda = \Lambda_o$. In the unstable region, as reasoned in the next paragraph, the system quickly evolves away from the region, ending up either at $\rho = 1$ and $\Lambda = \Lambda_{\text{sat}}$, the saturated case, or at $\rho \in [\rho', \rho_{\max})$ and $\Lambda = \Lambda_o$, the unsaturated case.

The offered load is the average packet arrival rate to the input queues over a long time period. Because of the statistical nature of packet arrivals and departures, the number of backlogged packets in the input queues varies over time. These backlogged packets in turn determine the instantaneous or near-term offered load to the network. Therefore, the instantaneous offered load fluctuates over time in accordance to the backlogged packets. Any increase in the number of nonempty input queues results in an increase in instantaneous offered load, which in turn raises the instantaneous link loading ρ of the network. By the same token, decreasing instantaneous offered load also decreases instantaneous ρ. In the unstable region, since $d\Lambda/d\rho$ is negative, the throughput decreases as a result of even a small increase in the instantaneous offered load. Therefore, the throughput Λ becomes smaller than the offered load Λ_o, and the negative balance between packet arrival and departure rates further increases the input-queue backlogs, resulting in an even higher instantaneous offered load. This creates a positive feedback effect that eventually brings Λ down to Λ_{sat} and the input queues are saturated. By the same token, in the unstable region any decrease in the instantaneous offered load will quickly bring the system back to the stable point at which $\Lambda = \Lambda_o$ through a positive feedback effect. Figure 5.10(b) shows Λ as a function of Λ_o. There is a bifurcation point at $\Lambda_o = \Lambda_{\text{sat}}$, and there are two operating regions between $\Lambda_o = \Lambda_{\text{sat}}$ and $\Lambda_o = \Lambda_{\max}$.

If the system is unsaturated with $\Lambda_o \geq \Lambda_{\text{sat}}$ (hence $\rho' \leq \rho < \rho_{\max}$), there is still the danger that the system will evolve to the unstable region and then become saturated. This can be caused by (1) a sudden increase in the number of packets arriving at the input queues, or (2) a series of bad luck in which packets already in the network are deflected more than the usual number of times so that there is a sudden increase in link loading ρ to beyond $\rho_{\max}$. Once that happens, the system may become saturated very quickly, making it impossible for the system to return to the original operating point. On the other hand, if $\Lambda_o < \Lambda_{\text{sat}}$ (hence $\rho < \rho'$), the system will eventually evolve back to the stable region even when there is a temporary increase in the number of packets in the system. This is because the saturation throughput Λ_{sat} can sustain the offered load over the long term. Such is not the case, however, when $\Lambda_o \geq \Lambda_{\text{sat}}$. In general, the system can only be safely operated at offered load below Λ_{sat} even though the maximum throughput $\Lambda_{\max}$ is higher. One can easily write a simulation computer program to verify this: when $\Lambda_o \in [\Lambda_{\text{sat}}, \Lambda_{\max}]$, even if the throughput and the offered load are balanced to begin with ($\Lambda = \Lambda_o$), the system eventually evolves to the saturation point $\Lambda = \Lambda_{\text{sat}} < \Lambda_o$.

In the unstable region, we have a typical network congestion scenario in which higher offered load brings about lower throughput and higher delay at the same time. Figure 5.10 also suggests that if some congestion control mechanism is exercised over the system, offered load of up to $\Lambda_{\max}$ may be sustainable without instability. For instance, we could introduce $2N\rho_{\max}$ tokens in the network. A packet is not allowed to enter the network until it has acquired a token from one of the input links, and the token will be released back to the network only when the packet has reached its destination. In this way, ρ is kept to be not more than $\rho_{\max}$. As with all networks operated with token access, one has to deal with the possibility of lost tokens due to transmission errors or malfunctioning nodes.

Another strategy is to use a contention-resolution policy that is more efficient than the random contention-resolution policy and it is hoped that it will reduce the deflection penalty enough to stabilize the system. It turns out that an arbitration policy that favors packets closer to their destinations will stabilize the system. When a packet in state i is deflected back to state n, $n - i + 1$ previous routing steps are wasted. The idea of favoring packets closer to their destinations is to minimize the amount of routing effort that is wasted given that deflection is inevitable.

The analysis of this strategy is more complicated because the deflection probability is state dependent. Instead of the uniform deflection probability q, we have q_i for the deflection probability at state i, $1 \leq i \leq n$. Although tokens are not used in reality, we can still imagine for analytical purposes that there are $2N$ fictitious tokens, one on each link, circulating around the network. A packet at an input queue must acquire an unused token before it can enter the network. A token is said to be active and in state i, $1 \leq i \leq n$, if the packet it carries is in state i. A token is said to be inactive or in state 0 if it is unused.

We trace the evolution of the state of a particular token. Let $P(i)$ be the probability that the token is in state i at the beginning of a time slot. We have

$$P(i-1) = P(i)(1 - q_i), \qquad 2 \leq i \leq n. \tag{5.14}$$

Define

$$\pi(i) = P(1) + \cdots + P(i), \qquad 1 \leq i \leq n,$$

to be the probability of the token either in state 1, 2, $\cdots$, or i. By definition, the probability of the token being active is $\pi(n)$. An active token implies the link it occupies is active. Since every link has a token on it at all time, the probability of an active token must also be the probability of an active link. Thus,

$$\pi(n) = \rho.$$

Assume that when two packets of the same state contend with each other, the winning packet will be chosen at random. The probability of a packet in state i being deflected is given by the sum of the probability of being deflected by a packet in state i and the probability of being deflected by a packet in state below i:

$$q_i = \frac{P(i)}{4} + \frac{P(1) + \cdots + P(i-1)}{2} = \frac{P(i)}{4} + \frac{\pi(i)}{2}, \qquad 1 \leq i \leq n.$$

$$\tag{5.15}$$

Substituting the above into (5.14) for each i, summing equations, and after some manipulation, we obtain

$$\pi(i-1) = [\pi(i) - \pi(1)]\left\{1 - \frac{\pi(1)}{2} - \left[\frac{\pi(i) - \pi(1)}{4}\right]\right\}, \quad 2 \leq i \leq n, \quad (5.16)$$

$$\pi(n) = \rho. \quad (5.17)$$

Equation (5.16) is subject to this interpretation: the probability that the token is active and in state $i-1$ or lower is equal to the probability that it was in the previous time slot an active token in state between 2 and i and that the packet it carries was not deflected. The above can be simplified to

$$\pi(i-1) = \pi(i) - \frac{\pi^2(i)}{4} - \pi(1) + \frac{\pi^2(1)}{4}, \quad 2 \leq i \leq n,$$

$$\pi(n) = \rho. \quad (5.18)$$

The throughput is related to $\pi(1)$ as follows. The probability of a token in state 1 not being deflected is $(1 - \pi/4)$, and if it is not deflected the packet carried by it will reach the final destination. Therefore,

$$\Lambda = 2N\pi(1)\left(1 - \frac{\pi(1)}{4}\right). \quad (5.19)$$

Unfortunately, (5.18) does not allow $\pi(1)$ to be solved in closed form in terms of ρ. Therefore, numerical method is needed for an exact solution. Equation (5.18) does let us show that the kind of instability associated with random contention-resolution policy does not occur here. The key is to show that $d\Lambda/d\rho > 0$ for all $0 \leq \rho \leq 1$. From (5.19),

$$\frac{d\Lambda}{d\pi(1)} = 2N\left[1 - \frac{\pi(1)}{2}\right] > 0. \quad (5.20)$$

From (5.18),

$$\frac{d\pi(i)}{d\pi(i-1)} = \frac{\left\{1 + \left[1 - \frac{\pi(1)}{2}\right]\frac{d\pi(1)}{d\pi(i-1)}\right\}}{[1 - \frac{\pi(i)}{2}]}, \quad 2 \leq i \leq n. \quad (5.21)$$

Substituting $i = 2$, we find that $d\pi(2)/d\pi(1) > 0$. It can be easily shown by induction using (5.21) that $d\pi(i)/d\pi(1) > 0$ for all $1 \leq i \leq n$. Thus,

$$\frac{d\Lambda}{d\rho} = \frac{d\Lambda}{d\pi(n)} = \frac{d\Lambda}{d\pi(1)}\frac{d\pi(1)}{d\pi(n)} > 0. \quad (5.22)$$

166 ADVANCED SWITCH DESIGN PRINCIPLES

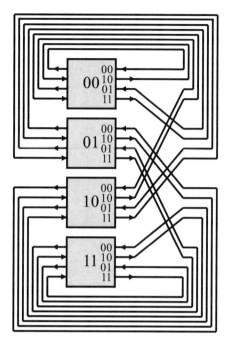

FIGURE 5.11 A feedback bidirectional shuffle-exchange network.

Although instability does not occur with shortest-distance priority contention resolution, the deflection penalty is still rather significant for large n. The next subsection considers a shuffle-exchange network in which the shuffle links are bidirectional. It is shown that the deflection distance can be reduced to 1.

5.1.4 Feedback Bidirectional Shuffle-Exchange Network

Let us consider a bidirectional shuffle-exchange network in which the shuffle links are bidirectional so that packets can travel in the reverse direction. In actual implementation, each node is 4×4 in dimensions, and there is a set of unshuffle links laying side by side with the set of shuffle links, as shown in Fig. 5.11. When a packet is deflected from node i to node j, thanks to the link in the reverse direction connecting node j to node i, the packet can travel back to node j from node i, correcting the deflection in one step. Implementation of self-routing is more complicated here, since a packet can be deflected a number of times in succession. We need to encode systematically in the packet header the routing steps required to correct deflections.

Since each node has four outputs, two routing bits are needed to indicate the outgoing link desired by a packet. In the original shuffle-exchange network, the two outputs of a node are labeled 0 and 1. Here, the four outputs are labeled 00, 01, 10, and 11: the most significant bit indicates whether the link is a shuffle or unshuffle link. Without loss of generality, let outputs 00 and 01 be unshuffle links 0 and 1, respectively, and outputs 10 and 11 be the shuffle links 0 and 1, respectively. The collections of shuffle and unshuffle links connected to all nodes are said to be in the shuffle and unshuffle *planes*, respectively.

Consider a packet at nodes $x_n \cdots x_1$. If it is routed to a shuffle link $1r$ ($r = 0$ or 1), by the cyclic left shift operation, the next node visited is $x_{n-1} \cdots x_1 r$. That is, the combination of switching and shuffling removes bit x_n from the left and attaches bit r to the right of the node label. If, on the other hand, the packet is routed to an unshuffle link $0r$, by the cyclic right shift operation, the next node visited is $rx_n \cdots x_2$: bit x_1 is removed from the right and bit r attaches to the left of the node label.

Let us first look at the routing of a packet from source $S = s_n \cdots s_1$ to destination $D = d_n \cdots d_1$ through the network without deflections. There are two possible ways of setting the initial routing tags of the packet. We can set it to $1d_n 1d_{n-1} \cdots 1d_1$, in which case the packet will travel to its destination via the shuffle plane. After each hop, the next two bits will be used for routing in the next node. The sequence of nodes traversed is then the same as that in the unidirectional shuffle-exchange network given by (5.7). Alternatively, we can set the routing tag to $0d_1 0d_2 \cdots 0d_n$, in which case the packet will travel through the unshuffle links (*note*: in the unshuffle network, the order in which the destination address bits are used is reverse, hence the way the routing tag is set). The sequence of nodes traversed is then

$$S = s_n \cdots s_1 \to d_1 s_n \cdots s_2 \to \cdots \quad (5.23)$$

$$\to d_i \cdots d_1 s_n \cdots s_{i+1} \to \cdots \quad (5.24)$$

$$\to d_n \cdots d_1 = D. \quad (5.25)$$

We now consider what happens when a packet is deflected. For illustration, let us assume that the routing tag of the packet is set to $1d_n 1d_{n-1} \cdots 1d_1$. The basic idea is to remove two routing bits from the routing tag whenever routing is successful in each stage, and add two error-correction routing bits whenever a deflection is encountered. After i hops without deflections, the packet arrives at an input of node $s_{n-i} s_{n-i-1} \cdots s_1 d_n d_{n-1} \cdots d_{n-i+1}$ with routing tag $1d_{n-i} \cdots 1d_1$. Suppose it is deflected at this node to the wrong shuffle output link. Instead of arriving subsequently at the correct node $s_{n-i-1} \cdots s_1 d_n \cdots d_{n-i}$, the packet will arrive at node $s_{n-i-1} \cdots s_1 d_n ... \bar{d}_{n-i}$, where $\bar{d}_{n-i}$ is the complement of d_{n-i}, as illustrated in Fig. 5.12.

In order to correct this deflection, we add two routing bits $0s_{n-i}$; this indicates that the packet wants to go back to the previous node where the deflection occurred. Thus, the routing tag at this point is $0s_{n-i} 1d_{n-i} 1d_{n-i-1} \cdots d_1$. Suppose routing at the node $s_{n-i-1} \cdots s_1 d_n \cdots \bar{d}_{n-i}$ is successful. Then the next node visited will be $s_{n-i} \cdots s_1 d_n d_{n-1} d_{n-i+1}$, the node at which the deflection occurred. At this point the routing tag will be $1d_{n-i} \cdots 1d_1$. We see that the packet returns to the *original state*, and the packet can proceed as before to its destination.

In general, a packet can be deflected to any of the three incorrect outputs at a node, and the basic idea is to add two routing bits to indicate the reverse link on which the

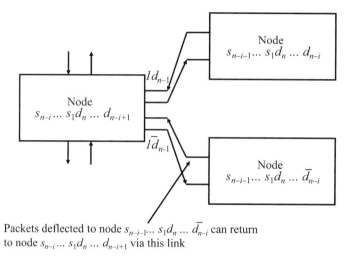

Packets deflected to node $s_{n-i-1}\ldots s_1 d_n \ldots \bar{d}_{n-i}$ can return to node $s_{n-i}\ldots s_1 d_n \ldots d_{n-i+1}$ via this link

FIGURE 5.12 Illustration showing how a routing error can be corrected via a reverse link.

packet can return to the node where the deflection occurred. Figure 5.13 summarizes the algorithm performed at the output of a node. Consider a packet at node $x_n \cdots x_1$ with routing tag $c_k r_k \cdots c_1 r_1$. If it is switched to the correct output indicated by the routing bits $c_k r_k$, then these two bits will be removed from its routing tag and it will

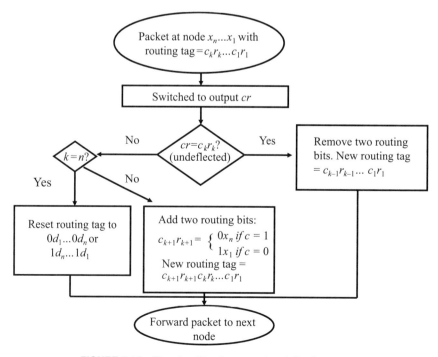

FIGURE 5.13 The algorithm for correcting deflection errors.

be forwarded to the next node with routing tag $c_{k-1}r_{k-1}\cdots c_1r_1$. If it is switched to an incorrect output cr, the routing bits $c_k r_k$ will not be removed and two routing bits $c_{k+1}r_{k+1}$ will be added to the routing tag:

$$c_{k+1}r_{k+1} = \begin{cases} 0x_n, & \text{if } c = 1, \\ 1x_1, & \text{if } c = 0. \end{cases}$$

The following explains why $c_{k+1}r_{k+1}$ are set in this manner. If $c = 1$, the packet has been deflected to the shuffle plane and will visit node $x_{n-1}\cdots x_1 r$ next. The two routing bits added, $0x_n$, will return the packet back to node $x_n\cdots x_1$ from node $x_{n-1}\cdots x_1 r$. If $c = 0$, the packet has been deflected to the unshuffle plane and will visit node $rx_n\cdots x_2$ next. The two routing bits, $1x_1$, will return the packet back to node $x_n\cdots x_1$ from node $rx_n\cdots x_2$.

Note that because of the addition of routing bits, the routing tag may grow in an unbounded fashion if the packet is deflected many times in successive nodes. To prevent this, whenever the distance is n and the packet is deflected, the routing tag is reset to either $1d_n\cdots 1d_1$ or $0d_1\cdots 0d_n$ so that the distance is always no more than n.

To see that successive deflections can be corrected with the algorithm in Fig. 5.13, refer to the finite-state machine representation in Fig. 5.14. The state of a packet is represented by a 2-tuple (*routing tag*, *node*). Since there is only a finite number of different combinations of routing tags and node labels, there is only a finite number of states. Suppose that a packet is deflected from the current state, say, state a, to one of the three possible states, say, state b. The error-correction algorithm guarantees that there is a transition through which the packet can return to state a later. But suppose

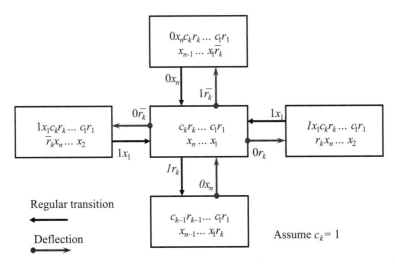

FIGURE 5.14 A state and its four adjacent states in the finite-state machine representation of packet state.

Pointer	Routing tag	Destination address
$\log n$ bits	$2n$ bits	n bits

(a)

Mode	Regular pointer	Regular routing tag	Error-correction pointer	Error-correction routing tag
1 bits	$\log n$ bits	$2n$ bits	$\log n$ bits	$2n$ bits

(b)

FIGURE 5.15 Two possible packet-header implementations.

that another deflection occurs immediately after the current deflection so that the packet reaches state c subsequently. By attaching two additional error-correction bits to the routing tag, we provide a return transition from state c to state b, from which the packet can eventually return to state a. The error-correction bits are added and used in a last-in-first-out fashion, and they guarantee a transition path back to state a no matter how many times the packet has been deflected. Thus, the algorithm is recursive and capable of correcting multiple deflections.

An alternative way to look at the problem that may appeal to readers who are familiar with Markov chain is as follows. Except for the absorption state that corresponds to packet reaching its destination, there is a transition path from each state to the other states. If at each state, the "correct" transition occurs with finite probability, then the packet will eventually reach the absorption state.

Figure 5.15 shows two possible implementations of the packet header. In (a), the header consists of a copy of the complete destination address, a routing tag, a pointer, and a destination indicator. The complete destination address is for the purpose of resetting the routing tag as described above. The routing tag contains the current remaining routing bits. The pointer points to the two routing bits to be used next. The removal of two routing bits upon successful switching can be implemented by decreasing the pointer value by 1 (*note*: pointer points to a pair of bits rather than one bit). When a deflection occurs, two routing bits are inserted at locations just beyond the position to which the pointer refers, and the pointer value increases by 1. Note that the pointer value is equal to packet distance minus 1, and it ranges from 0 to $n - 1$. The one-bit destination indicator is initially set to 0. After the last step in switching, it is set to 1 so that the next node recognizes the packet as being destined for it.

At any time, a packet in the bidirectional shuffle-exchange network is in one of two possible modes: it is either in the regular mode (when the packet is in its regular path to its destination), or in the error mode. In Fig. 5.15(b), a one-bit field is used to indicate the packet mode. There are two routing tags: the regular routing tag is used when the packet is in the regular mode and the error-correction routing tag is used when the packet is in the error mode. The regular routing tag is set to either $1d_n \cdots 1d_1$ or $0d_1 \cdots 0d_n$ initially and is never modified. The regular pointer points to the two routing bits in the regular routing tag that will

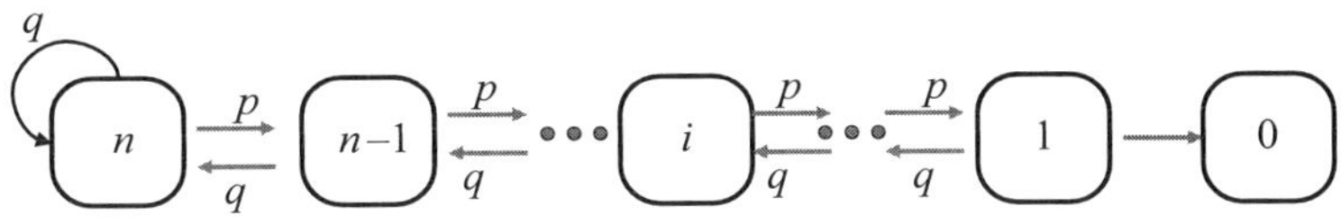

q = deflection probability

FIGURE 5.16 The state-transition diagram of a packet in the bidirectional shuffle-exchange network in which the distance is the state.

be used next. The error-correction routing tag consists of the error-correction routing bits. Whenever there is a deflection, two bits are added to this tag, and the error-correction pointer increases by 1. This field operates like a stack. When the bits in this tag are exhausted, the packet returns to the regular mode, and the regular routing tag will then be used for subsequent stages.

Let us now consider the performance of the network. As far as the distance to destination is concerned, the packet is performing a random walk associated with the state-transition diagram in Fig. 5.16. In actuality, the knowledge of distance is not enough for a complete analysis. A packet in general is not equally likely to be destined for the outputs in different planes. Therefore, for a fixed distance, the deflection probability for *cross-plane switching* is different from that for *in-plane switching*. The complete analysis is rather lengthy and involved.

We shall explore an approximate analysis that assumes that each packet is equally likely to be destined for any of the four outputs of a node regardless of whether the packet is in the error or regular mode. We shall also assume the random contention-resolution scheme.[2] As before, let T_i be the expected number of additional hops a packet in state i will experience before reaching the destination node. Then,

$$T_0 = 0, \tag{5.26}$$

$$T_i = 1 + pT_{i-1} + qT_{i+1}, \quad 1 \le i \le n-1, \tag{5.27}$$

$$T_n = 1 + pT_{n-1} + qT_n. \tag{5.28}$$

Let us define $M_i = T_i - T_{i-1}$. From (5.27) and (5.28), we have

$$pM_i = qM_{i+1} + 1, \quad 1 \le i \le n-1, \tag{5.29}$$

$$pM_n = 1. \tag{5.30}$$

---

[2] Since deflection distances are equal for all states (except stage n), favoring packets closer to their destinations in contention resolution will not yield as significant an improvement as in the shuffle-exchange network.

A particular solution to the linear difference equation (5.29) is $M_i = 1/(p-q)$, and the homogeneous solution to $pM_i - qM_{i+1} = 0$ is $(p/q)^i$. Therefore,

$$M_i = T_i - T_{i-1} = \frac{1}{(p-q)} + c\left(\frac{p}{q}\right)^i, \qquad (5.31)$$

where c is a constant that can be found from the boundary condition (5.30). Doing so yields

$$c = -\left(\frac{q}{p}\right)^n \left(\frac{1}{p-q} - \frac{1}{p}\right).$$

Now,

$$T_i = T_i - T_0$$

$$= \sum_{j=1}^{i} M_j$$

$$= \frac{i}{p-q} + c\frac{p}{q}\left[\frac{1 - (\frac{p}{q})^i}{1 - \frac{p}{q}}\right]$$

$$= \frac{i}{p-q} - \frac{\left(\frac{1}{p-q} - \frac{1}{p}\right)\left[\left(\frac{p}{q}\right)^i - 1\right]}{\left(\frac{p}{q}\right)^{n-1}\left(\frac{p}{q} - 1\right)}. \qquad (5.32)$$

To interpret the above equation, let us refer to the state-transition diagram in Fig. 5.16. In each step, with probability p, state i ($i < n$) will drift to the right (direction toward state 0) and with probability q, it will drift to the left. Therefore, $(p - q)$ is the expected drift toward state 0 at state i. Suppose that the state transitions were homogeneous without the reflection boundary at state n. That is, suppose that deflection distances in all states were equal to one, and a packet in state n would be deflected to state $n + 1$, a packet in state $n + 1$ would be deflected to state $n + 2$, and so on. It can be shown that $i/(p - q)$ would be the expected number of steps required to reach state 0 starting from any state i if $p > q$. In our case, thanks to the reflection boundary, the expected number of steps is reduced by the second term in (5.32). This reduction, however, is not significant for large n, and the system dynamics can be understood more readily by simply examining the first term.

To find p in terms of link loading ρ, consider the probability of at least one input packet desiring a particular outgoing link, which is $1 - (1 - \rho/4)^4$. If there is at least one input packet targeted for the output, there is an undeflected packet on it after the routing at the node. Thus, this is also the probability of finding an undeflected packet on an output link. By conservation, the probability must equal the probability of finding a packet on an input link and that it will not be deflected, ρp. Thus,

$$p = \frac{1 - (1 - \rho/4)^4}{\rho}. \tag{5.33}$$

Let us examine T_n for large n. Substituting $i = n$ into (5.32) and after some minor simplification, we obtain

$$T_n = \frac{n}{p-q} - \left(\frac{1}{p-q} - \frac{1}{p}\right)\left[\frac{1-(q/p)^n}{1-q/p}\right] \leq \frac{n}{p-q}. \tag{5.34}$$

In particular, T_n is dominated by the term $\frac{n}{p-q}$ for large n, and we can write

$$T_n \approx \frac{n}{p-q}. \tag{5.35}$$

Comparing the above with (5.11), the expected delay in the unidirectional shuffle-exchange network, we notice a very significant improvement: whereas the expected delay in (5.11) grows exponentially with n, the expected delay here grows only linearly with n. Since the expected delay is related to throughput, we should expect a marked improvement in throughput also.

Let us consider the system throughput Λ as a function of link loading ρ. There are four links to a node. By Little's law,

$$\Lambda = \frac{4N\rho}{T_n}. \tag{5.36}$$

It can be shown that $d\Lambda/d\rho < 0$ for ρ close to 1, and therefore the instability problem exists in this network. However, it can be shown that the instability problem is much less severe than the unidirectional shuffle-exchange network and that the maximum throughput $\Lambda_{\max}$ is very close to the saturation throughput Λ_{sat}. Problem 4.7 explores this issue in detail. With $\rho = 1$ in (5.33), (5.35), and (5.36), we get

$$\Lambda_{\text{sat}} \approx \frac{1.469N}{n}. \tag{5.37}$$

Instead of using a nonblocking 4×4 switch at each node, we could also use the simpler reverse banyan network shown in Fig. 5.17 (or any network within the Banyan class provided the associated routing algorithm is used) to perform switching at each node. The deflection probability will be higher since there is the possibility of internal conflict. For instance, as shown in the figure, although packet A and packet B are destined for different outputs, one of them will be deflected to the wrong output. In the reverse banyan network, bit x_1 of the routing bits x_2x_1 is used at the first column, and bit x_2 is used at the second column. Once a packet has been deflected in column 1, it does not matter how it is routed in the column 2 because it will reach the wrong

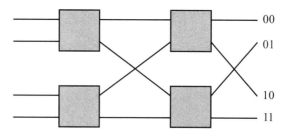

FIGURE 5.17 A 4 × 4 reverse banyan switch to be used in a node in the bidirectional shuffle-exchange network.

output link anyway. Therefore, priority at column 2 should be given to packets that have not been deflected in column 1 to reduce their deflection probability.

Let us consider the deflection probability. An incoming packet to a column-1 node will be deflected only if there is another packet on the other input link destined for the same outgoing link and it loses contention to this other packet. With the assumption that packets are equally likely to be destined for any of the four outputs of 4 × 4 switch, the probability that there is another input packet desiring the same outgoing link of the 2 × 2 column-1 node is $\rho/2$, where ρ is the link loading. With probability 1/2, this packet will win contention. Thus, the deflection probability at the column-1 node is $\rho/4$. The probability that an incoming link to a column-2 node has an undeflected packet is therefore $\rho(1 - \rho/4)$ and the deflection probability in column 2 is $1 - \rho(1 - \rho/4)/4$. Hence, the probability of an arbitrary packet not deflected in the overall 4 × 4 switch is

$$p = \Pr\{\text{not deflected in column 1}\}$$
$$\times \Pr\{\text{not deflected in column 2} \mid \text{not deflected in column 1}\}$$
$$= (1 - \rho/4)[1 - (\rho - \rho^2/4)/4]. \tag{5.38}$$

This results in a higher deflection probability than with 4 × 4 nonblocking switch elements, and therefore more number of hops is needed for the packet to reach its final destination (see Problem 5.8). However, the same random walk as depicted in Fig. 5.16 still applies and the deflection distance is still one. Therefore, we do not expect the use of simpler Banyan networks will result in significantly worse performance and the average number of hops needed is still proportional to n.

We close this subsection by pointing out that the performance of the bidirectional shuffle-exchange network can be further improved. With reference to Fig. 5.18, consider a packet that has not been deflected—the nodes in the figure are drawn in two separate columns for clarity. Suppose that it is at node a and wants to go to node c next. However, it is deflected to node d via the wrong forward link (forward links are shuffle links if the packet is to travel to its destination via the shuffle plane and are unshuffle links otherwise). In our original routing algorithm, we add routing bits so that the packet can return to node a via a reverse link from node d. This strategy attempts to go back to the original path. However, by the construction of the shuffle and unshuffle links, the packet can also travel via the other reverse link of node d to node

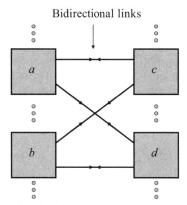

In bidirectional shuffle-exchange network, nodes a and b connected to node d implies they are also connected to another common node c

FIGURE 5.18 Illustration showing that both reverse links can be used to correct a deflection error.

b and then via a forward link in node b to node c; going back to node a is not strictly necessary. This is a simple observation of the shuffle and unshuffle interconnections: if nodes a and b are connected by shuffle (unshuffle) links to a common node d, then both nodes a and b are also connected by shuffle (unshuffle) links to another common node c.

By letting the packet use either of the reverse links, we reduce the deflection probability of the packet while its previous deflection is being corrected. In other words, we have a "don't care" situation in which the packet does not care which reverse link it acquires as long as one is available. In general, however, we do not always have a "don't care" situation when performing error correction. For example, if the packet were to be deflected to a reverse link instead of a forward link at node a, then it would be necessary for it to go back to node a to get to node c. When correcting successive deflections, it is also possible that some steps have a "don't care" situation while some do not. The general treatment is beyond the scope of this book. Suffice it to say that by considering the routing tags of packets and the nodes they occupy at each step, one can derive systematically the general "don't care" conditions and modify the routing strategy for further performance improvement. Further discussion can be found in Problem 5.10.

5.1.5 Dual Shuffle-Exchange Network

The idea of the feedback bidirectional shuffle-exchange network can be used in a feedforward dual shuffle-exchange network. Instead of feeding the outputs of the nodes back to the inputs of the same nodes through a set of shuffle links and a set of unshuffle links, the outputs are fed forward to a set of "new" nodes at the next stage. In this way, a packet traverses physically different nodes on route to its destination.

176 ADVANCED SWITCH DESIGN PRINCIPLES

This is the same idea as the shuffle-exchange network discussed in Section 5.1.2. The difference is that the deflection penalty in the dual shuffle-exchange network is much lower than that in the shuffle-exchange network.

One way to look at the dual shuffle-exchange network (DSN) is to consider it as being constructed of two subnetworks, a shuffle network (SN) and an unshuffle network (USN)—the mirror image of the shuffle network, as illustrated in Fig. 5.19. Of course, in order that packets can be transferred from one network to the other

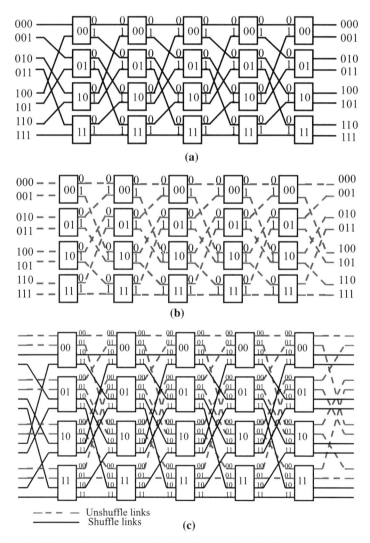

FIGURE 5.19 Construction of a dual shuffle network using shuffle-exchange and unshuffle-exchange networks: (a) Shuffle-exchange network, (b) unshuffle-exchange network, (c) dual shuffle-exchange network.

for correcting deflections, we need 4×4 nodes rather than 2×2 nodes. Conceptually, the routing and error-correction mechanisms of the DSN are the same as those in the feedback bidirectional shuffle-exchange network. There are, however, some differences relating to implementation and performance issues, and this subsection concentrates on these areas.

In the bidirectional shuffle-exchange network, the nodes are the packet sources and destinations; in the DSN, the input and output links are the sources and destinations. Consider a network with $N/2$ nodes at each stage. Since each node is 4×4, there are altogether $2N$ links connecting adjacent stages. We label the $N/2$ nodes at each stage with an $(n-1)$-bit number $x_n \cdots x_2$ (as opposed to $x_n \cdots x_1$ in the feedback network with N nodes). The four output shuffle and unshuffle links connecting to a particular node $x_n \cdots x_2$ are labeled as $(1, x_n \cdots x_1)$ and $(0, x_n \cdots x_1)$, respectively. The four input links are also similarly labeled. Routing of a packet in the DSN consists of first directing it to the node to which the desired output is attached—which is similar to routing in the feedback network—in the first $\log_2 n - 1$ steps, and then from the node to the destination output link in the last step. Therefore, the same error-correcting routing algorithm as depicted in Figs. 5.13 and 5.14 can be used, with the understanding that x_1 in the figures must be replaced by x_2 to take into account the new situation that x_2 is the least significant bit of the node label.

The DSN can be configured as an $N \times N$ switch or a $2N \times 2N$ switch, as shown in Fig. 5.20. Figure 5.20(a) depicts the $N \times N$ version. Both the SN and USN in the DSN are used for routing packets. In the $N \times N$ DSN, links $(1, d_n \cdots d_1)$ and $(0, d_n \cdots d_1)$ are associated with the same logical address $d_n \cdots d_1$. Packets reaching these two links as their final destinations will be multiplexed onto the same output link of the overall switch. Therefore, a packet with destination address $d_n \cdots d_1$ can

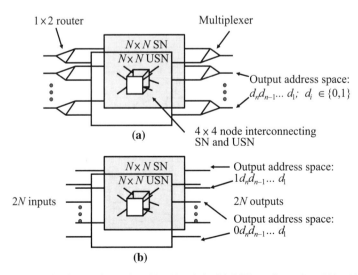

FIGURE 5.20 (a) DSN configured as $N \times N$ switch; (b) DSN configured as $2N \times 2N$ switch.

178 ADVANCED SWITCH DESIGN PRINCIPLES

have either of them as its destination. An incoming packet to the switch is routed either to an SN input or an USN input, and accordingly, its routing tag is set to either $1d_n \cdots 1d_1$ or $0d_1 \cdots 0d_n$. In the first case, the packet's *primary route* is in the SN, and in the second case, the primary route is in the USN. Any excursion to the companion network is for error-correction purposes only.

Figure 5.20(b) shows the $2N \times 2N$ version. Here, links $(1, d_n \cdots d_1)$ and $(0, d_n \cdots d_1)$ belong to two different logical addresses, $1d_n \cdots d_1$ and $0d_n \cdots d_1$, respectively. An incoming packet with destination $1d_1d_2 \cdots d_n$ is assigned the routing tag $1d_n 1d_{n-1} \cdots 1d_1$, and a packet with destination $0d_n d_{n-1} \cdots d_1$ is assigned the routing tag $0d_1 0d_2 \cdots 0d_n$. This setup has the advantage that the switch size is double that of Fig. 5.20(a) with essentially the same amount of hardware. For explanation purposes, however, we shall focus mainly on the design in Fig. 5.20(a) for the rest of this subsection.

Figure 5.21 shows an example of the routing of two packets A and B whose primary routes are both in the shuffle plane. The shuffle and unshuffle planes in the DSN are drawn separately for clarity, with the understanding that the two nodes with the same label at the same stage of the two planes are actually the same node in the DSN. As shown, the deflection of packet A to an incorrect shuffle link is corrected by forwarding it to an unshuffle link that returns it back to the previous node where the deflection occurred (node with the same label two stages later). In general, as in the bidirectional networks, deflections to shuffle links are corrected by routing to unshuffle links, and vice versa.

For further design simplification, let us reexamine the SN and USN shown in Fig. 5.19. We note that the first shuffle of the inputs to the SN is unnecessary and can

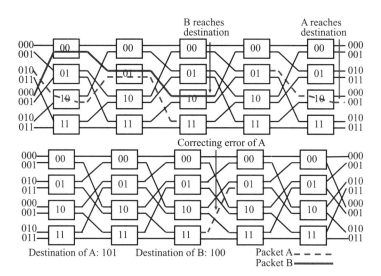

FIGURE 5.21 An example of deflection error in the shuffle plane being corrected in the unshuffle plane.

be removed. The links traversed by an undeflected packet with source $S = s_n \cdots s_1$ and destination $D = d_n \cdots d_1$ at the outputs of successive switching stages are then

$$S = s_n \cdots s_1 \to (1, s_n \cdots s_2 d_n) \to (1, s_{n-1} \cdots s_2 d_n d_{n-1})$$
$$\to \cdots \to (1, s_{n-i+1} \cdots s_2 d_n \cdots d_{n-i+1}) \to \cdots$$
$$\to (1, s_2 d_n \cdots d_2) \to (1, d_n \cdots d_1) = D.$$

From Fig. 5.19, we also see that whereas the shuffling of links precedes switching nodes in each stage of the SN, the switch nodes precede the unshuffling of links in the USN. Thus, a packet with primary route in the USN that has been switched successfully at the last switching stage must undergo another unshuffle before it can be considered as having reached its destination. The reader may verify this with an example, say, a packet destined for output 101, in Fig. 5.19. Alternatively, this can also be revealed by the sequence of links traversed by an undeflected packet at the outputs of successive switching stages:

$$S = s_n \cdots s_1 \to (0, s_n \cdots s_2 d_1) \to (0, d_1 s_n \cdots s_3 d_2)$$
$$\to \cdots \to (0, d_{i-1} \cdots d_1 s_n \cdots s_{i+1} d_i) \to \cdots$$
$$\to (0, d_{n-2} \cdots d_1 s_n d_{n-1}) \to (0, d_{n-1} \cdots d_1 d_n) \neq D.$$

One more unshuffle (cyclic right shift) is necessary to move the packet to the correct destination. To remove this inconvenience, the initial routing tag for a packet must be set to $0d_2 0 d_3 0 \cdots 0 d_n 0 d_1$. That is, a 2-bit cyclic right shift is performed on the original routing tag so that the two least significant bits $0d_1$ are used at the last stage of the new USN. The correctness of this strategy can be revealed by the sequence of output links traversed by an undeflected packet:

$$S = s_n \cdots s_1 \to (0, s_n \cdots s_2 d_2) \to (0, d_2 s_n \cdots s_3 d_3)$$
$$\to \cdots \to (0, d_i \cdots d_2 s_n \cdots s_{i+1} d_{i+1}) \to \cdots$$
$$\to (0, d_{n-1} \cdots d_1 s_n d_n) \to (0, d_n \cdots d_1) = D.$$

The main advantage of the above modification is that the bypass mechanisms within the SN and USN can both be implemented at the outputs of switch nodes; otherwise, the bypass mechanism of the USN must be implemented at the inputs of the next stage.

With the modification, the block diagram of a switch node is shown in Fig. 5.22. Figure 5.22(b) depicts the connection of bypass lines across two nodes of successive stages. The bypass lines are needed only for stages n and above. The postprocessors perform two functions: (1) process packets according to the error-correction routing algorithm; and (2) forward packets that have reached their destinations to the bypass lines. At the bypass lines, packets are multiplexed with other packets reaching the same destination in the previous stages. Note that although the multiplexers create

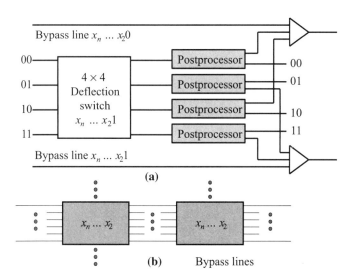

FIGURE 5.22 (a) Block diagram of a switch node; (b) interconnection of bypass lines between adjacent switch nodes of same labels.

physical queues internal to the network, the switch is logically an output-queued switch; the physical buffers for the output queues are distributed across many switch nodes. Instead of the above, we may also have lines leading out of the switch from the bypass locations and use a "large" multiplexer to multiplex packets for the same destination (see the configuration of the tandem-Banyan switch). In this case, there is only one physical queue for each output.

We now show that the complexity of the DSN is $N \log N$ using an approximate analysis. The validity of the analysis and the correctness of the routing algorithm have been verified with simulation.

Since routing requires at least $n = \log N$ stages, clearly the lower bound on the complexity of the DSN is $N \log N$. We shall now show that $N \log N$ is also an upper bound on the order of complexity for a given packet loss probability requirement P_{loss}. As a simplifying approximation, we assume that the input packets to a switch node are uncorrelated with each other and that they are equally likely to be destined for any of the four outputs. We further assume that the 4×4 switch elements are internally nonblocking. Let ρ_t be the load of an input at switch stage t, that is, ρ_t is the probability that there is an incoming packet on an input link. Let p_t be the probability that a packet is successfully routed at stage t. We have

$$p_t = \frac{1 - \left(1 - \frac{1}{4}\rho_t\right)^4}{\rho_t}. \qquad (5.39)$$

It is easy to show that $dp_t/d\rho_t \leq 0$ for $0 \leq \rho_t \leq 1$. Since the load cannot increase as t increases (number of packets cannot increase), ρ_t is a nonincreasing function of t.

Therefore, $p_t \geq p_{t-1} \geq \cdots \geq p_1$. This agrees with our intuition that the probability of success increases as more and more packets are removed from the network. To obtain an upper bound on L, the number of stages required to meet a given P_{loss}, we perform a worst-case analysis in which p_t, for all t, is replaced by $p_1 = p$. That is, the L required by our system is bounded above by the L required in a corresponding "time-invariant" random walk depicted by the Markov chain in Fig. 4.30, in which the state is the packet distance, and $q = 1 - p$ is the deflection probability.

For analytical convenience, let us adopt a version of the DSN similar to the one in Fig. 5.20(a), but in which each of the N incoming packets is randomly routed to one of the $2N$ input ports of the SN and USN, with at most one packet assigned to the same input port. The input load on each input is $\rho_1 = 0.5$. This gives

$$p = p_1 = \frac{1 - (1 - 0.5/4)^4}{0.5} \approx 0.828. \tag{5.40}$$

Let $g_i(k)$ be the conditional probability that a packet will reach its destination (or state 0) in k more steps given that its current state is i.

$$g_0(k) = \begin{cases} 1, & \text{if } k = 0, \\ 0, & \text{otherwise}, \end{cases} \tag{5.41}$$

$$g_i(k) = pg_{i-1}(k-1) + qg_{i+1}(k-1), \quad 0 < i < n, \tag{5.42}$$

$$g_n(k) = pg_{n-1}(k-1) + qg_n(k-1). \tag{5.43}$$

The generating function $G_i(z) = \sum_{k=0}^{\infty} g_i(k)z^k$ is given by

$$G_0(z) = 1, \tag{5.44}$$

$$G_i(z) = pzG_{i-1}(z) + qzG_{i+1}(z), \quad 0 < i < n, \tag{5.45}$$

$$G_n(z) = pzG_{n-1}(z) + qzG_n(z). \tag{5.46}$$

Equation (5.45) is a homogeneous linear difference equation in terms of i; (5.44) and (5.46) are the boundary conditions. The general technique for solving the difference equation (5.45) is to substitute $G_i(z) = S^i(z)$. This gives

$$S^2(z) - \frac{1}{qz}S(z) + \frac{p}{q} = 0. \tag{5.47}$$

The roots of the quadratic equation are

$$S_1(z), S_2(z) = \frac{1}{2qz}\left(1 \pm \sqrt{1 - 4pqz^2}\right). \tag{5.48}$$

The general solution of $G_i(z)$ is

$$G_i(z) = C_1(z)S_1^i(z) + C_2(z)S_2^i(z). \tag{5.49}$$

The constants, $C_1(z)$ and $C_2(z)$, can be found by matching the boundary conditions at $i = 0$ and $i = n$ using (5.44) and (5.46). This yields

$$G_n(z) = \frac{pz[S_2^{n-1}(z)S_1^n(z) - S_1^{n-1}(z)S_2^n(z)]}{(1 - qz)[S_1^n(z) - S_2^n(z)] - pz[S_1^{n-1}(z) - S_2^{n-1}(z)]}. \tag{5.50}$$

We can obtain a Chernoff bound on P_{loss} as follows:

$$P_{\text{loss}} \leq \sum_{k=L+1}^{\infty} g_n(k)$$

$$\leq \sum_{k=L+1}^{\infty} g_n(k) z^{k-(L+1)}, \quad \text{for some real } z \geq 1,$$

$$\leq z^{-(L+1)} \sum_{k=0}^{\infty} g_n(k) z^k$$

$$= z^{-(L+1)} G_n(z). \tag{5.51}$$

Thus, we are interested in $G_n(z)$ for real $z \geq 1$. It is clear from inequality (5.51) that to obtain a tight bound, $z^{-(L+1)}$ must be sufficiently small, or z sufficiently large. Let us determine *a priori* that we shall choose z large enough that $S_1(\cdot)$ and $S_2(\cdot)$ are both complex as a result. Then, it is more convenient to express them in the polar coordinates of the complex plane:

$$S_1(\theta), S_2(\theta) = \sqrt{p/q}\, e^{\pm i\theta}, \tag{5.52}$$

where $i = \sqrt{-1}$ and $\theta = \cos^{-1}\frac{1}{2z\sqrt{pq}}$. After some manipulation, we obtain

$$G_n(\theta) = \frac{(p/q)^{n/2} \sin\theta}{\sin(n+1)\theta - \sqrt{q/p}\sin n\theta}. \tag{5.53}$$

By inspection, substituting $\theta = 0$ appears to give a reasonably tight Chernoff bound. Doing so yields

$$G_n(\theta = 0) = \frac{(p/q)^{n/2}}{(n+1) - n\sqrt{q/p}}$$

$$\leq \frac{(p/q)^{n/2}}{(n+1)(1 - \sqrt{q/p})}. \tag{5.54}$$

Now, $\theta = 0$ implies $z = 1/2\sqrt{pq}$. Substituting the above $G_n(\cdot)$ and z into inequality (5.51) and taking logarithms on both sides yields an upper bound for L:

$$L \leq 2.793n - 3.554 \ln(n+1) + 3.554 \ln P_{\text{loss}}^{-1} + 1.162. \tag{5.55}$$

Since each stage consists of $N/2$ switch elements, the complexity of the DSN for a given P_{loss} is therefore of order $N \log N$.

Theoretically, this is the lowest complexity that can be achieved. We saw in Chapter 2 that the complexity order of a nonblocking switch must be at least $N \log N$. In a certain sense, this is also the least complexity order of any "reasonable" switch, not necessarily nonblocking. For example, the Banyan networks are the least complex switches that guarantee that there is at least one path from any input to any output, and they are also of order $N \log N$. Of course, the Banyan networks operated as loss systems would have very high loss probability, as has been demonstrated in Chapter 3. What we have shown for the DSN is that it can achieve the same order of complexity as the Banyan network while attaining arbitrarily small loss probability.

We close this subsection by observing that the complexity of the bypass mechanisms can be also reduced using a simple trick. The DSN operation assumed so far is associated with the random-walk model in Fig. 5.16. State n is different from other states because error in this state does not increment the distance further: the routing tag is simply reset to the original routing tag. Suppose that this state is treated no differently than other states. The corresponding random-walk model is shown in Fig. 5.23(a), in which there is no reflecting barrier.

The advantage of removing the boundary is that the number of bypass locations in the DSN can be decreased. With this new random walk, regardless of the state of a packet, each deflection means the packet will need two more steps to reach its destination. In other words, a packet that experiences a total of k deflections will exit at stage $n + 2k$ of the DSN. Therefore, only stages $n + 2k, k = 0, 1, 2, \cdots$, need to have the bypass mechanisms installed. The disadvantage is that the length of the routing tag may in principle grow in an unbounded fashion.

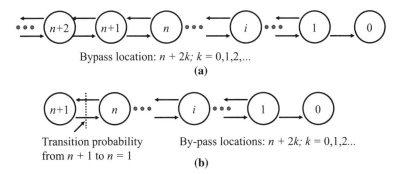

FIGURE 5.23 Two random-walk models realizable with DSN: (a) random walk without boundary; (b) boundary with one-step reflection to the right.

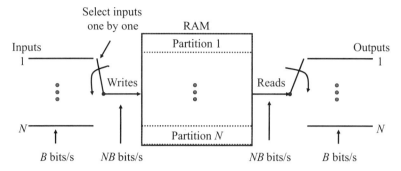

FIGURE 5.24 Switching by loading packets from N inputs into RAM locations based on their targeted outputs and outputting packets into the N outputs in a round-robin fashion.

The random walk in Fig. 5.23(b) also only needs bypass mechanisms at stages $n + 2k, k = 0, 1, 2, \cdots$, but bounds the routing tag to at most $2(n + 1)$ bits. To realize this random walk, if an error is made in state n, the routing tag is reset to the original destination address plus two dummy bits: $x_1 x_2 0 d_1 0 d_2 \cdots 0 d_n$ or $x_1 x_2 1 d_n 1 d_{n-1} \cdots 1 d_1$, where $x_1 x_2$ are the dummy routing bits.[3] This moves the packet to state $n + 1$. In state $n + 1$, routing will be considered successful regardless of the outcome and the two dummy routing bits removed. Consequently, a packet in state $n + 1$ will move to state n in one step with probability 1. Thus, we have a boundary in the random walk that reflects the walk one step to the right. Again, each deflection results in two additional steps before the packet can reach its destination.

5.2 SWITCHING BY MEMORY I/O

So far we have been mainly concentrating on packet switching in the space domain. In Chapter 2, it was shown that circuit switching in the time domain can be implemented using random access memory (RAM). Recall that in TDM, time is divided into frames, each of which consists of a number of time slots. The information on different time slots are associated with different circuits and are read into a RAM in the order of their arrival. By changing the order in which the information are read onto the output, the time slots occupied by the information can be interchanged.

A similar idea can be used in packet switching. Consider the $N \times N$ packet switch depicted in Fig. 5.24. An $N \times 1$ switch multiplexes the packets from the N inputs onto one channel connected to a RAM. The buffer space in the memory is divided into N partitions, one for each output. Packets are loaded into the partitions associated with their targeted outputs, and the N outputs take turns reading out their packets from the memory via a $1 \times N$ switch. In practice, small switches based on variants of this

---

[3] Alternatively, state $n + 1$ can be indicated using another extra bit in the header and the routing tag kept to length $2n$ bits.

scheme have been built. This switching approach, however, does not scale very well for large N: certainly, the $N \times 1$ and $1 \times N$ switches must work at N times the speed of input or output links; the memory must also have very high access speed. The reader is referred to Problem 4.15 for a discussion on the required memory access speed.

One way to reduce the required access speed is to use a parallelization approach in which several, say M, RAMs are used to store the packets. This strategy is similar to the switch speedup scheme by packet slicing discussed in Chapter 4. The information content of a packet is divided into M parts, with each part stored in one of the RAMs. The required memory access speed is then reduced M-fold. The trade-off is that there are more wires interconnecting the inputs and outputs to the memories. Thus, the architecture can only be parallelized to a certain extent, and the scaling problem is still not solved for large N.

If an output is congested and has many packets in its buffer, the buffering space allocated to it may overflow, leading to packet loss. Meanwhile, some other outputs may not be very busy and their buffers are underutilized. A simple extension, called *shared-buffer memory switching*, removes the partitioning of buffering space and allows it to be shared by all outputs. Buffers are allocated on an as-needed basis, and an arriving packet will be dropped only if the overall memory has been used. Therefore, for a given amount of memory, this scheme can achieve a lower packet loss probability.

The buffer management strategy is illustrated in Fig. 5.25. The buffer is organized as N linked lists, each implementing the queue of an output. A linked list has a *head* and a *tail*, corresponding to the first and last packets in the queue. In addition to the buffer required to store a packet, each entry of the linked list also has a pointer containing the address of the next entry. Thus, the first packet points to the second

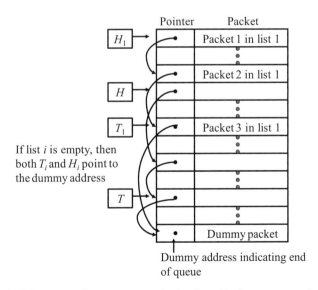

FIGURE 5.25 Buffer management in the shared-buffer memory switch.

packet, the second points to the third, and so on. The last entry, the tail, points to a specially designated *dummy address* (could be address 0, for example) to indicate the end of the linked list. The addresses of the head and tail of queue i are stored in buffers (or registers) H_i and T_i, respectively.

The N linked lists consist of all the used buffer locations. There is another linked list, called the *free* list, whose entries are the free buffer locations that could be allocated to the newly arriving packets. The addresses of the head and tail of this list are stored in buffers (or registers) H and T. An incoming packet destined for output i must first acquire a free location from this list. The operations on the linked lists are as follows, where we assume for the sake of brevity that the free list and list i are not empty (a few steps must be added if this is not the case):

$(H) \leftarrow$ PACKET	load packet into location pointed to by H,
(T_i).NEXT $\leftarrow H$	add packet to tail of list i,
$T_i \leftarrow H$	update the tail of list i,
$H \leftarrow (H)$.NEXT	update H to point to the next free location,
(T_i).NEXT $\leftarrow$ DUMMY	last entry of list i points to dummy address.

The N outputs take turns reading packets from the memory via the $1 \times N$ switch. After output i has successfully read out a packet, the free list and list i are updated as follows, where we assume for brevity that the free list is not empty before the update and list i is not empty after the update:

(T).NEXT $\leftarrow H_i$	return buffer to free list,
$T \leftarrow H_i$	
$H_i \leftarrow (H_i)$.NEXT	update list i,
(T).NEXT $\leftarrow$ DUMMY	last entry of free list points to dummy address.

Although it can be seen intuitively and shown quantitatively that the packet loss probability can indeed be reduced by the buffer-sharing approach, one must be careful in drawing the conclusion that the shared-buffer switch is therefore superior to unshared-buffer switch. There are several arguments against the shared-buffer switch that must also be considered.

In the unshared-buffer switch, output queues do not interfere with each other. In the shared-buffer switch, a misbehaving output (due to traffic congestion, network management errors, or hardware failures) can easily use up a large amount of the shared memory, to the detriment of the other outputs. The shared-buffer switch has an advantage over the unshared-buffer switch only to the extent that memory is an expensive resources. With the rapid advances in memory technology, this is fast becoming a weak argument. One might contend that in a real network, potential for traffic congestion always exists, and no matter how much buffer there is, there is a chance that it will fill up; therefore, having more buffer always helps. The problem with this argument is that more buffer does not help solve traffic congestion; in fact, it may delay the discovery of congestion. The shared-buffer switch allows an output queue to grow very large because of the sharing. If the carried traffic is real-time (e.g., video conference traffic) and must reach the receiver within certain time limit, excessive delay may have been incurred: the packets may as well be dropped at the switch. Furthermore, many network traffic management schemes use buffer overflow as an indication of congestion. For these schemes to be effective, the buffer cannot be too large. Overall, in designing a switch, we want it to have enough buffer to prevent excessive packet loss under normal traffic conditions; using more buffer than necessary, however, does not solve the congestion in heavy-traffic situations.

5.3 DESIGN PRINCIPLES FOR SCALABLE SWITCHES

Although various switches can theoretically be designed to large dimensions, technological and physical constraints often impose a practical limit on their maximum size, $N' \times N'$. If we want to construct a large switch system, say $2N' \times 2N'$, then two or more $N' \times N'$ switches have to be interconnected. Straightforward interconnections of these small switches create several stages of queueing delay. This may also result in severe performance degradation if congestion occurs at intermediate switches. Thus, increasing performance penalty for larger switch sizes seems unavoidable.

In this section we shall focus on switch scalability; we shall provide some key principles to construct large switches out of modest-size switches, without sacrificing overall switch performance.

5.3.1 Generalized Knockout Principle

As mentioned before, the knockout switch takes advantage of the fact that, with uncorrelated traffic among input ports and uniformly directed to all output ports, the probability of more than R packets destined for any particular output port in each time slot is very low. In this way, we can do away with the input queues and operate the switch as a loss system by simply dropping the very few packets that have lost contention.

The knockout principle can be implemented using very different switch architectures. In Chapter 4, we have presented two possible ways for its implementation. One is a design based on Batcher–banyan networks and the other is an implementation using broadcast buses and knockout concentrators. Although the complexity of the Batcher-banyan knockout switch is of order $N \log_2^2 N$, which is much lower than that of the second implementation for large N, the growing complexity of interconnection wiring between stages in this design makes it impossible to scale well. The knockout switch based on broadcast buses and concentrators has a uniform and regular structure and thus has the advantages of easy expansion and high integration density for VLSI implementation. It is one of the representative approaches to construct large-scale switches.

For large N (e.g., a few thousands), the complexity of a knockout concentrator at the output is still very high; the total number of switch elements in all N knockout concentrator is $N^2 R$. Although we can design VLSI chips to contain such large amounts of switch elements, it is still desirable to reduce the number of switch elements to

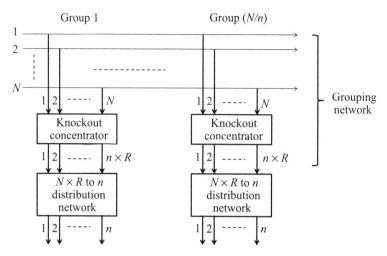

FIGURE 5.26 Partitioning outputs into multiple groups.

some extent by reducing the value of R. There are two ways to reduce the R value. One is to provide an input buffer at each input port, and an output buffer at each output port. Suppose that an eight-packet buffer is added at each input port; R can be reduced from 8 to 3 at the same packet loss probability of 10^{-6} for any switch size N with $\rho_o = 0.9$ (i.e., 90% load). Because the input buffer scheme is used, contention resolution must be performed among packets destined for the same output port. This will increase its complexity to the same extent as the pure input buffer scheme, which is the extreme case of $R = 1$.

The second way to reduce the value of R is to put a number of output ports into a group so that the routing links belonging to the same group can be shared by the packets that are destined for any outputs in this group. For example, as shown in Fig. 5.26, every n output ports are put together in one group, and $R \times n$ routing links are shared in this group. The center network, as shown in Fig. 5.26, used to partition output ports into a number of groups is thus called a "grouping network." If there are more than $R \times n$ packets in one time slot destined for the same output group, the excess packets will be discarded and lost while they are competing with others for available paths. Once packets are sent onto the $R \times n$ vertical routing links at the output of the knockout concentrator, they are routed to proper output ports through a distribution network, which can be implemented by any modest-size switch. This method generalizes the knockout principle from a single output to a group of outputs and is so-called *generalized knockout principle*. It is the combination of the knockout principle and channel grouping and provides a promising way to construct large switches out of small-scale switches.

Assuming the packets have independent, uniform destination address distributions, then the probability an input packet is destined to this group of outputs is simply n/N. If we only allow up to $R \times n$ packets to pass through to the group outputs, the

TABLE 5.1 Value of R for Different Group Sizes n and Loss Probability P_{loss} with $\rho_o = 0.9$

P_{loss}	$n=1$	$n=2$	$n=4$	$n=8$	$n=16$	$n=32$	$n=64$	$n=128$	$n=256$
10^{-6}	8	5.30	3.70	2.70	2.10	1.70	1.45	1.25	1.15
10^{-8}	10	6.45	4.40	3.15	2.40	1.90	1.55	1.35	1.20
10^{-10}	12	7.50	5.05	3.60	2.65	2.00	1.70	1.45	1.25

probability for a packet to be dropped is

$$P_{\text{loss}} = \frac{1}{n\rho_o} \sum_{k=Rn+1}^{N} (k-Rn) \binom{N}{k} \left(\frac{n\rho_o}{N}\right)^k \left(1 - \frac{n\rho_o}{N}\right)^{N-k}, \quad (5.56)$$

where ρ_o is the offered load of each input port. As $N \to \infty$, we have

$$P_{\text{loss}} = \left(1 - \frac{R}{\rho_o}\right)\left(1 - \sum_{k=0}^{Rn} \frac{(n\rho_o)^k e^{-n\rho_o}}{k!}\right) + \frac{(n\rho_o)^{Rn} e^{-n\rho_o}}{(Rn)!}. \quad (5.57)$$

The practical R and n combinations for three different packet loss probabilities are determined from (5.57) and listed in Table 5.1. For a given acceptable packet loss probability, for example, 10^{-10}, the required value of R decreases from 12 to 8 as the number n of output ports in each group increases from 1 to 2. The larger the size of the output group, the smaller the R value required. It is interesting to notice that as the R value increases from 1.25 to 1.45, the size n of the output group reduces considerably from 256 to 128 at 10^{-10} loss probability. Because it is practical to route packets to different groups based on some portion of the routing information carried in the packet header, the group size is therefore chosen as 2^i, where $i = 0, 1, 2, \cdots$. In actuality, it is always desirable to choose a smaller R for a large group size n to keep the hardware complexity smaller. However, to avoid the implementation difficulty resulting from the requirement of preserving correct packet sequence, every group is limited to a reasonable size.

We can also apply the generalized knockout principle to the distribution networks and further reduce the size n of the output groups by increasing the R value. For example, as the R value is increased from 1.25 to 2, the group size is reduced from 256 to 32 for the loss probability of 10^{-10}. This principle can be applied recursively until the n value becomes 1, where input packets are routed to an output port properly. In this way, many stages of the grouping network can be constructed recursively, which results in smaller hardware complexity and packet delay. Figure 5.27 illustrates one possible way to construct a switch with three stages of grouping network. In this switch design, the first stage grouping network has K groups ($K = N/n$), and each group has $R \times n$ fanouts. They are then fed to another grouping network at the second stage, which has $R \times n$ input lines and K' output ($K' = n/n'$), with each group's fanout equal to $R' \times n'$. At the third stage, the $R' \times n'$ input lines are then routed to n' groups with

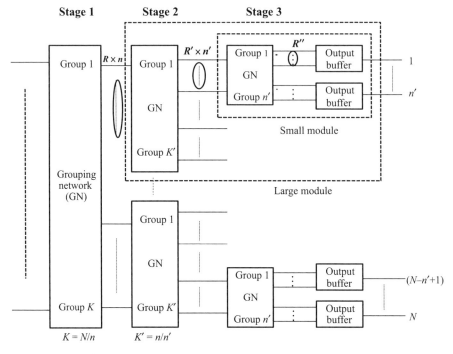

FIGURE 5.27 A switch with three stages of grouping networks.

R'' fanouts for each group. Packets on these R'' lines are statistically multiplexed and stored in a multiplexer with a logical FIFO to maintain packet sequence integrity.

As shown in Fig. 5.27, there are N/n knockout concentrators in the grouping network at the first stage. Thus, the total number of switch elements in the first stage is equal to

$N/n \times$ (number of input lines) $\times$ (number of output lines per concentrator)

$= N/n \times N \times Rn$

$= N^2 R.$ (5.58)

Similarly, we obtain the number of switch elements used in the second stage and the third stage; they are $N \times n \times R \times R'$ and $N \times n' \times R' \times R''$, respectively. The total number of switch elements in the switch fabric is, therefore, equal to the sum of them, or $N^2 R + N \times n \times R \times R' + N \times n' \times R' \times R''$. If we choose $n = 256$, $R = 1.25$, $n' = 32$, $R' = 2$, and $R'' = 12$ for the packet loss probability of 10^{-10}, the total number of switch elements will be equal to $1.25N^2 + 1048N \approx 1.25N^2$ for very large N. It is close to one order of magnitude reduction from $12N^2$, which is the number of switch elements required for building an $N \times N$ knockout switch with $R = 12$ and $n = 1$.

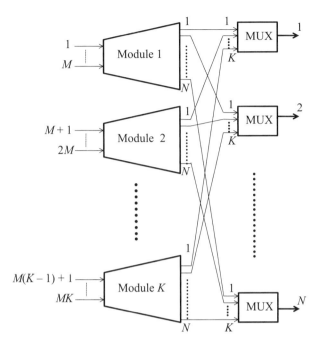

FIGURE 5.28 The decomposition of a large switch into modules.

5.3.2 Modular Architecture

The generalized knockout principle provides the flexibility and efficiency for increasing the switch size by equally partitioning all outputs into multiple groups and doing routing according to these groups. Here, we shall present a modular architecture, which does so by equally partitioning all inputs into multiple subsets and performing switching based on them.

Figure 5.28 illustrates the basic idea of the modular approach to construct a switch fabric with a large number of ports. The set of all inputs is equally partitioned into K subsets. Each subset of the inputs is then interconnected to all N outputs, forming the basic building block of the system called the switch module. The number of inputs of a switch module, $M = N/K$, is called the base dimension of the switch. In this architecture, each module is an autonomous nonblocking, self-routing packet switch. The outputs with the same index, one from each switch module, are multiplexed together and then fed to the output queue bearing that index as its port address.

Physically, as shown in Fig. 5.29, the modular architecture can be realized as an array of three-dimensional parallel processors. Each processor is a switch module consisting of a Batcher sorting network and an expansion routing network. The expansion network is a set of binary trees cross interconnected with a set of banyan networks. The set of concurrent modules is interconnected at the outputs by multiplexers. Thus, no interference between switch modules is possible, which is the key to simplifying

192 ADVANCED SWITCH DESIGN PRINCIPLES

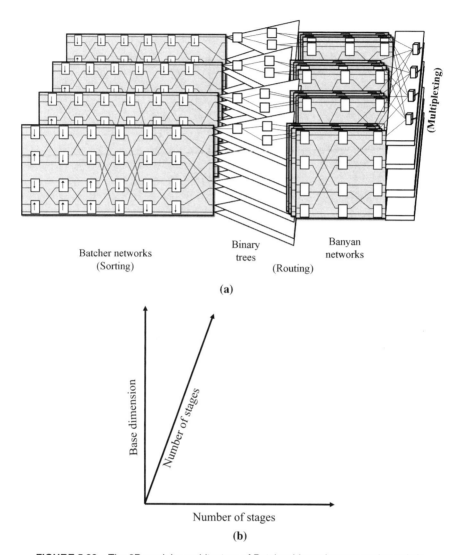

FIGURE 5.29 The 3D modular architecture of Batcher-binary-banyan packet switch.

the operation and maintenance of the whole system. This architecture also allows independent clocking of modules, which simplifies timing substantially. Within each module, the small physical size makes it fairly straightforward to synchronize, and the simpler hardware makes higher speed implementation possible.

5.3.2.1 Modular Batcher–Binary–Banyan Switch With reference to the previous discussion, each of the K modules of an $N \times N$ switch is a cascade of an $M \times M$ Batcher network, a stack of M binary trees and a group of K banyan

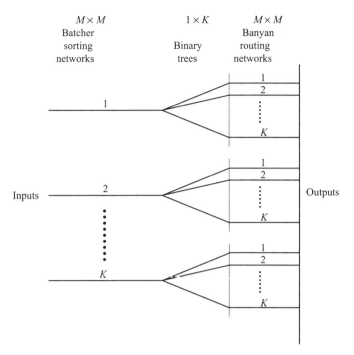

FIGURE 5.30 The divide-and-conquer modular approach.

networks where $M = N/K$. We will use $\sigma(M, N, K)$ to represent such a switch module. The Batcher network sorts the input packets based on their destination addresses. The succeeding binary trees and banyan networks then route the packets to the correct outputs.

The principle of this modular architecture is based on a common approach called divide-and-conquer, which has been used extensively in designing efficient algorithms. A well-known example is the fast Fourier transform. It simply splits a large problem into small parts, solves them, and then combines the solutions for the parts into a solution for the whole. This point is illustrated in Fig. 5.30. The set of inputs is first partitioned into K subsets. Each subset is sorted by a Batcher network. The sorted subset is then partitioned again by the binary trees into finer subsubsets. In each module, the ordered packets of these subsubsets are routed concurrently to their destinations by K parallel banyan subnetworks. Finally, these packets are buffered in respective output queues, waiting for transmission.

The combination of binary trees and banyan networks forms an expansion network, a network has larger fan-out than fan-in. As such, the output space of an expansion network can be arbitrarily enlarged by adding more banyan networks. Thus, each module may even have more outlets than the number of outputs. The structure and property of the switch modules are detailed below.

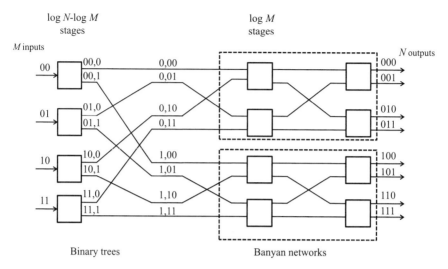

FIGURE 5.31 A 4 × 8 expansion network.

Binary–Banyan Expansion Network An n-stage expansion network with $M = 2^m$ inputs and $N = 2^n$ outputs is a combination of a set of M binary trees and a set of $K = N/M$ banyan networks. Figure 5.31 illustrates a three-stage expansion network with four inputs and eight outputs. Each $1 \times K$ binary tree has $k = \log K = \log N - \log M = n - m$ stages and each $M \times M$ banyan network has m stages. Every node of the network is either a 1×2 or a 2×2 switch element capable of performing the binary routing algorithm based on the n-bit destination address in the header of the packet. That is, a node at stage j sends the packet on link 0 (up) or link 1 (down) according to the jth bit of the header.

The cross interconnection of M binary trees and K banyan networks is similar to the link system of a multistage crossbar switch. The outputs of a binary tree can be labeled by two binary numbers $(x_1 \cdots x_m, y_1 \cdots y_{n-m=k})$, where $x_1 \cdots x_m$ is the top-down numbering of the binary tree and $y_1 \cdots y_k$ is the local address of each input within the binary tree. Similarly, the inputs of the banyan network can also be identified by two binary numbers $(a_1 \cdots a_{n-m=k}, b_1 \cdots b_m)$ where $a_1 \cdots a_{n-m}$ is the top-down numbering of the banyan network and $b_1 \cdots b_m$ is the local address of the input. The binary trees and banyan networks are transpose interconnected according to this numbering scheme; the two ports are interconnected if $(x_1 \cdots x_m, y_1 \cdots y_{n-m}) = (b_1 \cdots b_m, a_1 \cdots a_{n-m})$. This transpose cross interconnection results in a three-dimensional realization of switch modules as portrayed in Fig. 5.29.

The binary trees of the expansion network consist of 1×2 elements, which only allow one input packet at any instant of time. Packets will never collide in this stage of the network, but this may occur in the subsequent banyan subnetworks. In the following, we will discuss the condition on inputs that prevents any possible packet collisions in an expansion network.

DESIGN PRINCIPLES FOR SCALABLE SWITCHES

Nonblocking Property of Expansion Networks An interconnection network is nonblocking if the packet routing is not store-and-forward, and internal buffers are not needed within the switch nodes. It is known that if the incoming packets with distinct destination addresses are arranged in an ascending or a descending order, then the banyan network is internally nonblocking.

The cross interconnection defined by the above numbering scheme guarantees that the routing performed by the set of binary trees preserves the ordering of the destination addresses in input packet headers. Since a binary tree is actually a $1 \times K$ demultiplexer, at most one packet will be routed by a tree during any time slot. Packet collisions may only occur in the subsequent banyan subnetworks. This implies that the same nonblocking condition for banyan networks can still be applied to expansion networks. Formally, this property states the following.

[NB] *"If the set of destination addresses of input packets to the expansion network is monotone and concentrated, then so is every subset of input packets to each banyan subnetwork of the expansion network."*

This nonblocking condition can be best demonstrated by the following example. Consider two packets

A: from 01 to 1001 and B: from 10 to 1011

input adjacently to the 4×16 expansion network as shown in Fig. 5.32. According to the self-routing algorithm, they will emerge at outputs labeled 01,10 and 10,10 of the binary trees 01 and 10, respectively. Following the transpose interconnection, packet A and B will be routed to the inputs 10,01 and 10,10, respectively, of the banyan subnetwork 10. The two packets are still neighboring contiguously, no gap has been created, and in the same spatial order. The assertion in [NB] can be established by

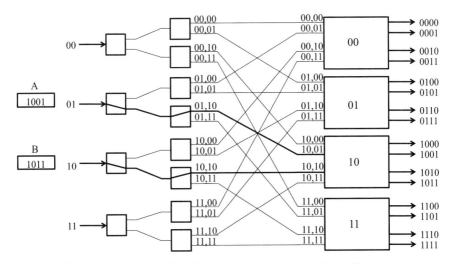

FIGURE 5.32 An example of input packets order preserving by binary trees.

TABLE 5.2 The statistics of a Switch Module $\sigma(M, N, K)$

	Number of networks	Dimension	Number of Stages	Number of Nodes
Batcher network	1	$M \times M$	$\frac{\log M(1+\log M)}{2}$	$\frac{M \log M(1+\log M)}{4}$
Binary tree	M	$1 \times K$	$\log K$	$K - 1$
Banyan network	K	$M \times M$	$\log M$	$\frac{M}{2} \log M$

applying the same argument to every adjacent pair of input packets. Consequently, a rectangular nonblocking self-routing switch with large fan-out than fan-in can be formed by combining a Batcher sorting network and an expansion network with arbitrary input/output ratio.

It should be noticed that the perfect shuffle preceding the inputs of banyan subnetworks is topologically equivalent to the perfect shuffle of inputs preceding to the stack of binary trees. This point can be demonstrated by the three-dimensional configuration shown in Fig. 5.29.

5.3.2.2 Complexity and Contention Resolutions The statistics of a switch module $\sigma(M, N, K)$ is tabulated in Table 5.2. The modularity only cuts down the complexity of Batcher networks. The total number of nodes is increasing with respect to K, the number of modules. However, these extra nodes are not entirely overhead. The modularity can improve switch throughput performance. Intuitively, this is simply because there are fewer input packets competing for outputs in each module. The analysis of this point will be elaborated later.

Theoretically, there is virtually no limit on the dimension of the modular switch that can be built from fixed-size Batcher and banyan networks. However, there is a trade-off between the base dimension M and the number of modules K for a given switch with dimension N, since $N = MK$. There are two notable special cases worth mentioning here. Namely, the Batcher–banyan switch for $M = N$ and the knockout switch for $M = 1$. In the case of $M = N$, it is obvious that the modular switch itself is a Batcher–banyan switch. On the other hand, in the case of $M = 1$, every module has only one input and each input is connected to every output directly, then the modular switch becomes a knockout switch, where the knockout function can be implemented by using a concentrator to regulate the maximum number of concurrent packet arrivals at the output during a time slot.

Therefore, the three-dimensional modular architecture is actually the unification of Batcher–banyan and knockout switches. Both switch architectures cannot be scaled to very large size N because either the base dimension $M = N$ or the number of modules $K = N$. That is, the decomposition of N into smaller factors does not apply. A choice for balancing these factors is $M = K = \sqrt{N}$, from which we can claim that this approach has at least the power to square the performance dimensions of these interconnection networks.

The throughput of the modular switch is limited mainly due to the head-of-line (HOL) blocking phenomena. This limitation can be relaxed and throughput can be increased by using look-ahead contention-resolution scheme described in the preceding chapter. In additional, when multiple packets simultaneously request the same output port of the same switch module, output conflict occurs. To resolve these conflicts, we can directly apply the contention-resolution algorithms developed for Batcher–banyan switches (see Chapter 4) without any modification. Also, we can engineer the switch throughput to meet any desired requirement. This can be achieved by adjusting the number of modules K and the maximum number R of packets that can be received by an output port simultaneously.

5.3.2.3 Performance Analysis
In the following, we analyze the throughput of a switch module $\sigma_R(M, N, K)$, which allows at most R packet arrivals at the output in a time slot. We assume that if there are $S(>R)$ packets waiting at the heads of input queues addressed to the same output, the selection of R to pass through the switch is done at random, and the $S - R$ that loses the contention will be ignored. Under these assumptions, the switch throughput is defined to be the carried load on an output line when all input lines are fully loaded.

Suppose the input line loading, or the probability that a slot contains a packet, is λ. Let M_i be the number of packets addressed to output i, we have

$$\Pr\{M_i = j\} = \binom{M}{j}\left(\frac{\lambda}{N}\right)^j\left(1 - \frac{\lambda}{N}\right)^{M-j},$$

$$j = 0, \cdots, M. \tag{5.59}$$

Suppose $A_i = \min(M_i, R)$ is the number of packets that actually arrived at output i in a time slot. For sufficiently large N, we can use Poisson approximation with parameter $M\lambda/N = \lambda/K$ to obtain

$$a_j = \Pr\{A_i = j\} = \left(\frac{\lambda}{K}\right)^j \frac{e^{-\lambda/K}}{j!},$$

$$j = 0, \cdots, R - 1. \tag{5.60}$$

and

$$a_R = 1 - a_0 - \cdots - a_{R-1}. \tag{5.61}$$

The carried load on an output line, denoted by λ', can be intuitively interpreted as the probability that a time slot on the output line is occupied by a packet. Because of the symmetricity of the K modules, we immediately obtain

$$\lambda' = K \cdot E[A_i] = K \cdot \left[R - \sum_{j=0}^{R-1}(R-j)a_j\right]$$

$$= K \cdot \left[R - e^{-\lambda/K}\sum_{j=0}^{R-1}\left(\frac{\lambda}{K}\right)^j \frac{R-j}{j!}\right]. \tag{5.62}$$

TABLE 5.3 Maximum Throughput of Modular Switches $\sigma_R(M, N, K)$

L	K = 1	K = 2	K = 4	K = 8	K = 16	K = 32
1	0.6321	0.7869	0.8848	0.9400	0.9694	0.9845
2	0.8964	0.9673	0.9908	0.9976	0.9994	0.9998
3	0.9767	0.9961	0.9994	0.9999	1.0000	1.0000

The maximum throughput of the switch is the carried load when all input lines are saturated, which can be obtained by substituting $\lambda = 1$ into (5.62). Table 5.3 lists the maximum throughput versus number of modules K for different values of R. For $K \geq 2$, maximum throughput $\lambda = 0.95$ can be achieved with $R = 2$. So, we can see that we can improve the switch throughput by increasing the value of K only.

PROBLEMS

5.1 Show formally that the shuffle link interconnection pattern corresponds to a cyclic left shift of the link label.

5.2 Consider a feedback shuffle-exchange network with nodes 00, 01, 10, and 11. Suppose that there is a packet at each of the nodes and the source–destination mapping is $00 \to 01, 01 \to 10, 10 \to 11$, and $11 \to 00$.
  (a) Show the sequence of nodes traversed by each of the packets on route to its destination node. Is there contention among packets?
  (b) Give an example in which there is contention among packets even though they are destined for different destinations. (*Hint*: Can there be contention if there is only one packet at each source?)

5.3 In the analysis of the feedback shuffle-exchange network in Section 5.1.3, we made the assumption that the packets on the incoming link are independently and equally likely to be destined for either of the output link. Do you expect the independence assumption to give rise to an optimistic or pessimistic performance estimate? Write a simulation program to verify your conjecture.

5.4 We want to study the relationship between the saturation throughput Λ_{sat} and the maximum throughput Λ_{max} of a feedback shuffle-exchange network with $N = 2^n$ nodes. From (5.13),
  (a) Show that $d\Lambda/d\rho \geq 0$ when $n \leq 4$.
  (b) Show that for each $n \geq 5$, there is a $\rho < 1$ beyond which $d\Lambda/d\rho < 0$.
  (c) Argue that $\Lambda_{\text{max}} > \Lambda_{\text{sat}}$ for $n \geq 5$ from parts (a) and (b).

5.5 Consider the input queue at a node in a feedback shuffle-exchange network with the random deflection scheme in which the winning packet is chosen at random under contention. This problem explores how the queueing delay incurred by a packet can in principle be derived, given an offered load to the queue $\rho_o = \Lambda_o/N < \Lambda_{\text{sat}}/N$. This is very similar to the waiting time derivation for the input-buffered internally nonblocking switch discussed in Chapter 3 except that the service time at the HOL is geometric.

(a) Let $P(0)$ be the probability of no active packet and $P(1)$ be the probability of a packet in state 1 at the beginning of a time slot. Assume for simplicity that the input queue may inject no more than one packet into the network per time slot, and when both input links are empty at the end of a time slot, one of the links will be chosen at random. Argue that the probability of finding an input link empty at the end of a time slot is $P(1)p + P(0)[P_0 + (1 - P_0)P(0)/2]$, where $p = (1 - q)$ is the probability of a packet not being deflected and P_0 is the probability of an input queue being empty.

(b) Show that $P(1)p = \rho_0/2$ and $P(0) = 1 - \rho$, where ρ is the link loading in the network.

(c) Show using Little's law that $P_0 = 1 - \overline{S}/\rho_o$, where S is the service time at the HOL of the input queue.

(d) From the above, derive a, the probability of finding neither of the input links empty at the end of a time slot in terms of only ρ_o, ρ, and $\overline{S}$.

(e) Show that $\Pr\{S = i\} = (1 - a)^{i-1}a$ and from that $\overline{S} = 1/a$. Thus, this and the previous part yield an equation governing ρ_o and the unknowns ρ and $\overline{S}$. Equation (5.13) gives the other equation governing ρ_o and ρ. Thus, in principle, $\overline{S}$ can be found given ρ_o. From this, a and therefore $S(z)$ can be found. Using the same approach as in the analysis of an input queue of an input-buffered switch (i.e., Eq. (4.31) can be used directly), the generating function of the waiting time $W(z)$ can then be found.

5.6 For the feedback shuffle-exchange network with the contention policy that favors packets closer to their destinations, write a computer program to solve numerically the link loading ρ as a function of the throughput Λ.

5.7 This problem investigates network stability in bidirectional shuffle-exchange network.

(a) Show from (5.33) that
$$\rho \frac{dp}{d\rho} = \left(1 - \frac{\rho}{4}\right)^3 - p.$$

(b) Show from (5.36) that
$$\frac{d\Lambda}{d\rho} = \frac{4N}{T_n}\left(1 - \frac{\rho}{T_n}\frac{dT_n}{d\rho}\right).$$

(c) Show from (5.35) that
$$\frac{\rho}{T_n}\frac{dT_n}{d\rho} = 1 + \frac{1 - 2(1 - \rho/4)^3}{2p - 1}.$$

(d) From the above, show that
$$\frac{d\Lambda}{d\rho} = \frac{4N}{T_n}\left[\frac{2(1 - \rho/4)^3 - 1}{2p - 1}\right] > 0$$

and that if $\rho < 0.825$, regardless of n, $d\Lambda/d\rho > 0$. Find the difference between Λ_{max} and Λ_{sat}.

(e) Write a computer program based on the more accurate formula for T_n, Eq. (5.34), and Eq. (5.36). By plotting graphs, show that Λ_{sat} and Λ_{max} are very close to each other for n up to 30 (3 billion nodes!). Argue therefore that the stability problem is not severe and can be ignored in this network.

5.8 Consider the feedback bidirectional shuffle-exchange network with the 4×4 switching nodes implemented using Banyan networks.

(a) From (5.33) and (5.38), show that the deflection probability is higher with the Banyan network than with a nonblocking switching network.

(b) Derive the mean delay and the saturation throughput of the bidirectional network when the Banyan networks are used.

5.9 Consider a network with four nodes 00, 01, 10, and 11 arranged as a square. There are eight links in the network interconnecting nodes whose labels differ in one bit only in both directions (e.g., there are two links between nodes 01 and 00 and no links between nodes 00 and 11). This is the two-dimensional hypercube network. Consider using deflection routing in this network. The nodes are the sources and destinations of packets. The distance of a packet inside the network at any time is either 2 or 1. There is zero probability of deflection when the distance is 2, since routing to either of the adjacent node decreases the distance by 1. A packet with distance 1 cannot be deflected by a packet with distance 2—priority is given to the packet with distance 1, since the packet with distance 2 is guaranteed not to be deflected. For this question, assume random contention resolution when two packets with distance 1 contend with each other. Furthermore, assume that a *new* packet just arriving to the network *externally* is equally likely to be destined for any of the three other nodes.

(a) Draw the state-transition diagram of a packet being routed in the network.

(b) Express T_1 and T_2, the mean delay given a packet is in states 1 and 2, respectively, in terms of the success probability at state 1, $p = 1 - q$, where q is the deflection probability. What is the mean delay for an arbitrary packet?

(c) Consider the saturation situation in which a packet leaving the network is immediately replaced by a new packet so that the link loading ρ is always 1. Express the saturation throughput Λ_{sat} of the overall network in terms of p.

(d) Let $\overline{N_1}$ be the expected number of packets in state 1 at the beginning of a time slot under saturation condition. Express $\overline{N_1}$ in terms of p and Λ_{sat}.

(e) Find p, hence the saturation throughput in terms of a *real number*.

5.10 Toward the end of Section 5.1.4, we discussed a strategy for reducing the deflection probability in bidirectional shuffle-exchange networks by considering certain "don't care" situation in routing. Let us examine the problem in more detail. Consider a packet in transit to its destination.

(a) Argue that in general, after a series of successes and deflections, the node at which the packet is currently residing has label $d_{n-k} \cdots d_{i+1} y'_i \cdots y_1$, $x'_k \cdots x_1 d_{n-k} \cdots d_{i+1}$, or $x'_k \cdots x_1 d_{n-k} \cdots d_{i+1} y'_i \cdots y_1$, where $d_n \cdots d_1$ is the destination address, and x_j's, y_j's, i, i' k, and k' are unknowns that depend on previous packet deflections. What are the physical meanings of i, i' k, and k'? Show that $i + k = i' + k'$.

(b) Consider the general case with node label $x'_k \cdots x_1 d_{n-k} \cdots d_{i+1} y'_i \cdots y_1$, what should the next few routing steps be in order to send the packet to the node $d_{n-k} \cdots d_{i+1} y'_i \cdots y_1 \cdots$? Do we have a "don't care" situation in these routing steps? What if we want to send the packet to node $\cdots x'_k \cdots x_1 d_{n-k} \cdots d_{i+1}$ instead?

(c) Consider sending the packet to node $d_{n-k} \cdots d_{i+1} y'_i \cdots y_1 \cdots$, and then to node $d_n \cdots d_{i+1} y'_i \cdots y_1$, then to node $\cdots d_n \cdots d_{i+1}$, and then to the destination node $d_n \cdots d_1$. How many undeflected steps are needed altogether? How many steps have "don't care" situations and how many do not, which steps involve routing to the shuffle plane and which steps to unshuffle plane?

(d) Answer the same questions for the other alternative: sending the packet to node $\cdots x'_k \cdots x_1 d_{n-k} \cdots d_{i+1}$, then to node $x'_k \cdots x_1 d_{n-k} \cdots d_1$, and then to node $d_{n-k} \cdots d_{i+1} \cdots$, and then to the destination node $d_n \cdots d_{i+1}$. Which alternative is more efficient?

(e) Show that by keeping the four parameters i, i', k, and k' and the destination $d_n \cdots d_1$ in the packet header, we have sufficient information for routing decision at each node.

5.11 Consider a variant of feedback shuffle-exchange network called the stay-or-shuffle network. Each node, in addition to the two regular input and two regular output links, has a third link that feeds back to the same node so that the node has three inputs and three outputs. The third link will be called a stationary link and the regular links, shuffle links.

(a) Suppose that at a node, only two of the input links have an incoming packet and both packets want to access the same shuffle output link. Is it better to deflect the losing packet to the other shuffle link or the stationary link? Why? Do you expect this network to perform better than the regular feedback shuffle-exchange network? Why?

(b) When all three inputs have an incoming packet, each output (stationary or shuffle link) will have a packet forwarded to it. When only two of the inputs have an incoming packet, assuming that they are equally likely to be destined for any of the two shuffle output links, what is the probability that the stationary output link will have a packet forwarded to it?

(c) Let the link loading of the stationary link be ρ' and the link loading of each of the shuffle link be ρ. Derive the relationship between ρ' and ρ drawing from clues in the previous part.

(d) Represent the state of a packet by a 2-tuple (i, j). State $(i, 0)$ means the distance of the packet is i and it is currently on a shuffle link, and state $(i, 1)$ means the distance of the packet is i and it is currently on a stationary link. Draw the state-transition diagram. Use p_i, q_i, and r_i to denote the probabilities of success, deflection to the wrong shuffle link, and deflection to the stationary link, respectively, at state $(i, 0)$. Use $p'_i, q'_i,$ and r'_i to denote the corresponding probabilities at state $(i, 1)$.

(e) Consider the random contention-resolution scheme in which contending packets are equally likely to be chosen as the winner. Derive $p = p_i, r = r_i,$ $q = q_i, p' = p'_i, r' = r'_i,$ and $q' = q'_i,$ in terms of ρ and ρ'.

(f) Let $T(i, j)$ denote the expected number of steps needed to reach the destination from state (i, j). Write down the dynamic equations of $T(i, j)$. Describe how you would solve for the throughput of the system.

5.12 This problem compares the configuration of a DSN as a $2N \times 2N$ switch and as an $N \times N$ switch.

(a) Given a fixed load per input, argue that a DSN configured as a $2N \times 2N$ switch has higher packet loss probability than the same DSN configured as an $N \times N$ switch. How would you modify the $2N \times 2N$ switch to achieve the same loss probability with the same load? Which switch has higher complexity now?

(b) Let us consider two DSNs, one with $N/2$ nodes and other with $N/4$ nodes in each stage. Both are configured as an $N \times N$ switch. For a given fixed input load and a fixed loss probability, intuitively, which network would have a higher number of nodes? Can you argue for your conclusion analytically?

5.13 Consider a $\log_2 N$-stage reverse shuffle-exchange network with the last unshuffling of links removed. The text explained that the routing tag of a packet should be set to $d_2 d_3 \cdots d_n d_1$ in order to route the packet to its destination $d_n \cdots d_1$ in this network. True or false? A set of input packets that can be routed through the original shuffle-exchange network (the one without the last unshuffling of links removed) without conflict can also be routed through this network without conflict.

5.14 In the text, contention is resolved in a random manner at the switch nodes of the DSN. This problem shows that deadlocks could occur when contention is resolved in a deterministic manner. Deadlock occurs when a group of packets contend with other in a periodic fashion and they take turns winning the contention.

(a) Give an example in which two packets keep contending with other, resulting in a deadlock situation in which the distance of neither packet progresses to 0.

(b) Argue that deadlocks could not occur if packets with the shorter distance are favored under contention.

5.15 Consider an unshared-buffer memory switch. Suppose that packets are 60 bytes in length and that the transmission rate of input and output ports is 150 Mbps. Assume that four bytes can be written into and read out of the RAM together each time and that the access times for read and write are the same. What is the required access time for the RAM for a 64×64 switch?

6

SWITCHING PRINCIPLES FOR MULTICAST, MULTIRATE, AND MULTIMEDIA SERVICES

A challenge of modern broadband digital networks is to efficiently support *multirate* and *multicast* connections for providing high-speed *multimedia* services. These services often have diverse quality-of-service (QoS) requirements. A key to the success of the future broadband network deployments, therefore, lies in the design of a high-speed packet switch to cope with these three *m*'s. In this chapter, we first present several fundamental switch design principles for multicasting. Then, we introduce the concept of path switching, which combines the advantageous elements of both dynamic and static routing. A cross-path switch is a Clos network operated under the principle of path switching. We shall show how cross-path switches can integrate various switch design principles together to support multirate, multicast, and multimedia traffic efficiently.

6.1 MULTICAST SWITCHING

So far we have been focusing on point-to-point switching in which an input packet is targeted for one and only one output. Many communication services, such as telephone calls, are point-to-point in nature and involve only two parties. At the other end of the spectrum, we have services such as television broadcast in which the same programs are received by all network users. The most general form of network connectivity is point-to-multipoint, in which information from one source is sent to a selected group of destinations. A three-party video conference call, for example, may be set up using three point-to-multipoint simplex (one-direction) connections, one from each source. Sending the same information from one source to a number of destinations

---

Principles of Broadband Switching and Networking, by Tony T. Lee and Soung C. Liew
Copyright © 2010 John Wiley & Sons, Inc.

206 SWITCHING PRINCIPLES FOR MULTICAST, MULTIRATE, AND MULTIMEDIA SERVICES

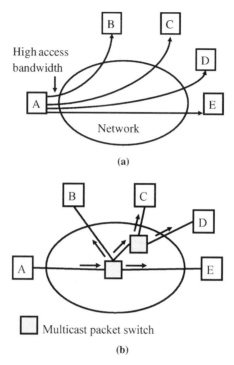

FIGURE 6.1 (a) Multicasting by separate point-to-point connections from source to destinations; (b) multicasting using multicast switches in network.

is called *multicasting*, and unicasting and broadcasting are the two extreme forms of multicasting.

While it is possible to perform multicasting by having the source establish a separate connection for each destination (see Fig. 6.1), this is usually not the most efficient way and is not viable when the number of destinations is large. Consider, for example, a video program vendor that wants to send its programs to hundreds of subscribers. This strategy would require very costly access facilities from the video source to the network. A better way is to set up just one connection from the source to some node (or nodes) in the communication network wherein the information is replicated and forwarded to the destinations. The node then requires a switch that can send information from one input to a number of selected outputs.

One version of the knockout switch architectures in the previous chapter uses N buses, one for each input, to broadcast input packets to the outputs. Filters at the outputs then select the packets destined for them. The multicast capability is inbuilt to such bus-based switch architectures. We may introduce, say, m bits in the packet header to indicate the multicast group. During connection setup, the filters of the outputs that belong to the multicast group are informed of the m-bit group identification. It is then possible for the filters to recognize and select the packets belonging to the group. The complexity of this switch architecture, as has been seen, is of order N^2, indicating that it does not scale well for large N.

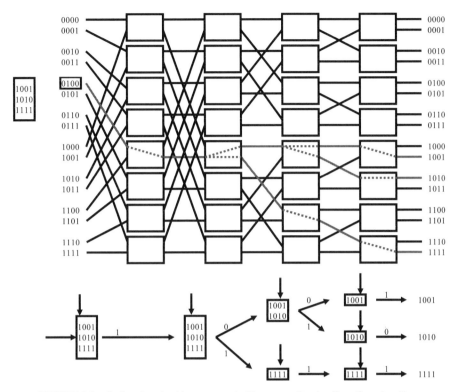

FIGURE 6.2 An input–output tree generated by generalized self-routing algorithm.

An alternative is to modify an interconnection network switch architecture, such as the banyan network, to handle packet replication and routing at the same time. Figure 6.2 shows the routing and copying of packets in a 16×16 banyan switch. Each 2×2 switch element has four states: in addition to the cross and bar states, the two other states correspond to connecting one of the inputs to both outputs. In the example, an input packet is destined for outputs 1001, 1010, and 1111. All these addresses are attached to the packet header. In stage i, bits i of all the addresses in the header are examined. If all of them are 0's (1's), the packet is forwarded to the upper (lower) output. On the other hand, if some bits are 0's and some bits are 1's, then the packet is duplicated with one copy sent to each output. The output addresses are split into two subsets: addresses with bit i equal to 0 attached to the upper packet, and addresses with bit i equal to 1 attached to the lower packet. It is easily seen that if there were only one packet being routed and replicated, a copy of the packet would eventually reach each desired output destination.

There are several problems with this approach. First of all, as we have seen in Chapter 3, the banyan network is highly blocking even for point-to-point connections when packets from many inputs are being routed. Adding the copying function to the routing function would only increase packet contention and exacerbate the situation. The second problem is more fundamental and applies to all switches (not necessarily

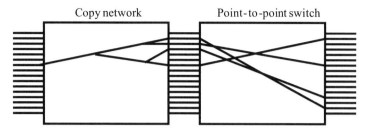

FIGURE 6.3 A multicast packet switch that consists of a copy network and a point-to-point switch.

banyan networks) that attempt to route and copy at the same time. With N outputs, there are $2^N - 1$ possible subsets of outputs for which a packet might be destined. To be able to specify all $2^N - 1$ possible subsets of outputs, there must be at least $\log_2(2^N - 1) \approx N$ bits in the packet header, which may be excessive for large N. Unless the fanout (number of outputs) of the connection is limited, this problem cannot be removed.

6.1.1 Multicasting Based on Nonblocking Copy Networks

A third approach that scales better is to separate the copying and the routing functions, as shown in Fig. 6.3. The copy network is responsible for making the exact number of copies required of the input packets without worrying about the outputs from which the copies emerge. A point-to-point switch, which could be any of those we have studied, then routes these copies to their respective targeted outputs.

Since the specific copy-network output reached by a packet copy is not important, we may introduce an additional restriction that all the copies of a packet must reach contiguous outputs. This restriction allows us to reduce the number of bits needed to specify the outputs in the packet header by using an output *interval*: (MIN, MAX). The output addresses MIN and MAX are the smallest and largest output addresses in the interval and the number of copies $CN = MAX - MIN + 1$. It turns out that we can design a nonblocking copy network using a broadcast banyan network, and this interval provides enough information for routing and duplication of packets at each stage within the network.

6.1.1.1 Packet Replication Using Boolean Interval Splitting Algorithm

The routing algorithm is called the *Boolean interval splitting algorithm*. The general algorithm is explained below; the reader is referred to Fig. 6.4 for a specific example. Suppose that a node at stage k received a packet with the header containing an address interval specified by two binary numbers: $MIN(k-1) = m_1 \cdots m_n$ and $MAX(k-1) = M_1 \cdots M_n$. The reason for introducing the argument $(k-1)$ is that the values of MIN and MAX are modified dynamically as the packet travels through the stages. The direction for packet routing is described as follows:

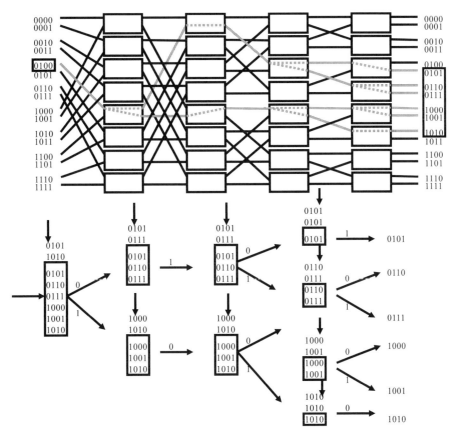

FIGURE 6.4 The Boolean interval splitting algorithm generates the equivalent input–output tree of a packet with interval addresses.

1. If $m_k = M_k = 0$ or $m_k = M_k = 1$, then send the packet out on link 0 (upper link) or 1 (lower link), respectively.
2. If $m_k = 0$ and $M_k = 1$, then duplicate the packet and modify the header (according to the scheme described below) and send both packets out on both links.
   (a) For the packet sent out on link 0
   $$\text{MIN}(k) = \text{MIN}(k-1) = m_1 \ldots m_n,$$
   $$\text{MAX}(k) = M_1 \ldots M_{k-1} 01 \ldots 1.$$
   (b) For the packet sent out on link 1
   $$\text{MIN}(k) = M_1 \ldots M_{k-1} 10 \ldots 0,$$
   $$\text{MAX}(k) = \text{MAX}(k-1) = M_1 \ldots M_n.$$

Physically, the above splitting means that the upper branch is responsible for replicating and routing packets to output interval $(m_1 \cdots m_n, M_1 \cdots M_{k-1} 01 \cdots 1)$, and the lower branch to output interval $(M_1 \cdots M_{k-1} 10 \cdots 0, M_1 \cdots M_n)$. These two intervals are contiguous to each other because the address $M_1 \cdots M_{k-1} 10 \cdots 0$ is next to the address $M_1 \cdots M_{k-1} 01 \cdots 1$. Therefore, the interval $(m_1 \cdots m_n, M_1 \cdots M_n)$ will be covered by both branches. Note from the above rules that $m_i = M_i, i = 1, \cdots, k-1$, holds for every packet that arrives at stage k (this can be easily argued by induction on k). Therefore, the event $m_k = 1$ and $M_k = 0$ is impossible due to the min–max representation of the address intervals.

6.1.1.2 Nonblocking Condition of Broadcast Banyan Network

A copy network is said to be *nonblocking* if it can produce all the packet copies requested provided the total number of copies required by all input packets is no more than N. In the broadcast banyan network, a set of input broadcast packets generates a set of embedded input–output trees. The network is nonblocking if the trees are link-independent (no common links among the trees). In Chapter 3, we showed that the banyan network is nonblocking for point-to-point connections if the input packets are sorted and concentrated. The following theorem is a generalization of the result. It specifies the conditions that must be met by the packets, hence the processing needed, before they can be forwarded to the broadcast banyan network for nonblocking replication. Let the input address of a packet i be x_i and the set of output addresses (i.e., addresses within the assigned output interval) to which the packet is targeted be Y_i.

Theorem 6.1 (Nonblocking Conditions of Broadcast Banyan Networks). A broadcast banyan network is nonblocking if the active inputs (inputs with arriving packets) $x_1, \ldots, x_m$ ($x_j > x_i$ if $j > i$) and their corresponding sets of outputs $Y_1, \ldots, Y_m$ satisfy the following:

1. *Distinct and monotonic outputs*: $Y_1 < Y_2 < \cdots < Y_m$ or $Y_1 > Y_2 > \cdots > Y_m$, where $Y_i < Y_j$ ($Y_i > Y_j$) indicates that every output address in Y_i is less than (greater than) every output address in Y_j.
2. *Concentrated inputs*: Any input between two active inputs is also active. That is, $x_i \le w \le x_j$ implies input w is active.

Proof. We know from Theorem 3.1 that the set of input–output paths $\langle x_1, y_1 \rangle, \ldots, \langle x_m, y_m \rangle$ is link-independent for an arbitrary choice of output addresses $y_1 \in Y_1, \ldots, y_m \in Y_m$. It follows that the set of input–output trees embedded in the broadcast banyan network must be link-independent. □

This theorem indicates that two tasks must be performed before the input packets enter the broadcast banyan network. First, the packets must be concentrated. This can be achieved by placing a concentrator before the broadcast banyan network. Second, the output addresses assigned to packets must be monotonic. One way is to assign the outputs to the active inputs from top to bottom as follows: for each active input i that requests CN_i copies, assign the next contiguous set of outputs Y_i, with $|Y_i| = CN_i$.

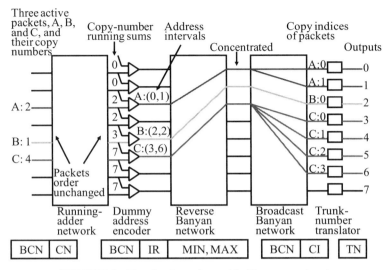

FIGURE 6.5 The structure of a nonblocking copy network.

Recall that a concentrator may be built by a running-adder network and a reverse banyan network. The running-adder network assigns a set of concentrated reverse banyan network's output addresses to the input packets by computing the running sums of their activity bits. It turns out that the running-adder address network can also be used to assign the broadcast banyan network's output intervals to the input packets, as will be described after the following overview.

The overall structure of the copy network is depicted in Fig. 6.5. As shown, it consists of a cascade of five components: a running-adder network, a set of N dummy address encoders, a reverse banyan network, a broadcast banyan network, and a set of N trunk-number translators. There are two additional fields in the header of each incoming packet: the broadcast channel number (BCN) identifies the multicast connection, and the copy number (CN) specifies the number of copies required. The BCN and CN are associated with a virtual connection and are determined during the call setup time. The different copies of a packet must reach different output destinations of the point-to-point switch that follows the copy network and they are distinguished by their copy indices (CIs). The CIs are generated inside the copy network after packet replication. The output addresses are referred to as the trunk numbers (TNs), and a trunk-number translator uses the packet's BCN and CI to look up its trunk number.

Packet replications are accomplished by two fundamental processes, an encoding process and a decoding process. The encoding process is performed by the running-adder network and dummy address encoders. The decoding process is performed by the concentrator, broadcast banyan network, and the trunk-number translator.

6.1.1.3 Encoding Process Two pieces of information must be generated for every packet during the encoding process: (1) its reverse banyan network output

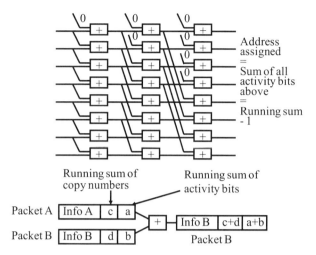

FIGURE 6.6 An adder in the running-adder network of the copy network: adding both the running sums of the activity bits and the copy numbers.

address and (2) its broadcast banyan network output intervals. The use of a running-adder network to compute the former has been described earlier in this chapter. To produce the latter, in addition to the running sums of the activity bits of packets, the running-adder network also computes the running sums of CNs, as depicted in Fig. 6.6. Similar to the computation of the activity bit running sums, it can be shown that the running sum

$$RS_j = \sum_{i=0}^{j} CN_i$$

is produced at output j, $0 \leq j \leq N-1$, of the network. Based on adjacent running sums, the dummy address encoders then generate the following sequence of address intervals:

$$(0, RS_0 - 1), (RS_0, RS_1 - 1), \ldots, (RS_{N-2}, RS_{N-1} - 1).$$

If the packet on output j is active, then $RS_j - 1 \geq RS_{j-1}$ and $(RS_{j-1}, RS_j - 1)$ is a legitimate interval and the output interval assigned to the packet is $(MIN, MAX) = (RS_{j-1}, RS_j - 1)$. If the packet is inactive, $RS_j = RS_{j-1}$ and the interval is illegitimate, which is of no adverse consequence anyway.

6.1.1.4 Decoding Process First of all, the reverse banyan network concentrates the input packets based on the running sums of their activity bits. The header of a packet as it enters the broadcast banyan network contains three fields important to the copying process: the BCN, address interval (MIN, MAX), and an index reference

(IR) that is set to MIN. The IR field is not used or modified in the broadcast banyan network, and all copies of the same master packet have the same IR value. Its purpose is the computation of copy indices (CI), which identifies the different copies at the outputs of the banyan network, as will be explained shortly.

We have already seen that packets can be replicated and routed in a nonblocking manner in the broadcast banyan network based on the Boolean routing algorithm. Some finer implementation details are described here. An activity bit must be posted in front of the address field to indicate whether the packet is active. At a switch node, if one of the input packets is active and one is inactive, routing will be performed based on the active packet header. Depending on the header, the switch element could be set to cross, bar, or the branching state. If both packets are inactive, we have a "don't care" situation and it does not matter to which state the switch is set. If both packets are active, neither packet needs to be duplicated. The switch element must be set to either the bar or cross state, and either packet header can be used to achieve the same setting.

When packets emerge from the broadcast banyan network, the address interval in their header contains only one address. That is, according to the Boolean interval splitting algorithm, we have

$$\text{MIN}(n) = \text{MAX}(n) = \text{output address}$$

and the copy index is computed by

$$\text{CI} = \text{output address} - \text{IR}.$$

In this way, the CN copies of a packet would have distinct CI, ranging from 0 to CN − 1. Based on BCN and CI, the trunk-number translator retrieves the trunk number (output address of the subsequent point-to-point switch) for which the packet copy is destined from a table. The trunk number corresponding to a BCN and a CI is determined during the setup of the multicast connection and is stored in the table throughout the duration of the session. Note that successive packets of the same BCN and CI may emerge from different broadcast banyan network outputs in different time slots. Therefore, it is important that the trunk number associated with the BCN and CI be stored in all the trunk-number translators.

6.1.2 Performance Improvement of Copy Networks

In the copy network discussed thus far, if the total number of copies requested by all input ports is not more than the network capacity N, all the requests are granted; otherwise, overflow of the copy requests occurs and some requests must be queued at the inputs for the next time slot.[1] The overflow problem gives rise to two issues:

---

[1] One may also design a loss system in which the overflow packets are dropped. As in the point-to-point switch, the issue then is how to design and engineer the copy network to have small packet loss probability.

1. How to remove the inherent bias of the running-adder network against the lower ports?
2. How to perform partial service of a copy request?

The first issue is related to *input fairness*. Since the address intervals of the broadcast banyan network are assigned to the copy requests by the running-adder network starting from top to bottom, lower input ports suffer when the available addresses run out. As a result, packets from lower input ports must be buffered much more often than those from upper ports under heavy-traffic situations. The solution is to allow computation of the running sums starting from any arbitrary input i so that port i is effectively the top port and port $(i-1)$ mod N the bottom port. Changing i from time slot to time slot based on the overflow condition allows input fairness to be achieved.

The second issue is related to the efficiency of the copy network. The utilization of the copy network would be improved if partial service of copy requests, called *call splitting*, can be implemented when overflow occurs. For argument's sake, suppose that the packet from the first input requests one copy and the packet from the second input requests N copies. If no partial service is allowed, the copy network would produce only one packet copy in the current time slot, resulting in much wastefulness. The solution is to allow $N-1$ copies to be made of the second packet in the current time slot, leaving a residual copy to be made in the next time slot. Call splitting allows the maximum number of packet copies (N) to be made whenever overflow occurs.

Before providing the solution, let us illustrate what we want to achieve with the 4×4 copy network example given in Fig. 6.7. In the current time slot, the total number of copies requested is seven, exceeding the capacity of the network. We note

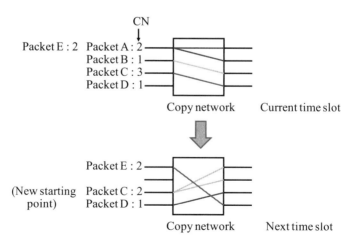

FIGURE 6.7 Illustration showing the principle for achieving input fairness by shifting the service priority order and the principle of achieving efficiency by call splitting in a copy network.

that although packet C requests three copies, the request cannot be satisfied due to overflow. However, call splitting allows one copy to be made in the current time slot. In the next time slot, the starting point for service will be packet C, because this is where overflow occurs in the current time slot.

6.1.2.1 Cyclic Running-Adder Network

To achieve the above, we introduce a modified running-adder network called the cyclic running-adder network. Figure 6.8 shows the basic structure and operation of the cyclic running-adder network. The structure is totally symmetrical across all ports. Recall that in the original running-adder network, partially computed running sums from the upper links are fed to the lower links, but not the other way round. In the cyclic structure, partially computed running sums from the lower links are also fed to the upper links, as shown in the figure, where the interconnections are indicated by letters $a, b, \ldots, g$. Wrapping the top of the figure to the bottom results in a cylindrically symmetrical structure.

In each time slot, a *starting point* for the computation of the running sums is defined. In Fig. 6.8, for example, running sums are computed starting from the fourth port. For explanation, we shall refer to the starting point as port 0, the next as port 1, and so on (see Fig. 6.8), with the understanding that the labeling changes from time slot to time slot. The partially computed running sums that cross the starting point are ignored by the adders to which they are fed: with reference to Fig. 6.8, the *darkened*

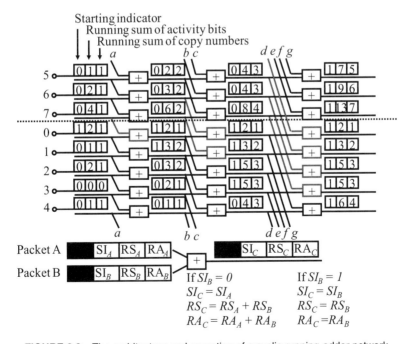

FIGURE 6.8 The architecture and operation of a cyclic running-adder network.

links that cross the *dashed* horizontal line are ignored by the *shaded* adders. In other words, the partial running sums on the darkened links are treated as having values of zero, effectively making port 0 the top port, port 1, the second port, and so on, in the running-sum computation (cf. Fig. 6.6).

The next question is how to inform the shaded adders to ignore their upper inputs, considering that the set of shaded adders changes from time slot to time slot with the starting point. We note that the paths from the starting input port to the shaded adders form a binary tree. Therefore, we introduce a one-bit starting indicator (SI) field, as shown in Fig. 6.8. Initially, only the starting port has this field set to 1; the other ports have it set to 0. The adders in the network simply add the SI fields on their two inputs. Notice that only the shaded adders have the SI field on their lower inputs set to 1. Therefore, this can be used as an indication that they must ignore their upper inputs in running-sum computation.

The starting point remains the same until overflow occurs, in which case the new starting point in the next time slot will be the first port that encounters overflow in the current time slot. By the end of each time slot, every input port must be informed of the effective starting point for the next time slot and the number of copies produced for its request in the current time slot—this is needed for implementing call splitting.

For this purpose, at the outputs of the running-adder network, the SI for the next time slot and the *served copy number* (SCN) for the current time slot are determined for every port based on its running sum and the one above. This information is then fed back to the corresponding input port via a backpropagating path. Let RS_i be the running sum computed at output i. The SI bit indicates whether the input port is the next starting point and is computed as follows:

$$SI_0 = \begin{cases} 1, & \text{if } RS_{N-1} \leq N, \\ 0, & \text{otherwise}, \end{cases}$$

$$SI_i = \begin{cases} 1, & \text{if } RS_{i-1} \leq N \text{ and } RS_i > N, \\ 0, & \text{otherwise}, \end{cases} \quad (6.1)$$

for port $i = 1, 2, \ldots, N - 1$. In Fig. 6.8, the next starting point is port 6 (the second port from the top) and its SI bit for the next time slot is set to 1 according to the above rule.

The SCN represents the number of copies allowed for the packet in the current time slot. It is computed as follows:

$$SCN_0 = RS_0,$$

$$SCN_i = \begin{cases} \min(N, RS_i) - RS_{i-1}, & \text{if } RS_{i-1} < N, \\ 0, & \text{otherwise}, \end{cases} \quad (6.2)$$

for input port $i = 1, 2, \cdots, N - 1$. We assume that the maximum allowable copies of each packet are bounded by the capacity of the copy network N. Therefore, the residual copy request from the current starting point can always be fulfilled within one cycle, and the delay difference between any two copies of the same master packet is at most one slot time. For any other input port i, its copy request will be fully satisfied if its running sum does not exceed N. Otherwise, the call will be split and only $\max(0, N - RS_{i-1})$ copies will be allowed.

In Fig 6.8, the SCN field returned to port 6 is 1 according to the above procedure, meaning that only one out of two requested copies is allowed. The residual copy request will be served in the next time slot. No copy is allowed for port 7, since its running sum and the one above both exceed $N = 8$. The activity bit of this overflow packet is set to 0, indicating that they become inactive in the current cycle.

6.1.2.2 Implication of Shifting Starting Point for Concentration
Recall that the packets that exit from the running-adder network must be concentrated by a concentrator before they can be replicated in a nonblocking manner in the broadcast banyan network. In the original design, the running sums of the activity bits are computed together with the running sums of the copy numbers in the same running-adder network. We may continue to do so in the cyclic running-adder network. However, shifting the starting point for computation running sums away from the top port presents a new problem: the reverse banyan network may not be nonblocking anymore. As shown by the example in Fig. 6.9, when the outputs of the reverse banyan network assigned to the packets are no more monotonic from top to bottom, the reverse banyan network could be blocking.

Figure 6.10 presents three ways to solve this problem. Problem 6.4 proposes two other methods. The approach in Fig. 6.10(a) uses an additional running-adder network for the computation of the running sums of SCNs and activity bits (RAs) starting

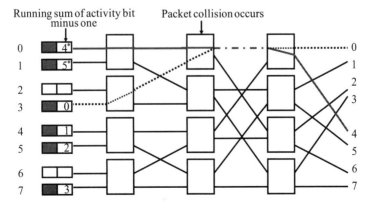

FIGURE 6.9 Cyclically monotone routing address gives rise to packet collisions in reversed banyan network. The packets at ports 2 and 6 are inactive. (*Note:* inputs connected to outputs of the cyclic running-adder network in Fig. 6.8.)

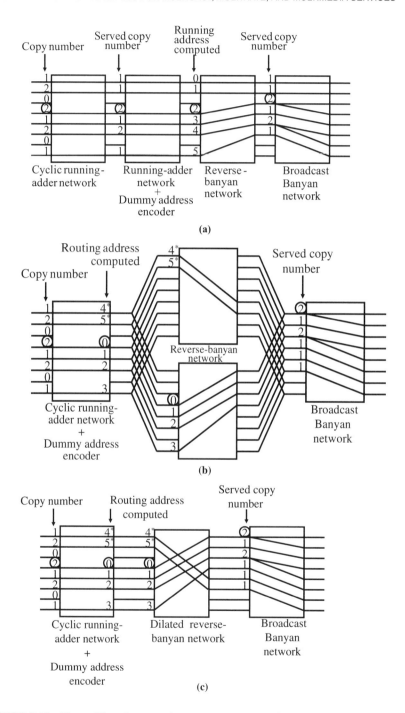

FIGURE 6.10 Three different approaches to concentrate active packets. The starting-point packet is marked by encircling its copy number, routing address in reverse banyan network, and served copy number.

from the top port. The preceding cyclic running-adder network makes sure that not more than a total N packet copies are allowed. Therefore, the additional running-adder network can perform exactly the same function as in the original design of copy network without encountering overflow. The cyclic network serves purely as a "contention-resolution" device.

The other approach is based on the observation that a cyclically shifted monotone sequence can be divided into two monotone subsequences. As shown in Fig. 6.10(b), the RA fields of star-marked packets form a monotone subsequence, and the RA fields of unmarked packets form the other one. The basic idea is to send these two monotone subsequences separately to two different reverse banyan networks. No blocking occurs in either reverse banyan network. In this design, packets above the starting point are forwarded to the upper reverse banyan network and packets at or below the starting point are forwarded to the lower one. For self-routing, a 1×2 switch at each output of the cyclic running-adder network must make this decision based on the packet header, and the knowledge of the starting point must be available.

To broadcast the starting point, let us reexamine Fig. 6.8. Notice that the SI fields of all packets are set to 1 at the outputs of the cyclic running-adder network. In other words, the SI bit of the starting port has propagated to all the other ports using the network as a broadcast device. Thus, for our purpose here, we may replace the SI field with a starting port (SP) field. At the inputs, the SP field of the starting port i is set to i; the SP fields of the rest are set to, say, -1 (assuming we label the ports using 0 to $N-1$ from top to bottom). The conditions "If $SI_B = 0$" and "If $SI_B = 1$" in Fig. 6.8 for the operation of an adder are replaced with "If $SP_B = -1$" and "If $SP_B \neq -1$," respectively. In this way, the starting port is encoded into the SP fields of all the packets at the outputs, and it is easy for a 1×2 switch to determine the position of the associated port with respect to the starting point.

The third approach, which is more elegant than the second, does not require physically separating the packets into two parts. Suppose we overlay one reverse banyan network in Fig. 6.10(b) on the other, we find that any internal link in this overlaid network will not be occupied by no more than two packets simultaneously. That is, any packet collision in Fig. 6.10 would only occur between a star-marked packet and an unmarked one. Therefore, as shown in Fig. 6.10(c), a *dilated* reverse banyan network with dilation degree 2 can be utilized as a concentrator network, and it will route the cyclically shifted monotone sequence in a nonblocking manner.

6.1.2.3 Implication of Call Splitting for Index Reference Recall that an index reference (IR) is needed to compute the copy index of a packet copy so that the trunk-number translator can distinguish it from the other copies of the same master packet. In the original design discussed in the preceding subsection, the IR field is always set to the minimum of the address interval (MIN, MAX). However, if call splitting is implemented, the IR field of the starting point should be adjusted according to the number of copies generated in the previous time slot to maintain the continuity of the index sequence. Specifically, for the port whose request is only partially fulfilled in the previous time slot,

$$IR = MIN - \text{previous SCN}. \tag{6.3}$$

It follows that the copy index (CI) of a packet that exits at output i of the broadcast banyan network would be

$$CI = i - IR = i - MIN + \text{previous SCN}. \tag{6.4}$$

In this way, the trunk-number translator can then assign the correct output address to the packet for routing in the subsequent point-to-point switch.

6.1.3 Multicasting Algorithm for Arbitrary Network Topologies

In the copy networks of the preceding subsections, the routes traversed by a packet are predetermined before they enter the copy networks. There is another category of copy networks that operate in an analogous way as deflection routing in that the routes are determined on the fly as packet traverses the networks.

In the previous approach, much preprocessing is performed before packets are launched into the broadcast banyan network for replication. The goal of the preprocessing is to avoid the conflict of packets inside the network. It can be easily seen that the complexity of the preprocessing hardware is actually more than that of the broadcast banyan network itself.

An alternative approach is to make no attempt to avoid packet conflict in the broadcast banyan network. Arriving packets are immediately launched into it. At any 2×2 node in the network, an input packet desiring replication will be duplicated and forwarded to both outputs provided there is no packet arriving on the other input. Otherwise, each of the input packets will simply be forwarded to one of the two outputs. In other words, a packet desiring replication will be duplicated when possible. Otherwise, the replication process will be delayed until a later stage. This strategy is rather general and does not depend on the nonblocking condition of the broadcast banyan network, since no attempt is made to avoid conflict in the network anyway. Therefore, it can be applied to arbitrary network topologies, including directed networks.

Let us first focus on the shuffle-exchange network before relating the replication algorithm to general network topologies. Instead of a $\log_2 N$-stage network, we have an L-stage network in which $L \geq \log_2 N$. In general, the larger the L value, the more likely that all the copy requests can be satisfied, since larger L implies more chances for packet replication attempts.

We shall generalize the Boolean splitting interval algorithm (Problem 6.5 considers an approach in which the copy number is encoded in the packet header rather than the splitting interval). Instead of using the interval to code the exact output addresses of the packet, we use it to code the *fanout* or the number of copies desired by the packet. For instance, the interval $(0, F - 1)$ can be assigned to every packet, where F is its fanout number. On which output a particular packet copy will end up depends on its interactions with other packets within the network and is not predetermined beforehand.

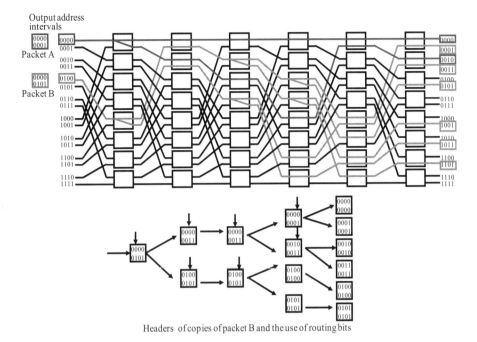

FIGURE 6.11 General replication scheme using interval splitting.

For illustration, an example of the replication process is given in Fig. 6.11. In the example, there are two packets, A and B, requesting for two and six copies, respectively. Let us track the replication of packet B.

Initially, its address interval is (0000, 0101). Unlike the original interval splitting algorithm, bits k are not necessarily used at stage k here. In our example, the first bits of 0000 and 0101 are both 0, so we progress directly to the second bits—in general, we skip the first bit position where MIN and MAX differ so that splitting can be performed immediately. After splitting, both copies of packet B find a copy of packet A at the other input of the same node at the second stage.

Let us first look at the upper copy of packet B. Since we do not want to buffer packets, splitting at this stage is not possible because each input packet must exit on one of the outputs. The splitting is therefore delayed until the third stage. This means that the third stage should examine the same bit position as the second stage to decide whether splitting is required. In the example, replication is possible at the third stage because the other input does not have a packet. The replication process is completed after five stages.

Now look at the lower copy of packet B at the second stage. Both bits examined are 0 and replication is not needed. Therefore, unlike the upper copy, we can progress to the next bit position at the third stage here. In general, only when replication is required and is denied will we examine the same bit position at the next stage.

The general algorithm can be described as follows:

Generalized Interval Splitting Algorithm

1. Initially, skip directly to the first bit position where MIN and MAX differ.
2. At any stage, examine the two routing bits.
   (a) If the bits are different and the other input of the switch node does not have a packet, perform splitting according to the original interval splitting algorithm.
   (b) If the bits are different and the other input of the switch node has a packet, do not perform splitting. At the next stage, the same two bits will still be examined.
   (c) If the bits are the same, progress to the next bits at the next stage.

Since the bits to be examined at stage k are not necessarily bits k, a way to indicate the effective routing bits is needed. This can be solved rather easily by positioning the effective routing bits at the front of the packet header at all times so that they are by default the first two bits, and this can be achieved by a logic design involving six-bit delays at each stage. The details of bit manipulation and logic implementation are beyond the scope of this book, which focuses only on the principles.

Several issues that must be addressed are discussed below.

6.1.3.1 Distinguishing Packet Copies
As discussed before, at the outputs of a copy network, packets of different multicast connections are distinguished using BCN and an IR. All copies of the same master packet have the same BCN, so the BCN can be incorporated into the header of the packet. For generation of distinct IRs, suppose that the interval $(0, F - 1)$ is assigned to all master packets. After successful replication, there will be F copies, all with MIN equal to MAX in their interval fields. Furthermore, the MIN (MAX) of the different copies are distinct and range from 0 to $F - 1$. Therefore, the MIN (or MAX) value can be used as the IR and no explicit IR field is needed in the header. If the replication process is not completed yet for a copy at an output of the copy network, then its MIN < MAX, and (MAX − MIN) copies will be considered as lost. The copy chosen to be the successful one can have an IR anywhere between MIN and MAX, inclusively.

6.1.3.2 Deadlock Prevention
Although larger L implies lower packet loss probability, there are two phenomena, however, that may prevent the completion of the replication process, no matter how large L is.

This first is obvious, and it is request overflow, which occurs when the sum of the fanouts of input packets exceeds N, the capacity of the network. The ways for dealing with overflow will be discussed later. The second phenomenon is deadlock, which occurs even when the total copies requested is less than N if certain deterministic routing policies are used.

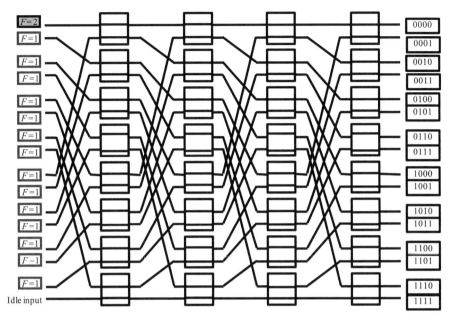

FIGURE 6.12 A deadlock example with a deterministic routing policy: set switch elements into bar state under "don't care" situations. A packet fails to be duplicated regardless of the number of stages while the bottom link is idle at each stage.

Figure 6.12 shows an example of a deadlock with the following policy: at a 2×2 node, if there are two input packets, set the switch element to *bar* state (i.e., forward the upper input packet to the upper output and the lower input packet to the lower); also set the switch element to bar state under the "don't care" situation when one input has a packet that does not need to be replicated while the other input does not have an active packet. In the example, the top input has a packet with $F = 2$, the bottom input does not have an active packet, and each of the rest of the inputs has a packet with $F = 1$. It is easily seen that with the above policy, no matter how large L is, the top packet will stay at the top link while the bottom link remains idle throughout all the stages. This is not a contrived example, and one can come up with many different deadlock scenarios.

There are two ways to prevent deadlocks. One is to use random routing policy. Whenever we have a "don't care" situation, set the switch element randomly either to bar or to cross state. One disadvantage with this approach is the need for implementing a random-number generator at each switch node. The alternative is to devise deterministic routing policies that are deadlock-free. We explain below a very simple deadlock-free policy.

For simple explanation, one can picture a token on each idle link. In order for a packet to be duplicated at a node, it must acquire and "consume" a token on the other input. Thus, the problem becomes that of devising a routing policy that will ensure the meeting of a token and the packet desiring replication. There are two "don't care" situations that must be made into "care" conditions: (1) whenever there is a token at one input and the other input does not have a packet that desires duplication, we forward the token to the upper output; (2) whenever the two inputs are occupied and at least one packet desires duplication, we forward that packet to the upper output.

To see why this strategy works, suppose we label the nodes in each stage from top to bottom in a binary fashion. As we have already learned, the construction of the shuffle-exchange network is such that a packet or a token residing at node $x_n x_{n-1} \cdots x_2$ will be forwarded via the upper outgoing link to node $0 x_n x_{n-1} \cdots x_3$ at the next stage. Therefore, the above policy can be viewed as attempting to route tokens and packets desiring replication to the top node $00 \cdots 0$. Thus, if there is a token and a packet that desires replication in the network, they will eventually meet at node $00 \cdots 0$, if not earlier. Consequently, deadlocks cannot occur.

In the worst case, a packet can be duplicated at node $00 \cdots 0$ every $\log_2 N$ stages. In actuality, packets desiring replication and tokens meet much more often. This becomes evident when we view the routing policy as attempting to concentrate tokens and packets desiring replications at nodes with many 0's (but not necessarily all 0's) in their labels. Packets whose replication has been completed are "pushed" to nodes with many 1's in their labels.

6.1.3.3 Overflow Problem Overflow occurs when the sum of the fanouts of input packets exceeds N. There are two classes of approaches to this problem. The first, such as that discussed in the preceding subsection, is to incorporate a reservation (contention-resolution) mechanism so that the packets allowed to enter the copy network desire at most a total of N copies. The overflow requests are buffered at the inputs so that replication can be attempted in the next time slot. One of the goals of the copy network in this subsection, however, is to eliminate this kind of preprocessing. Therefore, we examine the second approach, which is to increase the bandwidth (capacity) inside the network with respect to the external bandwidth.

One possibility is to employ expansion. An $M \times M$ ($M > N$) shuffle-exchange copy network could be used in such a way that only the upper N of its M input ports are connected to input links. Because only N of the input links are used, some switch elements at the initial stages are guaranteed not to be traversed by packets. An $N \times M$ expansion network results when these switch elements are removed. In this way, the capacity of the network is expanded to M and the internal load to the copy network is reduced, making the occurrence of overflow less likely.

Another possibility is to speed up the internal link with respect to the external link. With a speedup factor of 2, for example, each external time slot corresponds to two internal time slots, and by dividing the copy requests into two batches, one for each internal time slot, the overflow probability can be reduced. With this approach, if the point-to-point switch that follows the copy network does not employ speedup, buffers are needed in between them, since multiple copies may be delivered to the same link interconnecting them in the same time slot. In this sense, this scheme is compatible with point-to-point switches that employ input buffering. This scheme is also compatible with those point-to-point output-buffered switches that employ internal speedup with the same or higher speedup factor. The overall multicast switch will then be output-buffered.

6.1.3.4 Performance We consider a brief approximate analysis of the loss probability of packet copies. The analysis here applies to arbitrary network topologies. Consider an input packet requesting $F = 2^f$ copies in a large-size copy network. We can picture the input packet as having F embedded copies and focus on one of the copies. Each time splitting occurs, the copy becomes embedded in one of the output packets. The loss probability of the copy can be approximated by the probability that the copy experiences fewer than f splittings after traversing the L stages of the copy network.

For simplicity, we assume that a splitting attempt will be made as long as the replication process is not completed. Furthermore, we assume the probabilities of successful splitting at different stages to be independent. In reality, the probabilities of success at the earlier stages are higher than those at later stages because there are more packet copies at later stages. However, for analytical tractability, we assume that the probabilities are all equal to p, which will be set to the worst-case value later.[2] The situation is then the same as that in Bernoulli trials with success probability p and failure probability $q = 1 - p$. Denote the number of successes in L trials by S_L. The loss probability can be approximated by the probability that S_L is less than f, which can in turn be approximated by the error function when f is large:

$$P_{loss} \approx \Pr\{S_L < f\} \approx \frac{1}{\sqrt{2\pi}} \int_{-\infty}^{-y} e^{-x^2/2} dx, \qquad (6.5)$$

where

$$y = \frac{Lp - f}{\sqrt{pqL}} = \frac{\sqrt{L}(p - f/L)}{\sqrt{pq}}. \qquad (6.6)$$

The inequality

$$\frac{1}{\sqrt{2\pi}} \int_{-\infty}^{-y} e^{-x^2/2} dx \le \frac{1}{2} e^{-y^2/2} \qquad (6.7)$$

can be used to obtain a simpler bound. For L large enough that $(p - f/L) \approx p$, it can be seen that P_{loss} decreases exponentially with L.

Let us now take another viewpoint: what L is required to meet a given loss probability requirement? We note from (6.5) and (6.6) that f/L is an important quantity. Specifically, for a given constant $f/L < p$ the upper bound on P_{loss} approaches zero as f (and therefore L increases), whereas if $f/L > p$ the upper bound approaches one. Therefore, in general, the required L for a given loss probability requirement should be a constant times $f' = f/p$:

$$L = kf' = kf/p, \qquad (6.8)$$

where $k > 1$. For large f, L needs only be slightly more than f/p.

---

[2]This analysis has also ignored the overflow effect, which causes the success probabilities at different stages to be dependent. Thus, strictly speaking, our approximation is valid only if the overflow probability has been made very small by design.

Consider the following fact: when there is no contention and $p = 1$, f stages are needed to complete the replication process; but when $p < 1$, roughly $f' = f/p$ stages are required according to the above analysis. Therefore, f' can be interpreted as the *effective fanout* parameter when there is contention from other packets. For the worst possible fanout, $f = n$, the required L is proportional to n/p, where $n = \log_2 N$.[3]

To see how L relates to a given P_{loss} more specifically, write $k = 1 + \epsilon$ where $\epsilon > 0$. It is easy to derive from (6.5) and (6.7) that

$$\epsilon^2 \leq -\frac{2q}{f}\ln(2P_{loss}) - \epsilon < -\frac{2q}{f}\ln(2P_{loss}). \qquad (6.9)$$

Thus, the ϵ required to achieve a given loss probability goes to zero as f tends to infinity given a fixed q. This indicates that for large copy networks, in the worst case, the required L needs only be slightly more than n/p.

We now relate p to the network load. Let ρ denote the *output* offered load of the network. It is related to the input offered load ρ_o (probability of finding a packet at an input) by $\rho = \overline{f}\rho_o$, where $\overline{f}$ is the average fanout. Inside the network, the probability of a packet finding another packet at the same switch node, and therefore splitting is not possible, is bounded by ρ. In other words, $p = 1 - \rho$ and $q = \rho$. For a network that adopts expansion or speedup (see Section 6.1.3.3) with an expansion or speedup factor of S, $p = 1 - \rho/S$ and $q = \rho/S$.

Let us consider an example in which $\rho = 0.8$, $S = 2$, and $F = 32$. Suppose that a loss probability of 10^{-8} is required. It can be derived from (6.9) that the required $\epsilon < 1.7$, and the required $L < 23$. For a 256×256 copy network, $n = \log_2 N = 8$, this means $L < 2.8n$.

In reality, the required L should be much smaller, thanks to our generous simplifying assumptions in the analysis. The interested readers are referred to Ref. [Lie95] for simulated performance results.

6.1.3.5 Application to Arbitrary Topologies

Let us now illustrate the generality of the replication scheme by considering an entirely different network topology. The main observation is that the correctness of the algorithm does not depend on the shuffle-exchange pattern. The adjacent stages can be interconnected in any fashion.[4]

The same scheme also applies to undirected networks. An example is the shuffle-exchange network consisting of only one stage in which the output links of the stage are connected back to the inputs of the same stage via a shuffle pattern. For undi-

---

[3] In practice, the fanouts of multicast connections depend on the underlying services and applications, and they may not simply grow with the switch size. One would therefore expect the required L to be less than proportional to n/p as n increases.

[4] The performance, however, may be dependent on the interconnection pattern regularly or irregularly. For example, we can have an interconnection in which the $N \times N$ copy network is actually two unconnected $N/2 \times N/2$ shuffle-exchange networks in parallel. The performance will not be as good as a single $N \times N$ shuffle-exchange network.

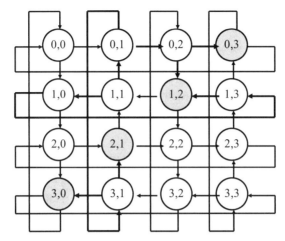

FIGURE 6.13 Packet replication in a Manhattan-street network using interval splitting: an example in which the source node is (1, 1) and the packet requests six copies.

rected networks, the same network can be used for both replication and routing: the replication algorithm is performed, and a copy is routed to its destination when no further splitting is required of it. Since the copies of the same master packet may be produced at different times, some copies may embark on journeys to their actual destination nodes while the others are still being replicated. Also, packets of different multicast connections may be replicated and routed simultaneously, and the network operation does not have to be divided into separate replication and routing phases.

Consider the application of the generalized interval splitting algorithm in the Manhattan-street network, a 16-node example of which is given in Fig. 6.13. In this example, the nodes are arranged as a 4×4 square grid. Each node has two inputs and two outputs. The data flows on the rows (columns) alternate between left-to-right (upward) and right-to-left (downward). The example adopts the interval splitting algorithm. We have a packet at node (1, 1) requesting for six copies, and its splitting interval is set to (0000, 0101). We end up with two copies each at nodes (0, 3) and (1, 2) and one copy each at nodes (2, 1) and (3, 0).

In general, when several packets are being replicated and routed, contention may occur and the replication process on some branches may be delayed. In addition, in undirected networks, copies of the same master packet may interfere with each other. For example, this happens in Fig. 6.13 if the two copies at node (0, 3) are to be split further (i.e., more than six copies are requested). Another situation that must be taken care of in undirected networks is when all the links have a packet and the splitting process of none of them is completed yet. This is an analog to the overflow problem. The copies circulate around the network indefinitely. In addition, there is also the possibility of deadlocks, in which even though overflow does not occur, packets desiring replication fail to do so no matter after how many steps. A mechanism must be installed to either prevent this from happening or break the deadlock. The details are beyond the scope of this book: Problem 6.6 explores this issue a little further based on a two-dimensional hypercube network.

The packet replication algorithm can readily be generalized to networks in which the nodes are not 2×2 (e.g., the hypercube network). The basic idea remains the same, except that now *partial* splitting is possible. For instance, when two packets enter a 4×4 node both wanting to split three ways, we may split both packets two ways.

6.1.4 Nonblocking Copy Networks Based on Broadcast Clos Networks

In this subsection, we extend the nonblocking properties of the Clos network to construct a nonblocking broadcast Clos network. We shall present a generalized nonblocking copy network architecture based on this broadcast Clos network.

A broadcast Clos network is a Clos network with switching nodes that can perform packet replication. Let the active inputs be $s_0, s_1, \ldots, s_{n-1}$ ($s_j > s_i$ if $j > i$). Each active input may go to a number of outputs. Let their corresponding *sets of outputs* be $D_0, D_1, \ldots, D_{n-1}$, respectively. The set of connection requests is monotonic if

$$s_i < s_j \Rightarrow d_i < d_j \quad \forall d_i \in D_i \text{ and } \forall d_j \in D_j$$

or

$$s_i < s_j \Rightarrow d_i > d_j \quad \forall d_i \in D_i \text{ and } \forall d_j \in D_j.$$

If the set of connection requests is monotonic, route assignment can be done by the rank-based assignment algorithm for broadcast Clos networks.

Theorem 6.2 (Rank-based Assignment Algorithm for Broadcast Clos Network). Let the set of connection requests $\pi = \{(s_0, D_0), \ldots, (s_{n-1}, D_{n-1})\}$ be monotonic. The assignment

$$f(s_i, D_i) = [m + i]_q, \qquad (6.10)$$

where q is the number of middle-stage modules, m is a constant integer, and i is the rank of the connection request, is nonblocking.

Proof. Construct the sequence $(d_0, d_1, \ldots, d_{n-1})$ by arbitrarily choosing a d_i from D_i for $0 \leq i \leq n - 1$. Since π is monotonic, the sequences $(s_0, s_1, \ldots, s_{n-1})$ and $(d_0, d_1, \ldots, d_{n-1})$ are monotonic. If two connection requests (s_i, d_i) and (s_j, d_j), where $i < j$, are assigned to the same middle-stage module, we have $[m + i]_q = [m + j]_q$ and this implies that $i + lq = j$, where $l \geq 1$. Now we have

$$s_j - s_i = (s_j - s_{j-1}) + (s_{j-1} - s_{j-2}) + \cdots + (s_{i+1} - s_i)$$
$$\geq \underbrace{1 + 1 + 1 + \cdots + 1}_{j-i} = j - i = lq \geq q$$

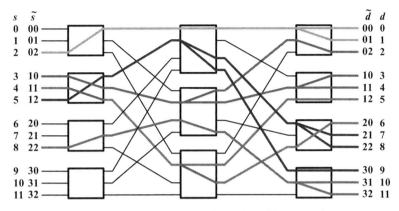

FIGURE 6.14 An example of broadcast Clos network.

and

$$d_j - d_i = (d_j - d_{j-1}) + (d_{j-1} - d_{j-2}) + \cdots + (d_{i+1} - d_i)$$
$$\geq \underbrace{1 + 1 + 1 + \cdots + 1}_{j-i} = j - i = lq \geq q.$$

This implies that $|s_i - s_j| \geq q$ and $|d_i - d_j| \geq q$. Packets from the same input-stage module will not be assigned to the same middle-stage module. Similarly, packets destined for the same output-stage module will not be assigned to the same middle-stage module. Since the choice of d_i from D_i is arbitrary, the broadcast Clos network is therefore non blocking. □

To illustrate this idea, consider a three-stage broadcast Clos network with $q = 3$ and $p = 4$ as shown in Fig. 6.14. The connection requests (which are monotonic) are shown in Table 6.1. Using the rank-based assignment algorithm (and setting $m = 0$), a nonblocking route assignment can be obtained. The connection request (s_i, D_i) is routed to middle-stage module $f(s_i, D_i)$. Packet replications are performed subsequently in the middle-stage modules and in the output-stage modules.

In general, each middle-stage module can be considered as a broadcast Clos network. Recall that the rank-based assignment algorithm preserves the order of the connection requests. Thus, we can apply it recursively to these broadcast Clos networks. Using the same notation as in Section 3.4.2, the rank $r(s_i)$ of a connection request (s_i, D_i) can be written as

$$r(s_i) = \sum_{i=1}^{n-1} \alpha_i \prod_{j=0}^{i-1} q_j + \alpha_0.$$

TABLE 6.1 Connection Requests to the Broadcast Clos Network

s_i	2	3	4	5	8
$\tilde{s}_i$	02	10	11	12	22
D_i	0,1	2,3,4	5,6	7,8,9	10,11
$\tilde{D}_i$	00,01	02,10,11	12,20	21,22,30	31,32
$m+i$	0	1	2	3	4
$f(s_i, D_i) = [m+i]_q$	0	1	2	0	1

The routing tag for the packet to go from input to the middle-stage module assigned to it is given by $(\alpha_0, \alpha_1, \alpha_2, \ldots, \alpha_{n-2})$. Once the packets reach the middle-stage modules, packet replications and routing are controlled by the generalized interval splitting algorithm that is discussed in the following.

6.1.4.1 Generalized Interval Splitting Algorithm for Broadcast Clos Network

We denote the switch modules from the middle stage to the output stage as broadcast switch modules due to their capability in performing replication. A packet arriving at a broadcast switch module can be routed to any one of the output links, or it can be replicated if necessary and sent out to a subset of the output links. For the nonblocking operation of the broadcast Clos network, assume that the set of connection requests is monotonic and the set of output destinations assigned to each input packet is consecutive. This assumption is justified because in a copy network, the replicated packets do not need to be routed to some specific output ports. Routing is done by a point-to-point switching network that is cascaded behind the copy network. We can therefore deliberately assign the sets of output destinations to the input packets such that the connection requests are monotonic. Under such assumption, the set of output destinations of each packet consists of an address interval that can be represented by two numbers, the *minimum* and *maximum*.

Let $N = q_0 \cdot q_1 \cdots q_{n-1}$. The broadcast Clos network consists of $2n - 1$ stages numbered $0, 1, \ldots, 2n - 2$. An address interval is a set of contiguous n-tuple. Suppose a node at stage i ($n - 1 \leq i \leq 2n - 2$) receives a packet containing an address interval specified by $\min(i - 1) = m_{n-1} \cdots m_{2n-2}$ and $\max(i - 1) = M_{n-1} \cdots M_{2n-2}$, where the argument $(i - 1)$ indicates that the packet comes from a node in stage $(i - 1)$. Since the size of the modules in different stages may be different, each tuple in an address can be of different bases. As discussed in Section 3.4.2, m_j and M_j, where $n - 1 \leq j \leq 2n - 2$, are of base q_{2n-2-j}. When a node in stage i receives a packet, the following procedure is performed:

1. If $m_i = M_i$, then send the packet out on link m_i.
2. If $m_i \neq M_i$, then $(M_i - m_i + 1)$ copies of the packet are required. Replicate the packet, modify the headers and send the packets out on links m_i to M_i.

The purpose of header modification is to split the original address interval into several subintervals, each contained in a replicated packet. It is easy to see that

$m_j = M_j$ for $j = n-1, \ldots, i-1$ holds for every packet that arrives at stage i, where $n-1 \le i \le 2n-2$. The event $m_i > M_i$ is impossible. For the packet sent out on link m_i, the header becomes

$$\min(i) = \min(i-1) = (m_{n-1}, \ldots, m_i, m_{i+1}, \ldots, m_{2n-2}),$$
$$\max(i) = (m_{n-1}, \ldots, m_i, q_{(2n-2)-(i+1)} - 1, \ldots, q_0 - 1). \quad (6.11)$$

This is due to the fact that the bases of $m_{i+1}, m_{i+2}, \ldots, m_{2n-2}$ are $q_{(2n-2)-(i+1)}, q_{(2n-2)-(i+2)}, \ldots, q_0$, respectively. For the packet sent out on link M_i, the header becomes

$$\min(i) = (M_{n-1}, \ldots, M_i, 0, \ldots, 0),$$
$$\max(i) = \max(i-1) = (M_{n-1}, \ldots, M_i, M_{i+1}, \ldots, M_{2n-2}). \quad (6.12)$$

For those packets going to output j where $m_i < j < M_i$, the header becomes

$$\min(i) = (m_{n-1}, \ldots, m_{i-1}, j, 0, \ldots, 0),$$
$$\max(i) = (m_{n-1}, \ldots, m_{i-1}, j, q_{(2n-2)-(i+1)} - 1, \ldots, q_0 - 1). \quad (6.13)$$

Figure 6.15 shows the operations of the generalized interval splitting algorithm on a five-stage broadcast Clos network. An input or output address is represented by a 3-tuple. The first and the second components are of base 2 while the third one is of base 3. By recursively applying the rank-based assignment algorithm, the packet is routed to one of the middle-stage modules. The packets are then subsequently replicated and routed to the destinations using the generalized interval splitting algorithm. The address intervals contained in the header of each packet are shown in the shaded boxes.

6.1.4.2 Decomposition and Degeneration
The broadcast Clos network is in fact the cascade combination of a reverse omega network and a broadcast omega network. As mentioned in Chapter 3, the omega network belongs to a class of multistage interconnection networks (MINs) and possesses the nonblocking and self-routing properties. A broadcast omega network is an omega network with switching nodes that can perform packet replications. The sufficient conditions for the broadcast omega network to be nonblocking are exactly the same as that for broadcast banyan network. They are given as follows.

Theorem 6.3 (Nonblocking Conditions of Broadcast Omega Networks). A broadcast omega network is nonblocking if the active inputs $s_0, s_1, \cdots, s_{n-1}$ ($s_j > s_i$ if $j > i$) and their corresponding sets of outputs $D_0, D_1, \ldots, D_{n-1}$ satisfy the following:

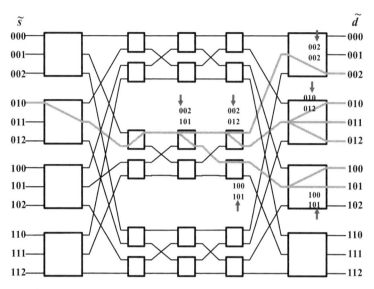

FIGURE 6.15 Packet replication process in a five-stage Clos network.

1. *Concentrated outputs*: Any input between two active inputs is also active.
2. *Monotonic outputs*: $D_0 < D_1 < \cdots < D_{n-1}$ or $D_0 > D_1 > \cdots > D_{n-1}$. The inequality $D_i < D_j$ indicates that every output address in D_i is less than every output address in D_j.

It can also be shown that the broadcast-omega network can be recursively constructed.

Consider the five-stage network formed by cascading a reverse omega network and a broadcast omega network as shown in Fig. 6.16. This five-stage network is a nonblocking copy network given monotonic connection requests. To see this, suppose the set of monotonic connection requests defined in Table 6.1 is to be fulfilled. In the reverse omega network, the input packets can be routed to a set concentrated ports in stage 2 so that the nonblocking conditions for the reverse omega network are satisfied. In this example, the packets are routed to the set of ports from 10 to 21, but any other sets of concentrated ports can be used, including those that are cyclically concentrated. In the broadcast omega network, since the inputs are concentrated and their sets of outputs are monotonic, the packets can be routed and replicated in a nonblocking manner. The routing and replications of packets in the broadcast omega network can be performed by using the generalized interval splitting algorithm in a similar manner as in the broadcast Clos network.

It is not difficult to realize that this five-stage network is in fact equivalent to a broadcast Clos network. Note that stage 2 is dummy and may be eliminated. After that, stage 1 can be combined with stage 3 and the resulting network is a broadcast Clos network. In the five-stage network, routing the input packets to a set of concentrated

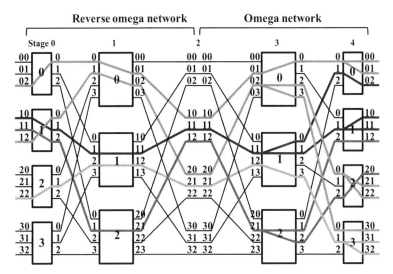

FIGURE 6.16 Broadcast Clos network decomposed into a reverse omega network and a broadcast omega network.

ports in stage 2 is equivalent to routing consecutive input packets to consecutive modules in stage 1 in a modulo fashion. In our example, the first input packet is routed to module 0, the second routed to module 1, and so on. In fact, this is the basis of the rank-based assignment algorithm used in the broadcast Clos network, which performs route assignment based on the rank of each connection request.

When the reverse omega network and the broadcast omega network in this five-stage network are recursively constructed using 2×2 switching elements, we obtain a cascade combination of the reverse banyan network and the broadcast-banyan network, which is the structure of the copy network presented in Section 6.1.1. The reverse banyan network, together with a running-adder network, performs the function of packet concentration and the broadcast banyan network is responsible for packet replication and routing. Therefore, we can see that the copy network based on broadcast Clos network is in fact the generalization of the structure based on broadcast banyan network.

6.1.4.3 Generalized Connection Networks
A *generalized connection network (GCN)* is a switching network with N inputs and N outputs in which each output may be connected to any one of the inputs for a total of N^N different connection patterns. In other words, a mapping $\pi = \{(s_0, \pi(s_0)), \ldots, (s_{n-1}, \pi(s_{n-1}))\}$, where $\pi(i)$ is a set of output destinations (not necessarily continuous), can be realized if

$$\pi(i) \cap \pi(j) = \emptyset \quad \text{for} \quad i \neq j \quad \text{and} \quad 0 \leq i, j \leq n-1.$$

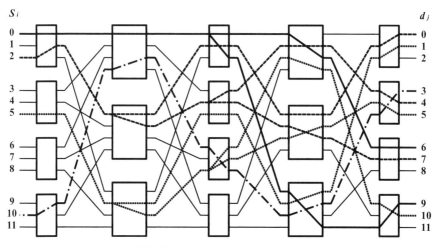

FIGURE 6.17 A five-stage GCN.

Let $|\pi(s_i)|$ denote the number of elements in $\pi(s_i)$. Here, we show by an example that a five-stage GCN can be constructed using a broadcast Clos network and the concepts discussed in this subsection. Consider the network formed by cascading a 12×12 broadcast Clos network and a 12×12 Clos network. Let $q = 3$ and $p = 4$ in both networks. The resulting five-stage network is a GCN. Suppose the following set of connection requests is to be satisfied:

$$\pi = \begin{pmatrix} 0 & 2 & 5 & 10 \\ \{6,9\} & \{0,4,7\} & \{1,5,10\} & \{3\} \end{pmatrix}.$$

Using the previous notations, we have $\pi(0) = \{6, 9\}$, $\pi(2) = \{0, 4, 7\}$, $\pi(5) = \{1, 5, 10\}$, and $\pi(10) = \{3\}$. The broadcast Clos network is responsible for broadcasting the inputs based on the number of output destinations of each input. Given π, the number of output destinations of each input is determined and π is transformed to a set of connection requests for the broadcast Clos network. In this example, input s_i is assigned a set of consecutive outputs D_i in stage 2 based on $|\pi(s_i)|$ (Table 6.2). By using the rank-based assignment algorithm and the generalized interval splitting algorithm, the inputs are broadcasted to the outputs in stage 2 in a nonblocking manner.

When the calls emerge from the broadcast Clos network, they subsequently enter the Clos network. Since the calls appear consecutively at the output of stage 2, the set of connection requests for the Clos network can be easily determined. In our example, let $\pi_j(s_i)$ denote the jth element of the set $\pi(s_i)$ and let π' be the set of connection

TABLE 6.2 Connection Requests to the Broadcast-Clos Network.

s_i	0	2	5	10
D_i	0,1	2,3,4	5,6,7	8
$m+i$	0	1	2	3
$f(s_i, D_i) = [m+i]_q$	0	1	2	0

requests in the Clos network. We can obtain π' as follows:

$$\pi' = \begin{pmatrix} 0 & 1 & 2 & 3 & 4 & 5 & 6 & 7 & 8 \\ \pi_0(0) & \pi_1(0) & \pi_0(2) & \pi_1(2) & \pi_2(2) & \pi_0(5) & \pi_1(5) & \pi_2(5) & \pi_0(10) \end{pmatrix}$$

$$= \begin{pmatrix} 0 & 1 & 2 & 3 & 4 & 5 & 6 & 7 & 8 \\ 6 & 9 & 0 & 4 & 7 & 1 & 5 & 10 & 3 \end{pmatrix}.$$

This is because there are two output destinations from input 0, three from inputs 2 and 5, and one from input 10, respectively. Based on π', a nonblocking route assignment can be determined and the connection requests can be established in a nonblocking manner.

Obviously, no switching is performed between modules in stage 2 and modules in stage 3. We can therefore combine them together to form a five-stage rearrangeably nonblocking multiconnection network. Note, however, that this network is only suitable for circuit switching applications since centralized route assignment is required in the Clos network.

6.2 PATH SWITCHING

Large packet switch architectures can be built based on the modular Clos network. As discussed in Chapter 2, the original Clos network was proposed for circuit switching systems, in which a path from an input to an output is established and dedicated to a connection during its whole lifetime. In contrast to these systems, packet switches carry information in multiplexed format. Often, each connection only requires a portion of the capacity of the internal link carrying it, and the bandwidth needed for the connection may be time varying. One way to operate such systems is to use static route assignment, called *multirate circuit switching*, where a path is established for each connection such that a portion of the path capacity is allocated to the connection during its lifetime. The capacity allocated along the path can be the peak bandwidth requirement of the connection in order to ensure low delay and packet loss probability. Static routing, however, may result in low utilization of the system, because it does not exploit the full advantage of the statistical multiplexing gain offered by packet switching. The other extreme is the dynamic routing scheme, called *cell switching*, where the connection pattern of the Clos network is rearranged in every time slot according to the targeted outputs of the arrived packets. Since paths for incoming packets must be assigned on a slot-by-slot basis, the collecting and processing of global traffic information for path hunting on the fly would become a bottleneck severely limiting the growth of the switch size and speed. Unlike multirate circuit switching, the dynamic routing scheme usually does not guarantee the bandwidth for each individual connection, thus only desired overall system performance, such as total packet loss probability or throughput, is attainable. In general, the system utilization of the dynamic scheme is expected to be higher than the static scheme, but the latter provides better guarantee of QoS for each connection.

In this section, we shall present a novel quasi-static routing scheme called *path switching*. It is a unification of circuit switching and packet switching and is a compromise of the dynamic and the static routing schemes. In actuality, static switching is similar to the railway system, where path scheduling and reservation for trains are such that the railway is idle most of time, resulting in low utilization. On the other hand, although dynamic routing i packet switching can raise utilization, its complexity places a limit on the size of the switch. Path switching is an analog to the traffic signals governing the routing of cars at the crossroad. It does not require any complicated or special hardware, but still yields high utilization. Because of its simplicity and its flexibility, this quasi-static routing scheme provides distributed control in the Clos network.

The routing of path switching is based on the concept of virtual path within the Clos network. We consider that there is a virtual path between an input module and an output module, which comprises all virtual circuits interconnecting any incoming port and any outgoing port on this pair of modules. The scheduling of path switching consists of two parts. First, the capacity required for each virtual path is determined by the traffic statistics among all pairs of input and output modules so that the path-level QoS of the traffic on each pair can be satisfied. Then, a finite number of regular bipartite multigraphs are generated according to the capacity assignment matrix. If the switch is operated repeatedly according to a set of connection patterns, predetermined by the edge coloring of the bipartite graph, then the capacity requirement on each virtual path can be satisfied in the long run, and the computation of route assignment on the fly can be avoided. Path switching is a quasi-static routing scheme, and it is a compromise of the static scheme and the dynamic scheme.

The path-switched Clos network will be called *cross-path switch*, a statistical Clos network. Instead of using a central controller to process, schedule, and route all incoming packets simultaneously. The cross-path switch makes use of a routing algorithm that is distributed over the three stages of the Clos network. The route assignment in central modules, predetermined by the bipartite multigraphs, will be used repeatedly. Storing the routing table in the local memory of every input module, the connection pattern of the central stage is known in every time slot. Each connection pattern specifies exactly how many packets can be delivered to a particular output module through which central modules in that time slot. Based on this routing specification, each input module will select those packets waiting in the local input buffers according to their destinations and priorities. The selection process will match the destination addresses with the desired output modules only, such that each output module would have to handle the output port contentions.

To construct the cross-path switch, any nonblocking, self-routing interconnect fabrics can be used as the building blocks of the first two stages. The switch modules of the third stage must be capable of resolving the output port contentions; either the Batcher–banyan network with extended outputs or the knockout switch can be employed to perform this task. The implementation of path switching is completely distributed. Although the computation of the capacity assignment and route assignment by the central controller still requires global information, it is not a slot-by-slot task. The routing tables stored in the local memory of input modules would be updated only if the traffic matrix changes significantly and the switch performance becomes unacceptable.

6.2.1 Basic Concept of Path Switching

A three-stage Clos network is shown in Fig. 6.18. The first stage consists of k input modules, each of dimensions $n \times m$. The dimensions of each central module in the middle stage are $k \times k$. As illustrated in Fig. 6.19, the routing constraints of Clos network are briefly stated as follows:

1. Any central module can only be assigned to *one* input of each input module, and *one* output of each output module.
2. Input i and output j can be connected through any central module.
3. The number of alternate paths between input i and output j is equal to the number of central modules.

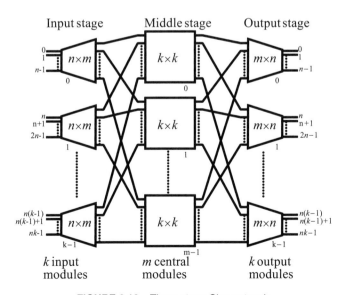

FIGURE 6.18 Three-stage Clos network.

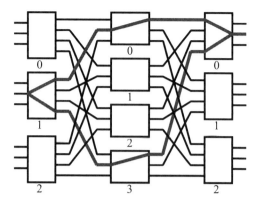

FIGURE 6.19 Routing in Clos network.

If we consider each input module and each output module as a node, a particular connection pattern in the middle stage of the Clos network can be represented by a regular bipartite multigraph with node degree m as illustrated in Fig. 6.20, where each central module corresponds to a group of k edges, each connecting one distinct pair of input–output nodes (modules). Suppose the routing algorithm of the Clos network is based on dynamic cell switching, and the amount of traffic from input module I_i to output module O_j is λ_{ij} cells per time slot. The connection pattern will change in every time slot according to arrived packets, and the routes will be computed on a slot-by-slot basis. Let $e_{ij}(t)$ be the number of edges from I_i to O_j of the corresponding bipartite multigraph in time slot t. Then the capacity C_{ij} of the virtual path between

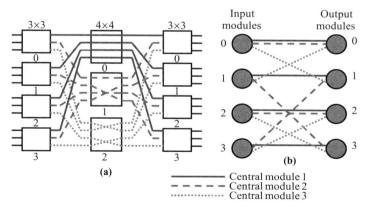

FIGURE 6.20 Correspondence between the middle-stage route scheduling in a Clos network and the edge coloring of the regular bipartite multigraph: (a) three-stage Clos network; (b) the equivalent regular bipartite graph.

I_i and O_j must satisfy

$$C_{ij} = \lim_{T \to \infty} \frac{\sum_{t=1}^{T} e_{ij}(t)}{T} > \lambda_{ij}. \qquad (6.14)$$

On the other hand, the routes of circuit-switched Clos network are fixed, and the connection pattern will be the same in every time slot. The capacity must satisfy

$$C_{ij} = e_{ij}(t) = e_{ij} > \lambda_{ij}, \qquad (6.15)$$

which implies that the peak bandwidth C_{ij} is provided for each virtual circuit at call setup time, and it does not take statistical multiplexing into consideration at all. We implement the idea of quasi-static routing, called path switching, using finite number of different connection patterns in the middle stage repeatedly, as a compromise of the above two extreme schemes. For any given λ_{ij}, if $\sum_i \lambda_{ij} < n \leq m$, and $\sum_j \lambda_{ij} < n \leq m$, we can always find a finite number f of regular bipartite multigraphs such that

$$\frac{\sum_{t=1}^{f} e_{ij}(t)}{f} > \lambda_{ij}, \qquad (6.16)$$

where $e_{ij}(t)$ is the number of edges from node i to node j in the tth bipartite multigraph. A cross-path switch is said to be *statistically stable* if and only if the capacity requirement (6.14) can be satisfied by providing connections repeatedly according to the coloring of these F bipartite multigraphs. These finite amounts of routing information can be stored in the local memory of each input module to avoid the slot-by-slot computation of route assignments. Path switching becomes circuit switching if $f = 1$, and it is equivalent to cell switching if $f \to \infty$.

The scheduling of path switching consists of two steps, the capacity assignment and the route assignment. The capacity assignment is to find the capacity $C_{ij} > \lambda_{ij}$ for each virtual path between input module I_i and output module O_j; it can be carried out by optimizing some objective functions subject to $\sum_i C_{ij} = \sum_j C_{ij} = m$. The choice of the objective function depends on the stochastic characteristic of the traffic on virtual paths and the quality of service requirements of connections.

The next step is to convert the capacity matrix $[C_{ij}]$ into edge coloring of a finite number f of regular bipartite multigraphs; each of them represents a particular connection pattern of central modules in the Clos network. An edge coloring of a bipartite multigraph is to assign m distinct colors to m edges of each node such that no two adjacent edges have the same color. It is well known that a regular bipartite multigraph with degree m is m-colorable. Each color corresponds to a central module, and the color assigned to an edge from input module i to output module j represents a connection between them through the corresponding central module.

Suppose that we choose a sufficient large integer f such that fC_{ij} are integers for all i, j and form a regular bipartite multigraph, called *capacity graph*, in which the number of edges between node i and node j is fC_{ij}. Since the capacity graph is regular with degree fm, it can be edge colored by fm different colors. Furthermore, it is easy to show that any edge coloring of the capacity graph with degree fm is the superposition of the edge coloring of f regular bipartite multigraphs of degree m. Consider a particular color assignment $a \in \{0, 1, \ldots, fm - 1\}$ of an edge between input node I_i and output node O_j of the capacity graph. Let

$$a = r \cdot f + t, \qquad (6.17)$$

where $r \in \{0, 1, \ldots, m - 1\}$ and $t \in \{0, 1, \ldots, f - 1\}$ are the quotient and the remainder of dividing a by f, respectively. The mapping $g(a) = (t, r)$ from the set $\{0, 1, \ldots, fm - 1\} \to \{0, 1, \ldots, f - 1\} \times \{0, 1, \ldots, m - 1\}$ is one-to-one and onto, that is,

$$a = a' \iff t = t' \text{ and } r = r'.$$

That is, the color assignment a, or equivalently the assignment pair (t, r), of the edge between I_i and O_j indicates that the central module r has been assigned to a route from I_i to O_j in the tth time slot of every cycle. Adopting the convention in TDMA system, each cycle will be called a *frame* and the period f *frame size*. As illustrated by the example shown in Fig. 6.21(a)–(c), where $m = 3$ and frame size $f = 2$, the decomposition of the edge coloring into assignment pairs guarantees that route assignments are either space interleaved or time interleaved. Thus, relation (6.17) will be called the *time–space interleaving principle*.

For uniform traffic, where the distribution of traffic loading between input modules and output modules is homogeneous, the fm edges of each node can be evenly divided into k groups, where k is the total number of input (output) modules. Each group contains $g = fm/k$ edges between any I/O pair, where the frame size f should be chosen properly to make the group size g an integer. The edges of this capacity graph can be easily colored by the *Latin square* given in Fig. 6.22, where each A_i, $0 \leq i \leq k - 1$, represents a set of distinct colors, for example,

$$A_0 = \{0, 1, \ldots, g - 1\}; A_1 = \{g, g + 1, \ldots, 2g - 1\}; \ldots; A_{k-1}$$
$$= \{(k - 1)g, (k - 1)g + 1, \ldots, kg - 1\}.$$

Since each number in the set $\{0, 1, \ldots, fm - 1\}$ appears only once in any row or column in the table, it is a legitimate edge coloring of the capacity graph. The assignment $a = (t, r)$ of an edge between the I_i/O_j pair indicates that the central module r will connect the input module i to output module j in the tth slot of every frame. As an example, for $m = 3$ and $k = 2$, we can choose $f = 2$ and thus $g = 3$. Then, the groups of colors are $A_0 = \{0, 1, 2\}$ and $A_1 = \{3, 4, 5\}$, respectively. The procedure described above is illustrated in Fig. 6.22, and the correspondence between the route assignments and the connection patterns in the middle stage is shown in Fig. 6.23.

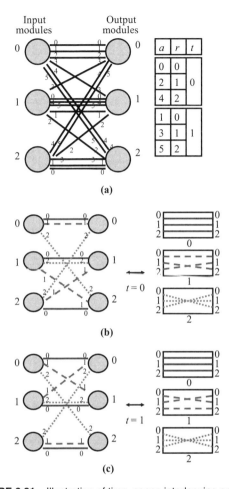

FIGURE 6.21 Illustration of time–space interleaving principle.

	O_0	O_1	O_2	...	O_{k-1}
I_0	A_0	A_1	A_2	...	A_{k-1}
I_1	A_{k-1}	A_0	A_1	...	A_{k-2}
⋮	⋮	⋮	⋮	⋱	⋮
I_{k-1}	A_1	A_2	A_3	...	A_0

FIGURE 6.22 Latin square assignment.

Central module	Connected I/O pairs at time slot	
	0	1
0	$I_0/O_0, I_1/O_1$(bar)	$I_0/O_0, I_1/O_1$(bar)
1	$I_0/O_0, I_1/O_1$(bar)	$I_0/O_1, I_1/O_0$(cross)
2	$I_0/O_1, I_1/O_0$(cross)	$I_0/O_1, I_1/O_0$(cross)

(a)

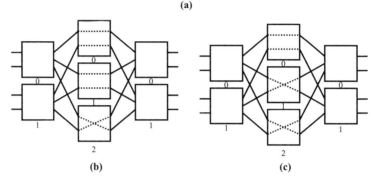

(b)　　　　　　　　　　(c)

FIGURE 6.23 Route scheduling in the middle stage for uniform traffic: (a) connection pairs in central modules; (b) time slot 0; (c) time slot 1.

In the above example, since the number of central modules m is greater than the number of input modules k, it is possible that more than one central modules are assigned to some I/O pairs in one time slot. In the case that $m < k$, there are not enough central modules for all I/O pairs in one time slot assignment. Nevertheless, the total number of central modules assigned to every I/O pair within a frame should be the same, for uniform input traffic to fulfill the capacity requirement, and it is equal to $g = fm/k$. This point is illustrated in the following example. For $m = 4$ and $k = 6$, we choose $f = 3$ and $g = 2$. The same method will result in the connection patterns shown in Fig. 6.24. It is easy to verify that the number of central modules (paths, edges) assigned for each I/O pair is equal to $g = 2$ per $f = 3$ slots.

6.2.2 Capacity and Route Assignments for Multirate Traffic

In general, the traffic loadings among different I/O pairs may not be homogeneous. The uniform capacity assignment would result in nonuniform distribution of offered load on different virtual paths, and packets may suffer dissimilar delays and losses. Thus, for nonuniform traffic, the distribution of the limited capacity among virtual paths should depend on their loadings, and it can be determined by optimizing some objective functions. The procedure of converting quantitative capacity assignments to route assignments is the same as before: coloring the edges of the capacity graph

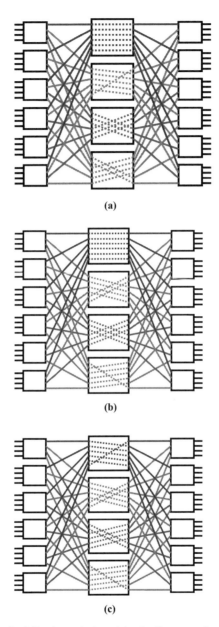

FIGURE 6.24 Route scheduling in central modules for the second example of uniform traffic: (a) time slot 0; (b) time slot 1; (c) time slot 2.

first and then decomposing them into assignment pairs according to the time–space interleaving principle.

6.2.2.1 Capacity Allocation
In this part, we shall discuss the allocation of the limited capacity in the cross-path switch to meet diverse QoS requirements for individual calls. The capacity in the cross-path switch can be classified into three levels, namely, the switch level, the call level, and the packet level. The capacity of switch level can be viewed as the overall switching resources in the cross-path switch. It is defined to be the number of packets that can be routed from the input stage to the output stage in one slot time. The switch-level capacity is fixed and determined by the number of central modules. The call-level capacity allocation is constrained by the capacity of the switch level. At the call level, the cross-path switch will allocate a certain amount of capacity, called *effective bandwidth*, to individual calls such that their statistical QoS constraints in terms of delay or loss at the packet level can be satisfied. The effective bandwidths can be estimated through the direct approach, model fitting, or the virtual-buffer approach. An example of calculating effective bandwidths based on–off source model is provided in the appendix.

Let α_{uv}, in unit of cells/slot, be the effective bandwidth required to accommodate the aggregate traffic between an input link u and an output link v. We assume that the following link capacity constraints are observed by call admission procedure:

$$\begin{cases} \sum_u \alpha_{uv} \leq 1, \; \forall \, 0 \leq v \leq N-1, \\ \sum_v \alpha_{uv} \leq 1, \; \forall \, 0 \leq u \leq N-1. \end{cases} \quad (6.18)$$

The capacity of each link may not be fully utilized, and its remaining capacity can be distributed to available bit rate (ABR) traffic, which is not bounded by any delay constraints, as specified in the following chapter. The ABR service can be supported by feedback flow control mechanisms to provide rapid access to unused network bandwidth whenever available. Let ρ_{uv} be the mean rate of the admitted ABR traffic between input link u and output link v, satisfying the following constraint of remaining link capacity:

$$\begin{cases} \sum_u \rho_{uv} \leq 1 - \sum_u \alpha_{uv}, \; \forall \, v, \\ \sum_v \rho_{uv} \leq 1 - \sum_v \alpha_{uv}, \; \forall \, u. \end{cases} \quad (6.19)$$

As shown in Fig. 6.25, let $\lambda_{ij} = \sum_{(u,v)|u \in I_i, v \in O_j} \rho_{uv}$ be the total rate of ABR traffic on the virtual path between input module I_i and output module O_j, and we assume that the following $k \times k$ ABR traffic matrix is given:

$$\mathbf{T} = \begin{bmatrix} \lambda_{0,0} & \lambda_{0,1} & \cdots & \lambda_{0,k-1} \\ \lambda_{1,0} & \lambda_{1,1} & \cdots & \lambda_{1,k-1} \\ \vdots & \vdots & \ddots & \vdots \\ \lambda_{k-1,0} & \lambda_{k-1,1} & \cdots & \lambda_{k-1,k-1} \end{bmatrix}.$$

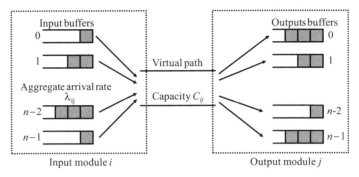

FIGURE 6.25 Virtual path between input module i and output module j.

Then, the capacity allocation is to find a matrix

$$\mathbf{C} = \begin{bmatrix} c_{0,0} & c_{0,1} & \cdots & c_{0,k-1} \\ c_{1,0} & c_{1,1} & \cdots & c_{1,k-1} \\ \vdots & \vdots & \ddots & \vdots \\ c_{k-1,0} & c_{k-1,1} & \cdots & c_{k-1,k-1} \end{bmatrix},$$

subject to the following constraints:

$$\begin{cases} c_{ij} \geq \lambda_{ij}, & \forall i, j, \\ \sum_j c_{ij} = m - \sum_{u \in I_i, v} \alpha_{uv}, & \forall i, \\ \sum_i c_{ij} = m - \sum_{v \in O_j, u} \alpha_{uv}, & \forall j. \end{cases} \quad (6.20)$$

The goal of capacity allocation is to distribute limited resources fairly among all ABR calls. In the one-dimensional case, a fair allocation is to distribute the capacity proportional to individual arrival rates as follows:

$$c_i = \frac{\lambda_i \cdot c}{\sum_j \lambda_j}, \quad \forall i, \quad (6.21)$$

where c is the total bandwidth. This is equivalent to minimizing the total weighted offered load λ_i / c_i function

$$z = \sum_i \frac{\lambda_i^2}{c_i},$$

subject to

$$\sum_i c_i = c.$$

The offered load is defined to be the ratio of mean arrival rate λ_i to the allocated capacity c_i. In the cross-path switch with two-dimensional constraints (6.20), we can minimize the following similar objective function:

$$z = \sum_{i,j} \frac{\lambda_{ij}^2}{c_{ij}}. \tag{6.22}$$

Alternatively, each virtual path can be modeled as an independent M/M/1 queue with arrival rate λ_{ij} and service rate c_{ij} for all i, j; then the average delay for the packets from input module i to output module j is given by

$$T_{ij} = \frac{1}{c_{ij} - \lambda_{ij}}, \quad \forall (i, j). \tag{6.23}$$

The objective is to minimize the total weighted delay [BeG92]

$$z = \sum_{i,j} \frac{\lambda_{ij}}{c_{ij} - \lambda_{ij}}. \tag{6.24}$$

The total capacity allocated for the virtual path between I_i and O_j is then equal to $C_{ij} = c_{ij} + \sum_{(u,v)|u \in I_i, v \in O_j} \alpha_{uv}$.

There are several efficient methods that can be used to solve the optimization problem. The objective functions (6.22) and (6.24) are convex, and each can be minimized subject to the linear constraints (6.20) by using the sequential linear approximation algorithm [Frank], or the generalized reduced gradient (GRG) method [Lasdon] for convex programming. Furthermore, since it is a sum of some single-variable functions, and each of them can be approximated by a linear function, the problem can then be transformed into a linear programming problem and solved by the modified simplex method to reduce the computation complexity [Hillier].

6.2.2.2 Route Assignment

In this part, we shall discuss the issue of converting the overall capacity allocation matrix $[C_{ij}]$ into time–space interleaved route scheduling. Consider the cross-path switch with parameters $n = 3$, $k = 3$, and $m = 4$ and assume that the traffic is of ABR type only. If the traffic matrix is given by

$$\mathbf{T} = \begin{bmatrix} 1 & 1 & 1 \\ 2 & 1 & 0 \\ 0 & 1 & 1 \end{bmatrix}, \tag{6.25}$$

the capacity assignment matrix calculated by the minimization of the total weighted delay (6.24) is

$$\mathbf{C} = \begin{bmatrix} 1.3366 & 1.2774 & 1.3859 \\ 2.6564 & 1.3366 & 0 \\ 0 & 1.3859 & 2.6044 \end{bmatrix} \approx \begin{bmatrix} \frac{4}{3} & \frac{4}{3} & \frac{4}{3} \\ \frac{8}{3} & \frac{4}{3} & 0 \\ 0 & \frac{4}{3} & \frac{8}{3} \end{bmatrix}. \qquad (6.26)$$

In general, the resulting capacity assignment matrix is non-integer such as the example shown above. However, the capacity assignments in a frame of size f, $f \cdot C$, can be rounded off into integers, and the round-off error is inversely proportional to f. That is, the error can be arbitrarily small if the frame size is sufficiently large. However, since the amount of routing information stored in the memory is linearly proportional to f, the frame size is limited by the access speed and the memory space of input modules. In practice, the choice of frame size f is a compromise between the round-off error and the memory requirement. We assume, without loss of generality, that $e_{ij} = f \cdot c_{ij}$ are all integers for a sufficiently large f. Then,

$$\sum_j e_{ij} = \sum_i e_{ij} = f \cdot m \qquad (6.27)$$

and

$$\mathbf{E} = f \cdot \mathbf{C} = \begin{bmatrix} e_{0,0} & e_{0,1} & \cdots & e_{0,k-1} \\ e_{1,0} & e_{1,1} & \cdots & e_{1,k-1} \\ \vdots & \vdots & \ddots & \vdots \\ e_{k-1,0} & e_{k-1,1} & \cdots & e_{k-1,k-1} \end{bmatrix}.$$

In the above matrix $\mathbf{E}$, each element e_{ij} represents the number of the edges between the I_i/O_j pair in the $k \times k$ capacity graph, in which each node has degree of fm. As mentioned in Section II, this capacity graph can be colored by fm colors, and each color represents one distinct time–space slot based on the time–space interleaving principle (6.17). Coloring can be found by complete matching, which is repeated recursively to reduce the degree of every node one by one. One general method to search for a complete matching is the so-called Hungarian algorithm or alternating path algorithm [Leighton, McEliece]. It is a sequential algorithm with the worst time complexity $O(k^2)$, or totally $O(fm \times k^2)$ because there are fm matchings. For the above example, the route scheduling is shown in Fig. 6.26. If each of fm and k is a power of 2, an efficient parallel algorithm proposed in Ref. [LeLi02] for conflict-free route scheduling in a three-stage Clos network with time complexity of $O(\log^2(fmk))$ can be used.

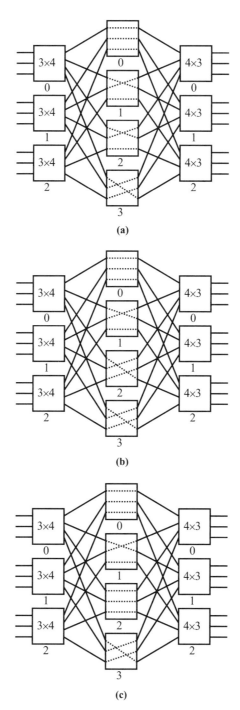

FIGURE 6.26 Route scheduling for the example of nonuniform traffic: (a) time slot 0; (b) time slot 1; (c) time slot 2.

6.2.3 Trade-Off Between Performance and Complexity

Although the core issue of the cross-path switch is to incorporate the concept of virtual path in the switch architecture and provide a coherent network management paradigm, we still need to investigate the trade-off between its cell-level performance and switch complexity to provide information for switch design. First, we shall study various trade-offs between performance and complexity in terms of the parameters n, m, and k of an $N \times N$ Clos Network. Since a packet arriving at its destined output module of $m \times n$ still needs to be routed to one of the n output ports, the complexity of output modules will grow with respect to n. On the other hand, a large value of n means a small value of k for fixed switch size $N = nk$, and the traffic stream on each virtual path would be much smoother, resulting in better statistical multiplexing gain. This is because the superposition of n point processes will approach a Poisson process for large n. Also, the capacity assignment and route assignment will be simpler for a smaller value of k, since both the optimization procedure and edge-coloring algorithm are based on the $k \times k$ traffic matrix. In practice, the determination of the precise module size n should take the overall loss probability and the throughput into consideration.

A fair comparison of systems with different values of n and m/n must be based on the same amount of overhead. There are three types of overhead in implementing the cross-path switch.

1. *Input arbitration:* For better throughput performance, the look-ahead contention-resolution scheme described in Section 4.2.1 is implemented at the input modules. Recall that the depth of search in each queue is called *window size*. The larger the window size, the better the throughput, but the processing of look-ahead selection is limited by the transmission time of one packet. For window size w and module size n, the maximum number of packets that can be examined is wn. This product reflects the processing time required by the look-ahead scheme, and therefore we assume that it is fixed in making the throughput comparison. For example, the throughput of the 1024×1024 switch with module size $n = 8$ and window size $w = 8$ will be compared with that of $n = 16$ and $w = 4$, because they have the same product $wn = 64$. Notice that the module size n should not be too large, otherwise the internal processing speed may not be fast enough to support the look-ahead scheme.

2. *Input memory:* In a cross-path switch, the predetermined central-stage connection patterns are stored in the memory of each input module. The total number of memory space needed to store the addresses of all central modules is exactly m, and the overall memory requirement is $mkf = m/n \cdot (Nf)$ for k input modules with frame size f. We assume that the frame size f is a constant in all cases, and use the expansion factor m/n to represent this memory requirement. This point is illustrated by the example shown in Fig. 6.27.

3. *Expansion factor:* The expansion factor m/n represents the overall capacity expansion, and it is the most critical parameter in the performance of

250 SWITCHING PRINCIPLES FOR MULTICAST, MULTIRATE, AND MULTIMEDIA SERVICES

Destined output module	Through which central modules at time slot		
	0	1	...
0	a_0, a_1	b_0, b_1	...
1	a_2	b_2, b_3	...
⋮	⋮	⋮	⋮
$k-2$	a_{m-2}	b_{m-2}, b_{m-1}	...
$k-1$	a_{m-1}	Nil	...

(a)

Destined output module	Central modules at time slot		
	0	1	2
0	0	0	0, 2
1	1, 2	1	1
2	3	2, 3	3

(b)

FIGURE 6.27 Storage of the middle-stage connection patterns at an input module: (a) a general format and (b) a specific storage for input module 1 in Fig. 6.26.

cross-path switch. Intuitively, the system will perform uniformly better with larger expansion factor at the cost of extra overhead $m/n - 1$ due to the increased number of modules in the central stage. In the following analysis, however, we will show that the packet loss probability at output module will also increase with respect to the expansion factor. That is, arbitrary large expansion factor does not necessarily result in better system performance as one might expect.

In a cross-path switch, the switching of a packet in the first two stages is performed according to the address of its destined *output module*, while the switching in the last stage only depends on the local address of its destined *output port*, and these two "addresses" are independent. To simplify the analysis, the performances of the first stage and the last stage of the cross-path switch will be evaluated separately. For an $N \times N$ switch, once the *module size n*, number of incoming (outgoing) links to an input (output) module, and the *expansion factor m/n* are given, the other parameters $k = N/n$, and $m = n \cdot m/n$ are all fixed. We will focus our discussion on the effects of the variation of these two parameters on the throughput and the packet loss probability of the switch.

We shall consider a 1024×1024 switch with uniform traffic as an example and make performance comparisons among five different cases of $n \times k = 4 \times 256 = 8 \times 128 = 16 \times 64 = 32 \times 32 = 64 \times 16$. For the sake of simplicity, we assume that the output modules can achieve nearly 100% throughput at the last stage, while there is no packet loss in the first stage for sufficiently large input buffers. Thus, the system performance is mainly characterized by the loss probability in the output modules and the throughput limitation in the input modules.

We first estimate the loss probability of an $m \times n$ output module with evenly distributed input traffic. This is a straightforward extension of the loss probability analysis in the knockout switch in Chapter 4. Suppose each output port can accept up to L packets in one-slot time. The loading on an input link of an output module, denoted by ρ_{out}, is related to the loading on an input port of an input module, denoted by ρ_{in}, as follows:

$$\rho_{out} = \frac{n}{m} \cdot \rho_{in}. \tag{6.28}$$

It follows that the packet loss probability at the last stage, under the assumption that a packet arrives at each link independently with probability ρ_{out}, is given by

$$P_{loss} = \frac{1}{\rho_{in}} \cdot \sum_{l=L+1}^{m} (l-L) \cdot \binom{m}{l} \left(\frac{\rho_{out}}{n}\right)^{l} \cdot \left(1 - \frac{\rho_{out}}{n}\right)^{m-l}. \tag{6.29}$$

Figure 6.28 shows the loss probability as a function of L, and Fig. 6.29 shows the loss probability versus the expansion factor m/n, for different n with $\rho_{in} = 0.8$. These figures reveal the fact that the loss probability will increase with respect to both parameters n and m/n.

As shown in Figs. 6.30–6.33, however, it is obvious that larger values of n and m/n will always result in better throughput of input modules with the look-ahead selection scheme in the first stage. We assume in the simulation that packets arrive independently at each input link, and they are equally likely to be destined for each output module. Figure 6.30 shows the throughput versus the window size for different

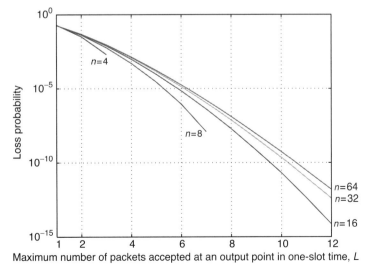

FIGURE 6.28 Loss probability versus L with 80% loading.

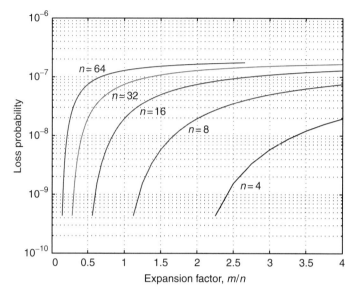

FIGURE 6.29 Loss probability versus expansion factor (m/n) with fixed $L = 8$ and 80% input loading.

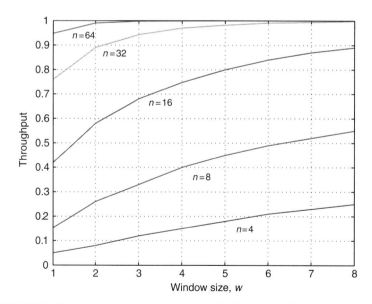

FIGURE 6.30 Throughput versus window size w with fixed expansion factor (m/n) = 1.5.

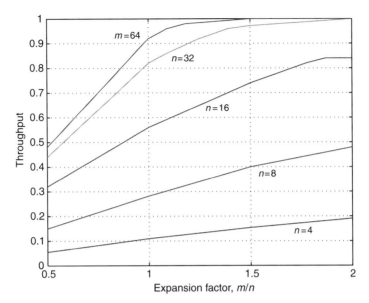

FIGURE 6.31 Throughput versus expansion factor (m/n) with fixed window size $w = 4$.

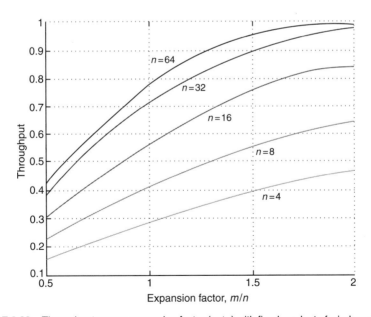

FIGURE 6.32 Throughput versus expansion factor (m/n) with fixed product of window size and module size $wn = 64$.

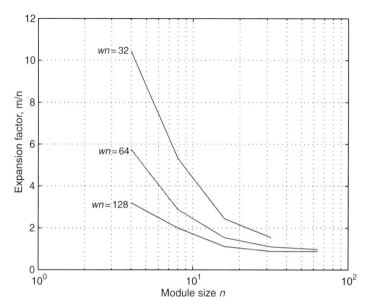

FIGURE 6.33 Expansion factor (m/n) versus module size n to achieve 80% throughput for several fixed products of window size and module size.

module sizes n with constant expansion factor $m/n = 1.5$. Figure 6.31 illustrates that the throughput is increasing with respect to the expansion factor m/n. However, larger values of m/n and w will increase the complexity of the central stage and the processing time at the input stage, respectively. To have a fair comparison, we fix the input-arbitration overhead $wn = 64$ in Fig. 6.32 when plotting the throughput for different module sizes n. It shows again that the larger the module size, the higher the throughput.

Figure 6.33 shows how large the expansion factor m/n should be for some particular module size n to achieve 80% throughput with several fixed products of the window size and the module size. Again, the larger modules behave better. The trade-off between the loss probability of output modules and the throughput of input modules is summarized in Table 6.3. It shows that a proper choice of the module size and the expansion factor should be $n = 16$ and $m/n = 2$, respectively, for switch size $N = 1024$. With window size $w = 4$, the maximal throughput that can be achieved is above 80%, while the loss probability at the output modules is kept below 10^{-7} with $L = 8$.

6.2.4 Multicasting in Path Switching

In this subsection, we shall study two multicast schemes to enhance the cross-path switch to support multicasting. The first scheme, namely scheme 1, performs replication in both input and output modules. As shown in Fig. 6.34(a), a packet in this case is

TABLE 6.3 Summary of Trade-Off in Module Size n

Module Size n	Expansion Factor $m/n = 1$, $P_{loss}(L = 8, \text{Load} = 80\%)$	Throughput	Expansion Factor $m/n = 2$, $P_{loss}(L = 8, \text{Load} = 80\%)$	Throughput
4	0	0.29	0	0.47
8	0	0.41	$O(10^{-8})$	0.64
16	$O(10^{-8})$	0.56	$O(10^{-7})$	0.84
32	$O(10^{-7})$	0.72	$O(10^{-7})$	0.97
64	$O(10^{-7})$	0.78	$O(10^{-7})$	0.99

first replicated at the input stage. The number of copies replicated at this stage is equal to the number of destined output *modules* of the packet. These copies will be routed by the central modules to their destined output modules. Each copy arriving at the last stage will be further replicated if it is destined for more than one output port within that output module. In contrast to the scheme 1, the second scheme, namely scheme 2, only performs packet replication in input modules. In this scheme, a packet that has different destined output ports within the same output module needs to be replicated in the first stage and the resulting copies will be routed independently through different central modules (see Fig. 6.34(b)). We do not consider multicasting at the middle stage, because the capacity provided by some point-to-multipoint connection pattern in the central modules will be affected by the time-varying replication requests and it is not appropriate for the capacity assignment of the cross-path switch.

With the same number of central modules, one would intuitively expect higher throughput in the first scheme. However, as it requires packet replication at the last stage, the complexity of output modules will be higher. On the contrary, by using more central modules in the second multicast scheme, the same throughput can also be attained. Since both schemes have their own advantages, it is necessary to evaluate the trade-off between the performance and the complexity.

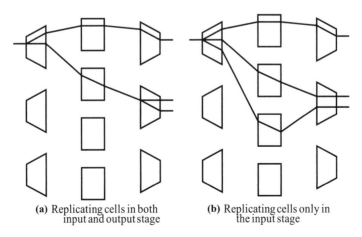

(a) Replicating cells in both input and output stage

(b) Replicating cells only in the input stage

FIGURE 6.34 Two multicast schemes in cross-path switch.

6.2.4.1 Capacity Allocation for Multicast Traffic

In the original unicast cross-path switch, the capacity allocation is based on the traffic load and the effective bandwidth of each virtual path at the middle stage to guarantee the QoS requirement. For multicast traffic, the middle-stage traffic load will be affected by the packet fanout distribution and the multicast schemes used. The modified capacity assignment procedure is described below.

Let S_u, S_v, S_{I_i}, and S_{O_j} be the sets of the call requests on input link u, output link v, input module I_i, and output module O_j, respectively. We assume that the traffic load (before replication) ρ_r, in unit of packet/slot, the effective bandwidth (before replication) α_r, and the packet fanout number y_r of call request r are given at the call setup, and they satisfy the following constraints to fulfill the packet-level QoS requirement:

1. *Input link load constraint:*

$$\sum_{r \in S_u} \rho_r \leq 1, \; \forall u. \qquad (6.30)$$

2. *Output link capacity constraint:*

$$\sum_{r \in S_v} \alpha_r \leq 1, \; \forall v. \qquad (6.31)$$

We say that $r \in S_u$ when the call request r comes from input link u, and $r \in S_v$ when one of the destinations of r is output link v.

Define λ_{ij} to be the total traffic load of the virtual path connecting input module I_i and output module O_j. For scheme 1, in which a call request has only one copy routed in the middle stage for each destined output module,

$$\lambda_{ij}^{(1)} = \sum_{r \in S_{I_i}, \; r \in S_{O_j}} \rho_r. \qquad (6.32)$$

In contrast, the traffic load of call request r at the middle stage of scheme 2 is expanded y_r times, and $y_r(i, j)$ of them will be routed through the virtual path from I_i to O_j. Then the total traffic load of this virtual path is given by

$$\lambda_{ij}^{(2)} = \sum_{r \in S_{I_i}, \; r \in S_{O_j}} \rho_r \cdot y_r(i, j). \qquad (6.33)$$

Similarly, the aggregate effective bandwidths β_{ij} of a virtual path connecting I_i and O_j in two multicast schemes are given, respectively, as follows:

$$\beta_{ij}^{(1)} = \sum_{r \in S_{I_i}, \; r \in S_{O_j}} \alpha_r, \quad \beta_{ij}^{(2)} = \sum_{r \in S_{I_i}, \; r \in S_{O_j}} \alpha_r \cdot y_r(i, j). \qquad (6.34)$$

The capacity assignment is to find the capacity C_{ij} for the virtual path connecting I_i and O_j such that the following constraints are satisfied:

$$\begin{cases} C_{ij} \geq \beta_{ij}, & \forall i, j, \\ \sum_i C_{ij} = m, & \forall j, \\ \sum_j C_{ij} = m, & \forall i, \end{cases} \tag{6.35}$$

where m is the number of central modules.

Like the case of unicast traffic, an optimal assignment requires establishing an objective function such as minimizing total weighted delay z after each virtual path is modeled as an $M/M/1$ queue:

$$z = \sum_{i,j} \frac{\lambda_{ij}}{C_{ij} - \lambda_{ij}}. \tag{6.36}$$

6.2.4.2 Performance Evaluations In the following, we shall study how the multicast traffic affects the throughput and the packet loss probability in a 1024 × 1024 cross-path switch. The performance analysis is based on two assumptions. First, the traffic distribution among input ports and output ports is homogeneous, and a packet is equally likely to be destined for any output port. Second, the arrival process at each input link is Bernoulli, that is, the probability that there is a packet arriving in each time slot is identical and independent.

Fanout Distribution and Middle-Stage Traffic Load To characterize the effect of the multicast traffic on the cross-path switch, we consider three fanout distributions to make comparisons. Let Y be the random variable of the fanout number, and M be its maximum.

1. *Constant distribution:*

$$\Pr\{Y = M\} = 1. \tag{6.37}$$

   Mean $E[Y] = M$ and variance $\text{Var}[Y] = 0$.

2. *Uniform distribution:* Suppose the requested fanout is uniformly distributed from 1 to M, which is the maximum. In other words,

$$\Pr\{Y = y\} = 1/M, \quad 1 \leq y \leq M. \tag{6.38}$$

   Thus, $E[Y] = M + 1/2$ and $\text{Var}[Y] = M(M+1)/6$.

3. *Truncated geometric distribution:*

$$\Pr\{Y = y\} = \frac{(1-q)q^{y-1}}{1 - q^M}, \quad 1 \leq y \leq M, \tag{6.39}$$

where q is used to control the shape of the distribution. The mean is given by

$$E[Y] = \frac{1}{1-q} - \frac{Mq^M}{1-q^M}. \tag{6.40}$$

This distribution is often used in the literature for modeling the fanout distribution. By fixing M equal to the switch size $N = 1024$, the parameter q can be calculated for a given mean fanout $E[Y]$.

Multicast traffic load in input stage is different from that in the output stage because packets are replicated inside the switch. Let ρ_{in} and ρ_{out} denote the traffic load of each *input* link at *input* stage and the traffic load of each *output* link at *output* stage, respectively. The traffic in output stage is more than that in input stage, and the ratio is given by

$$\frac{\rho_{out}}{\rho_{in}} = E[Y], \tag{6.41}$$

which is the mean packet fanout number.

Here, we define the middle-stage traffic load ρ_{mid} such that ρ_{mid}/ρ_{in} is equal to the ratio of the amount of traffic in *middle stage* to that in *input stage*. It can be viewed as a reference load of the switch and reflects some difference between two multicast schemes when comparing their throughput and loss performance.

To calculate the middle-stage traffic load ρ_{mid}, we let X be the number of copies that are replicated from a packet at input stage, that is, the number of distinct output modules that the packet is destined for, then $\rho_{mid} = \rho_{in} \cdot E[X]$.

For scheme 1 in which there is packet replication at both the input and the output stages, $E[X]$ varies for different fanout distributions. To evaluate $E[X]$, we first notice that the probability that a packet has no copy destined for a specific output module is given by

$$p_0 = \sum_{y=1}^{M} \frac{\binom{n}{0}\binom{N-n}{y}}{\binom{N}{y}} \cdot \Pr\{Y = y\}. \tag{6.42}$$

Then the average of the total number of copies replicated from a packet at the input stage is $k \cdot (1 - p_0) = E[X]$. For scheme 2 where there is no replication in output modules, $E[X] = E[Y]$; thus, $\rho_{mid} = \rho_{out}$.

To facilitate the comparison of the two schemes in the following, we provide some numerical figures of ρ_{mid} by fixing $\rho_{out} = 1$ in Table 6.4.

Throughput Since output queueing implemented in output modules ensures 100% throughput at the last stage, the throughput of the whole switch is limited only by the head-of-line blocking in input modules with look-ahead selection. Here, we shall only focus on the throughput at the input stage.

TABLE 6.4 The Middle-Stage Traffic Load in Multicast Cross-Path Switches for Different Fanout Distributions

Module Size	Multicast Scheme	Fanout Distribution	Mean Fanout		
			$E[Y]=2$	$E[Y]=4$	$E[Y]=8$
n=16	S1	Geometric	0.9855	0.9578	0.9064
		Uniform	0.9902	0.9712	0.9345
		Constant	0.9927	0.9782	0.9501
n=16	S2	G/U/C		1	
n=32	S1	G	0.9706	0.9165	0.8242
		U	0.9800	0.9416	0.8712
		C	0.9849	0.9554	0.9000
n=32	S2	G/U/C		1	
n=64	S1	G	0.9419	0.8438	0.6978
		U	0.9596	0.8857	0.7620
		C	0.9692	0.9113	0.8087
n=64	S2	G/U/C		1	

S1: replicating packets in both input and output stages; S2: replicating packets only in input stage.

In a cross-path switch, the throughput of an input module is mainly characterized by the window size w of the input queues and the group size m/k. Recall from Chapter 4 that the window size in look-ahead selection is defined to be the number of packets waiting at each input queue that will be checked in one time slot, and it is limited by the module size n and the processing speed. For multicast packets, the window size is defined to be the number of copies instead of packets. On the other hand, the group size represents the maximum number of packets that can be delivered to an output port in one time slot. For an $n \times m$ input module in cross-path switch, the real destination of a packet is one of the k *output* modules; thus, the average group size is m/k. The window size is used to release the head-of-line blocking, while the group size can be regarded as the output capacity provided by each destination. Increasing either of them will certainly result in better throughput, but there is a trade-off. The processing speed limits the product of the window size w and the module size n, thus a smaller n is desirable. On the contrary, when n is larger, the statistical multiplexing gain is higher. We choose the median size $n = 32$ in this analysis.

For clear presentation, we itemize the factors affecting the throughput as follows:

1. *The expansion factor m/n:* Recall that in a cross-path switch, the ratio m/n is called the expansion factor. One can think of it as the average capacity per input link. As shown in Fig. 6.35, the throughput is increasing with the expansion factor. Nearly 100% throughput can be achieved by a sufficiently large expansion factor. The saturation speed is relatively faster in scheme 1 or with a larger mean fanout.

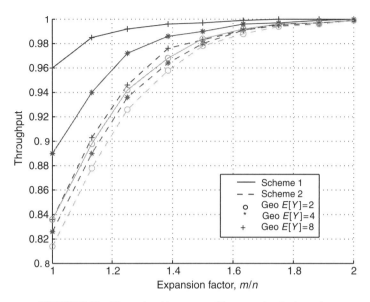

FIGURE 6.35 Throughput increase with expansion factor m/n.

2. *The mean fanout $E[Y]$:* To see the multicast effect, we first focus on the first moment of multicast traffic, that is, the mean fanout. The throughput versus $E[Y]$ curve is plotted in Fig. 6.36. It can be observed that the throughput is higher when the mean fanout is larger. The reason is that the copies from the same packet have distinct destinations, thus the output contention at the head of line is reduced, compared with the case of having the same total number of unicast packets waiting. In conclusion, the look-ahead throughput performance is more favored to multicast traffic.

3. *The fanout distribution:* Even with the same mean fanout number, the throughput still varies for different fanout distributions as shown in Figs. 6.37 and 6.38. The throughput discrepancy is larger in scheme 1 because the middle-stage traffic load ρ_{mid} is different for different fanout distributions, as shown in Table 6.4. The larger the ρ_{mid}, the more congested the switch and the smaller the throughput. It helps to explain why the highest throughput is attained when the packet fanout is geometrically distributed.

   Conversely, the throughput discrepancy in scheme 2 is smaller because the middle-stage traffic load ρ_{mid} is the same as ρ_{out} for all fanout distributions. However, since the geometric fanout distribution has the largest variance, its performance is the worst. This is similar to other switching systems that the performance is probably better in handling traffic with less fluctuations.

Loss Probability The overall loss probability of the switch can be well approximated by the knockout loss at the output stage, provided the buffer size of each input link is sufficiently large and thus the packet loss at the input stage is negligible.

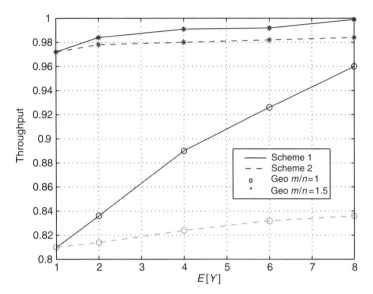

FIGURE 6.36 Throughput increase with mean fanout number $E[Y]$.

Consider an output module of size $m \times n$. Since the packets of the m input links may come from different input modules and the pattern is time varying, we need to simulate all k input modules and totally $N = 1024$ links in order to obtain the loss probability. However, we devise a procedure in order to obtain this value more easily.

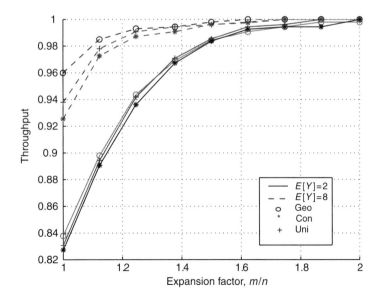

FIGURE 6.37 Throughput varying with fanout distribution in scheme 1.

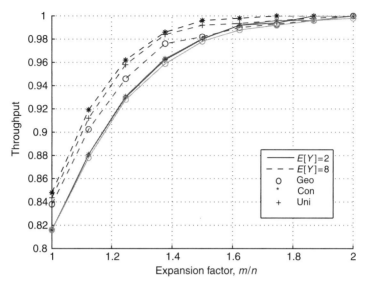

FIGURE 6.38 Throughput varying with fanout distribution in scheme 2.

Since there is no buffering at the input links of the output modules, the time correlation between the arrival process of the input links need not be taken into consideration. Although we still need the simulation data to capture the "space" correlation among the active neighboring input links connected from the same input module in order to calculate the loss probability, it is sufficient to simulate only one input module and collect the statistics of the packets delivered to the specific output module. The procedure is presented as follows:

Procedure for Calculating Loss Probability

1. Consider all m input links of an output module, and let A be the random variable of the number of active links. To calculate the loss probability, we first need to obtain the probability distribution of A. Define $G_A(z) = \sum_{a=0}^{m} \Pr\{A = a\} \cdot z^a$ to be the generating function. Suppose m_i of the m input links are connected from input module i, $0 \leq i \leq k-1$; $\sum_{i=0}^{k-1} m_i = m$. Similarly, we define A_i to be the number of corresponding active links among the group of m_i links. Since the traffic from different modules is independent, it follows that

$$G_A(z) = \Pi_{i=0}^{k-1} G_{Ai}(z). \tag{6.43}$$

Also since the traffic statistics from the input modules are independent and identically distributed, we can obtain the probabilities $\Pr\{A_i = a\}, 0 \leq a \leq m_i$, $\forall i$, by simulating only one input module and collecting the departure statistics.

TABLE 6.5 The Statistics of the Number of Packets Leaving from an Input Module for an Output Module

Time Slot	C	Pr {x Links Are Active}			The Corresponding Generating Function
		$x = 0$	$x = 1$	$x = 2$	
0	2	0.1589	0.2474	0.5937	$G0(z) = 0.1589 + 0.2474z + 0.5937z^2$
1	1	0.2286	0.7714	–	$G1(z) = 0.2286 + 0.7714z$
2	1	0.2109	0.7891	–	$G2(z) = 0.2109 + 0.7891z$
3	1	0.1973	0.8027	–	$G3(z) = 0.1973 + 0.8027z$
4	1	0.1850	0.8150	–	$G4(z) = 0.1850 + 0.8150z$
5	1	0.1795	0.8205	–	$G5(z) = 0.1795 + 0.8205z$
6	1	0.1710	0.8290	–	$G6(z) = 0.1710 + 0.8290z$
7	1	0.1682	0.8318	–	$G7(z) = 0.1682 + 0.8318z$

C: number of links connected to the specific output module.

As an example, we demonstrate the calculation of $G_A(z)$ for the case that $m = 36$ and $E[Y] = 2$. The frame size is equal to $f = 8$, and $g = mf/k = 9$, indicating that there are totally nine links connected from an input module to an output module during eight time slots. The statistics of the packet delivered from any input module to the specific output module is summarized in Table 6.5. The $m = 36$ input links of an output module can be partitioned into four groups, and each has nine links with statistics as given in Table 6.5. By convolution, the generating function of the probability distribution of the number of active packets arriving at an output module is given by

$$[G0(z)G1(z)G2(z)G3(z)G4(z)G5(z)G6(z)G7(z)]^4. \tag{6.44}$$

2. Let Y_o be the random variable of the fanout number of a packet arriving at the output module. Consider a packet (at the input stage), the average number of the copies destined for the specific output module is obviously $E[Y]/k$, including the case that no copy of the packet is destined for the output module. Then

$$\frac{E[Y]}{k} = p_0 \cdot 0 + (1 - p_0) \cdot E[Y_o], \tag{6.45}$$

where p_0 is the probability that no copy of the packet is destined for the output module. Thus,

$$E[Y_o] = \frac{E[Y]}{k(1 - p_0)} = \frac{E[Y]}{E[X]}. \tag{6.46}$$

3. Consider the probability that one copy of a packet arriving at the output module is destined for a particular output port, which is simply $p = E[Y_o]/n$. In scheme 1, this probability is independent from link to link, because no two packets arriving at the output module come from the same original packet. In scheme 2,

since there may be the case that more than one packet simultaneously arriving at the output module is generated from the same original packet, this probability is dependently distributed over the links from the same input module. This dependence here is beneficial to having less loss because the copies generated from an original packet will not be destined for the same output port, resulting in less output port contention. We make the independence assumption in the following computation to produce an upper bound for the loss probability of scheme 2. When k is sufficiently large or m/n is small, the bound is expected to be tight.

4. Consider a particular output port. Let L be the random variable that represents the number of packet copies destined for the port in one time slot. Its generating function is given by

$$G_L(z) = \sum_{a=0}^{m}(1 - p + pz)^a \Pr\{A = a\}, \qquad (6.47)$$

where $p = E[Y_o]/n$ and $\Pr\{A = a\}$ is the probability that a of the m links are active. Thus,

$$\Pr\{L = l\} = \sum_{a=l}^{m} \Pr\{A = a\} \cdot \binom{a}{l}(1-p)^{a-l} p^l. \qquad (6.48)$$

The loss probability is

$$\frac{1}{\rho_{out}} \sum_{l=G+1}^{m}(l - G)\Pr\{L = l\}, \qquad (6.49)$$

where G is the group size of an output module.

The numerical results are presented below.

1. *The group size G:* First, we demonstrate the function of group size G in output modules in Figs. 6.39 and 6.40. The group size G is defined to be the number of packet copies that can be delivered to each output port during one time slot. The larger the group size, the smaller the loss probability but the higher the complexity. Arbitrarily small loss probability can be achieved by a sufficiently large G. In addition, the discrepancy in the loss probability is not so significant with different parameters such as the number of central modules m and the mean fanout number $E[Y]$. In other words, it is mainly characterized by the group size G.

2. *The expansion factor:* As shown in Fig. 6.41, the loss probability is also varying with the expansion factor m/n. When the number of central modules is large, the number of active packets allowed to be simultaneously delivered to an

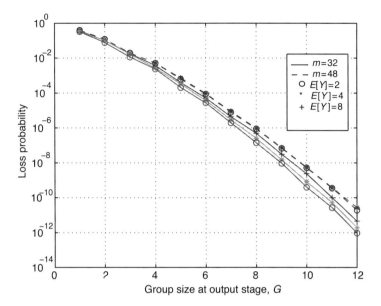

FIGURE 6.39 Loss probability decrease with group size G in scheme 1.

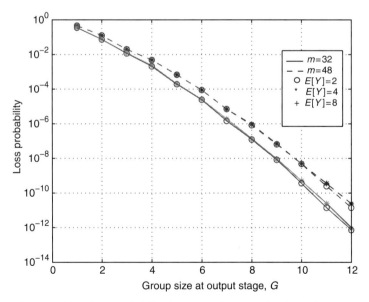

FIGURE 6.40 Loss probability decrease with group size G in scheme 2.

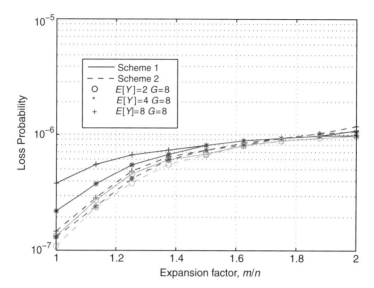

FIGURE 6.41 Loss probability increase with expansion factor m/n.

output module tends to be more uniformly distributed, and thus a higher loss probability is expected.

3. *The mean fanout number:* The effect of multicast traffic on the loss probability is not so significant. As shown in Fig. 6.42 with fixed group size $G = 8$, the expansion factor m/n dominates the trend of the loss probability. It changes only a little when the mean fanout number $E[Y]$ increases, except when the expansion factor is small ($m/n = 1$) in scheme 1 due to the dramatic increase in throughput. Please refer to Fig. 6.36.

6.2.4.3 Overall Complexity Measurements and Comparisons
We have considered feasible designs for each multicast schemes to implement the multicast cross-path switch. In the following, we shall estimate their complexity in terms of the number of switching nodes used under certain performance requirement.

The input modules and the central modules in both schemes will be the same. Since the central modules are required to be rearrangeably nonblocking only, so that every connection pattern of the route assignment can be realized, Benes networks are suitable and sufficient. Its complexity is given by

$$C_{cm} = k \left(\log_2 k - \frac{1}{2} \right), \tag{6.50}$$

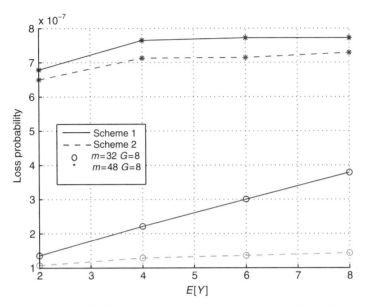

FIGURE 6.42 The little change of loss probability with mean fanout $E[Y]$.

where k is the size of a central module, that is, the number of input modules. Then the complexity of the middle stage consisting m central modules is

$$C_{mid} = mk \left(\log_2 k - \frac{1}{2} \right). \tag{6.51}$$

Since packet replications are necessary at the input stage, a copy network cascaded with a point-to-point switch is sufficient. As mentioned in the previous section, the copy network is estimated as two banyan interconnection fabrics, each with complexity $C_{cn} = \frac{m}{2} \log_2 m$, while the point-to-point switch consists of a Batcher–banyan and its complexity is given by

$$C_{pp} = \frac{m}{4} \log_2 m (\log_2 m + 1) + \frac{m}{2} \log_2 m. \tag{6.52}$$

Thus, the overall complexity of the input stage is $C_{in} = k(C_{cn} + C_{pp})$.

At the last stage, two different switches are used. For scheme 1 where packet replications are still needed in the output modules, a knockout switch with broadcast buses can be used. Since each concentrator with group size G has the complexity

$$m + (m-1) + (m-2) + \cdots + (m-G) = mG - \frac{G(G+1)}{2},$$

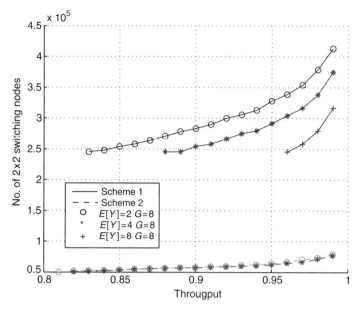

FIGURE 6.43 The complexity comparison between two multicast schemes for different throughput requirement.

we need $C_{\text{knock}} = n[mG - G(G+1)/2]$ switching nodes to build up one knockout switch and totally $C_{\text{out1}} = kC_{\text{knock}}$ for the whole output stage. For scheme 2, a Batcher–banyan switch with channel grouping can be used in the output module. Recall from Chapter 4 that this switch contains an $m \times m$ Batcher sorting network, an $m \times m$ concentrator, and G parallel banyan networks. The total complexity of the output stage for scheme 2 is given by

$$C_{out2} = k \cdot \left[\frac{m}{4} \log_2 m (\log_2 m + 1) + \frac{m}{2} \log_2 m + G \cdot \frac{n}{2} \log_2 n \right]. \quad (6.53)$$

The complexity measures of the switch architectures are shown in Figs. 6.43 and 6.44, respectively. The trade-off between performance and complexity of scheme 2 is superior to scheme 1 because of the huge complexity in establishing the broadcast knockout switches. Therefore, it might not be worth implementing multicast switching at the output stage of the cross-path switch.

6.A APPENDIX

6.A.1 A Formulation of Effective Bandwidth

The basis of call-level capacity allocation is the QoS requirement at the packet level. In general, the QoS requirement could be the loss rate, the mean delay, the delay jitter,

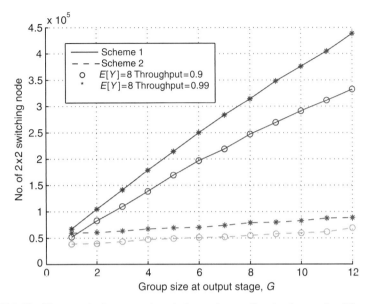

FIGURE 6.44 The complexity comparison between two multicast schemes for different order of loss probability requirements.

or some combinations of these performance indicators. The effective bandwidth is the minimum capacity required to satisfy the prerequisite QoS of each call. We formulate the packet-level QoS requirement by a pair of constraints: the delay and the loss. The delay constraint is defined to be

$$-\log_{10} \Pr\{W > \tau\} > \delta_D, \qquad (6.54)$$

where $\Pr\{W > \tau\}$ is the probability that the packet waiting time W exceeds a certain accepted amount τ. The loss constraint [Guerin, Veciana] can be defined in a similar manner:

$$-\log_{10} \Pr\{X \geq B \mid \text{seen by arrivals}\} > \delta_L, \qquad (6.55)$$

where X is the random variable of queue length and B is the required buffer size. Thus, the required delay and loss QoS constraints can be satisfied by the resources α and B allocated for the connection request.

6.A.2 Approximations of Effective Bandwidth Based on On–Off Source Model

We provide the widely used Markovian on–off source model here to demonstrate the approximate estimation of the effective bandwidth of a call request. A two-state Markovian process can be fully characterized by the following three parameters [AMS81]:

$$\begin{cases} \text{mean arrival rate,} & \lambda & \text{cell/slot,} \\ \text{peak arrival rate,} & \lambda_p & \text{cell/slot,} \\ \text{average "on" period,} & T_{on} & \text{slot.} \end{cases}$$

With service rate (capacity) μ, the queue length distribution given in Ref.[Anick] is as follows:

$$\Pr\{X > x\} = \frac{\lambda}{\mu} e^{-\frac{\lambda_p(\mu-\lambda)x}{\mu T_{on}(\lambda_p-\mu)(\lambda_p-\lambda)}}. \tag{6.56}$$

The overflow probability seen by arrivals is given by

$$\Pr\{X > x \mid \text{seen by arrivals}\} = e^{-\frac{\lambda_p(\mu-\lambda)x}{\mu T_{on}(\lambda_p-\mu)(\lambda_p-\lambda)}}. \tag{6.57}$$

The waiting time distribution $\Pr\{W > \tau\}$ is just a scaled version of (6.57) due to the relation $x = \mu\tau$, and given as follows:

$$\Pr\{W > \tau\} = e^{-\frac{\lambda_p(\mu-\lambda)\tau}{T_{on}(\lambda_p-\mu)(\lambda_p-\lambda)}}. \tag{6.58}$$

The effective bandwidth α can be calculated from the delay constraint (6.54), and it is given by

$$\alpha = \lambda + \frac{T_{on}(\delta_D + \log_{10} e)(\lambda_p - \lambda)^2}{\lambda_p \tau + T_{on}(\delta_D + \log_{10} e)(\lambda_p - \lambda)}. \tag{6.59}$$

Similarly, the required buffer size B given below can be obtained from (6.55) through (6.57).

$$B = \frac{\alpha T_{on}(\lambda_p - \alpha)(\lambda_p - \lambda)(\delta_L + \log_{10} e)}{\lambda_p(\alpha - \lambda)}. \tag{6.60}$$

PROBLEMS

6.1 For the broadcast banyan network, argue that two bits are the minimum number of bits required to perform routing at each switch element, and therefore, the Boolean interval splitting algorithm is a "minimal" algorithm.

6.2 Consider the following broadcast banyan network that attempts to perform copying and routing at the same time with a $2n$-bit routing tag. Two bits are used in each stage. Bits 00 and 11 mean the packet is to be forwarded to upper and lower outgoing links, respectively. Bits 01 mean the packet is to be duplicated and forwarded to both outgoing links. Not all the $2^N - 1$ subsets of outputs can be reached this way. Describe the subsets of outputs that can be reached.

6.3 Consider an 8×8 copy network based on the Boolean interval splitting algorithm in the broadcast banyan network operated as a waiting system such that overflow requests that cannot be fulfilled will be queued at inputs.

(a) Show the encoding and decoding processes for the following copy requests: two copies for input 0, one copy for input 3, and three copies for input 5; the other inputs are inactive. Show the values adopted by the various fields in the packet headers as they travel through the five components of the copy network. Assume the running-adder network is not cyclic.

(b) Now, do the same assuming a cyclic running-adder network and a dilated reverse banyan concentrator, with input 5 being the starting point.

(c) Give an example showing that fewer than N packet copies are allowed even though there are more requests still waiting at the input queues (*Hint:* consider HOL blocking).

(d) Show how the efficiency of the copy network can be improved by placing a shifting concentrator before the copy network so as to implement a logical FIFO queue out of the N input queues.

(e) Consider an input–output pair of a multicast communication session of the overall multicast switch. Show that packets arriving at successive time slots to the input to the shifting concentrator may have their copies delivered simultaneously to the output of the point-to-point switch, assuming that the switch is an output-buffered switch. However, it is not possible for a later arriving packet to have its copy delivered to the output before an earlier arriving packet.

(f) Part (e) means that the switch output must be able to uncover the underlying sequence of the simultaneously delivered packets in order to avoid out-of-sequence packet transmission on its transmission links. Propose a scheme that enables the output to do so.

6.4 The text explained how a reverse–banyan network fails to concentrate the packets between the cyclic running-adder network and the broadcast–banyan network in a copy network. Three approaches to solving this problem were proposed. This question considers two other approaches.

(a) Show that a Benes network can be used to concentrate the packets. In particular, show how you would set the $2n - 1$ bits for routing in the Benes network.

(b) A three-phase scheme analogous to the sort-banyan point-to-point switch could be used. In the first two phases, only the headers of packets enter the cyclic running-adder network. The first phase calculates the running sums and determines the served copy numbers. The second phase acknowledges

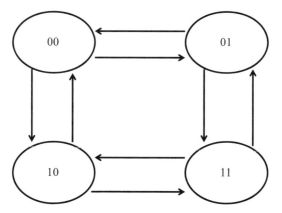

FIGURE 6.45 Hypercube network.

the inputs of the served copy numbers and the current starting port. The third phase performs a cyclic shift on all the winning input packets so that the starting-port packet appears on the uppermost output of the adder network. In this way, a reverse–banyan network can then be used to concentrate the packets in a nonblocking manner. Show how to use the adder-network structure to perform the cyclic shift.

6.5 For the probabilistic copy algorithm for arbitrary network topologies, the text considered the splitting interval to keep track of the replication process. An alternative is to directly encode the copy number, CN, into the header.
  (**a**) Explain how you would modify CN at each stage whenever it is possible to duplicate a packet.
  (**b**) An index reference (IR) field is also needed in the packet header. How would you modify the IR field whenever duplication occurs so that the IR fields of packet copies of the same master packet are distinct?

6.6 Consider applying the probabilistic copy algorithm on a two-dimensional hypercube network with four nodes, labeled as 00, 01, 10, and 11, as shown in Fig. 6.45. The links are bidirectional and simultaneously carry one packet in each direction.
  (**a**) Suppose that there is a packet at node 00 desiring four copies and there are no other packets. Show the packet replication process.
  (**b**) Show that a deadlock occurs if the packet desires eight copies.
  (**c**) Suppose that there is a buffer that can hold just one packet at each node. In other words, when there are no incoming packets to a node, both desiring replication, we can hold one of the packets in the buffer while duplicate the other packets on the two outgoing links. Show that this eliminates the deadlock in part (b).

6.7 Consider multicasting in a large *feedback* shuffle-exchange network. A master packet to be routed to F destinations will be replicated first before the individual copies are routed to their destination nodes. That is, the F copies will be made according to the following algorithm. Each packet has a CN field that keeps track of the number of copies to be made from it. The CN field of the original master packet is set to F initially. Each time a packet with a CN field greater than 1 arrives at a node, it will be duplicated and forwarded to both outputs if there is no incoming packet on the other input. Otherwise, it will be forwarded to one of the outputs without being replicated, and the replication process is deferred until the packet visits a node with no packet on the other input.

(a) Describe how the CN fields of the two packet copies will be updated each time a packet is duplicated.

(b) Consider a master packet with $F = 2^f$, where f is a positive integer (i.e., F is a power of 2). Focus on a particular copy to be made. We can picture that it is embedded in a packet with CN copies yet to be made during replication process. Let the state of the copy be $\log_2$ CN. Draw the state-transition diagram that corresponds to the state evolution of the copy during replication, where $p = $ probability of successful duplication and $q = 1 - p$.

(c) Let ρ be the link loading in equilibrium. Express the transition probabilities in the previous part in terms of ρ.

(d) Let $T(i)$ be the expected number of additional steps needed before the replication of the copy is completed (i.e., before its CN value becomes 1). Write down the dynamic equation and boundary condition for $T(i)$.

(e) Calculate $T(f)$ in terms of ρ.

7

BASIC CONCEPTS OF BROADBAND COMMUNICATION NETWORKS

This chapter discusses various elementary concepts of broadband communication networks. The focus is on networks in which information is transported in fixed-length packets called cells. The asynchronous transfer mode (ATM) standardized worldwide, for example, assumes such a transport mechanism. The coverage is brief but should provide enough background information for the more detailed study in Chapter 8.

7.1 SYNCHRONOUS TRANSFER MODE

To understand the motivations behind ATM, let us consider the limitations of the synchronous transfer mode (STM) that lead to the proposal of ATM in the first place. Simply put, STM is time-division multiplexing (TDM) and ATM is fixed-length packet switching.

Recall from Chapter 1 that in TDM, information from several sources is multiplexed onto one physical transmission medium in which the time is slotted. Time slots are grouped into frames. Each time-slot position in a frame is dedicated exclusively to a particular source. Thus, time slot i of frames 1, 2, 3, ... can carry information from one and only one source once the connection has been set up. There is no sharing of time slots among different connections.

While this exclusive dedication of bandwidth works well for sources that produce traffic in a continuous fashion (e.g., voice and video), it is wasteful for sources with bursty traffic (e.g., computer data): when the sources are idle, their associated time slots do not carry any information. This limitation of STM is rather obvious and has long been recognized as one of the disadvantages of circuit switching when compared

---

Principles of Broadband Switching and Networking, by Tony T. Lee and Soung C. Liew
Copyright © 2010 John Wiley & Sons, Inc.

with packet switching. It turns out that this efficiency consideration, although valid, was not the main motivation for ATM when it was first proposed. ATM is advantageous in many respects even if all sources had more or less constant-rate traffic. In fact, some people view the bandwidth efficiency of ATM for bursty traffic as icing on the cake.

The ATM transport mechanism was originally proposed by the telephone network community, although the technology is also aggressively being adopted by the computer communication community. To understand the original motivation for ATM, let us briefly review the STM technology the way it is adopted in telephone networks. There are several fixed transport rates in the telephone networks. A voice conversation, after digitization, requires 64 kbps of bandwidth, and this forms the basic transport rate. The Digital Signal Level 0 (DS-0) of the TDM hierarchy carries information at this rate. The next higher rate is the DS-1 rate, which is about 1.5 Mbps. Thus, 24 telephone conversations, or DS-0 channels, can be multiplexed onto a DS-1 channel. Each time slot is 1 byte and each frame consists of 24 bytes, one for each DS-0 channel.

Instead of subscribing communication channels from the network operator at the basic 64 kbps rate, a customer may also lease a T-1 line that carries a 1.5 Mbps DS-1 data signal. Note that the term "DS-1" refers to the data signal whereas the term "T-1" refers to the carrier system, that includes the transmitter, the transmission medium, and the receiver for the DS-1 data signal. The T-1 line can be used to send either voice or computer data, and the choice is up to the subscriber. If it is used to carry data, the whole 1.5 Mbps can be used as an entire chunk of bandwidth without the need for a preliminary stage of multiplexing DS-0 signals.

A number of DS-1 signals can be multiplexed onto a high-speed data-signal level. The DS-2 signal consists of 4 DS-1 signals and the DS-3 signal consists of 28 DS-1 signals. The DS-3 rate is about 45 Mbps. For the purpose of TDM, instead of multiplexing from DS-1 to DS-2 and then to DS-3, it is a common practice to multiplex directly from DS-1 to DS-3. In this case, each time slot in a DS-3 is dedicated to data from a particular DS-1 signal.

With this transport-rate hierarchy, the next question is how to switch these channels. The STM switches are typically designed to switch at one of the basic rates (i.e., at 64 kbps, 1.5 Mbps, or 45 Mbps). Switching at rate r means that all the data in an input channel of rate r will be switched to an output channel of rate r; that is, one cannot switch some of the data within the input channel across different output channels. For instance, a switch that switches at DS-1 rate does not take apart the underlying DS-0 signals before switching is being performed, and the whole input DS-1 is switched to an output DS-1.

The switching rate may not be the same as the data rate of the input lines. For example, as illustrated in Fig. 7.1, a switch that switches at the 64 kbps DS-0 rate may have input lines at the DS-1 rate. It is more complicated to switch at the lower rate for two reasons. First, the switch must be able to take apart the data at the higher rate to obtain the data at the lower rate: this is essentially the demultiplexing function. Second, there are more data channels at the lower rate and their switching is more complicated simply because of their large number.

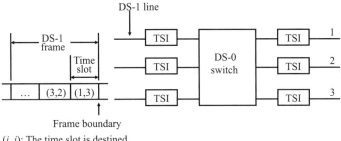

FIGURE 7.1 A T-S-T DS-0 switch with DS-1 line interface.

It should be noted that implementation-wise, it is not necessary to *spatially* demultiplex the DS-1 channel so that each of the underlying channels is carried on a physical channel. In practice, the demultiplexing function and part of the switching function are often performed in the time domain. Returning to the example in Fig. 7.1, the DS-0 data on each time slot of an input DS-1 line are destined for a particular time slot of a particular output DS-1 line. Each input DS-1 line is fed into a time-slot interchanger (TSI), which is essentially a time-domain switch (see Chapter 2). The TSI switches the time-slot positions taken by the DS-0 data channels and forwards the time-slotted data in a one-by-one fashion to a next stage of space-domain switch. The space-domain switch operates in a time-slotted fashion, and the switching configuration changes from time slot to time slot. The outputs of the space-domain switch are connected to yet another stage of TSI. By controlling the TSI at both the outer stages and the connectivites of the center-stage space-domain switch, it is possible to switch any input DS-0 channel to any output DS-0 channel provided the switch is nonblocking (see Chapter 2).

One limitation of the STM transport-rate hierarchy is that a subscriber can only acquire a channel of 64 kbps, 1.5 Mbps, or 45 Mbps (ignoring the DS-2 rate that is less popular in practice). In fact, the difficulty of offering data-transport services other than the basic rates was the reason why ATM was first proposed. We discuss the issues in the following.

If a subscriber needs, say, a bandwidth of 15 Mbps, the straightforward option is to lease a 45 Mbps DS-3 channel. However, this is more than that is needed and can be rather expensive. Alternatively, the subscriber may acquire ten 1.5 Mbps channels to make up the 15 Mbps bandwidth and disperse the 15 Mbps source traffic over the ten 1.5 Mbps channels. A difficulty is that the network does not recognize these 10 channels as being logically related to each other and may switch them over different paths from source to the destination. Since different paths may introduce different delays on the data, the receiving end needs to compensate for the longer-delay paths by introducing additional delays in the shorter-delay paths after the data are received so that the original information can be reassembled in the correct sequence. The end equipment required at the source and destination for such synchronization is commercially available from third-party equipment vendors. The process of dispersing a channel over several parallel subchannels is called *inverse multiplexing* for obvious reason. The network operator is unaware of (and has no need to be aware of) the use of inverse multiplexing by the subscriber. The inverse-multiplexing equipment, however, introduces additional cost to the subscriber.

Instead of inverse multiplexing, one may also modify the switches so that they can perform multirate switching so that channels of rate $n \times$ basic rate, n an integer, can be supported. For example, if a switch has input and output lines of DS-3 rate and can switch at the DS-1 rate, we can modify it so that it can switch a channel of $n \times$ DS-1 rate, where $n = 1, 2, \ldots, 28$. The switch must recognize the n DS-1 channels as logical subchannels of the actual channel during both call setup and switching of time slots. There are two ways to go about this and both introduce additional complexity to the switch design.

The first method is to simply perform some sort of inverse multiplexing at the switch. A subscriber sends to the network a channel of $1.5 \times n$ Mbps bandwidth. After being multiplexed with data from other subscribers, the channel occupies n time slots of a DS-3 input line to a switch. During call setup, the switch controller must be informed of the n time-slot positions taken by the input channel. One by one, the controller sets up n DS-1 connections over the switch. Although these connections have the same input and output, they may be switched over different paths within the switch. This is illustrated in Fig. 7.2(a), which shows many T-S-T switch modules (each T-S-T module is like that in Fig. 7.1) connected in three stages. These n time slots, if switched over different internal paths, may suffer different processing and switching delays. Consequently, it is possible that the n time slots within a frame period at the input may arrive at the output in different frame periods. Hence, it is necessary to introduce artificial delays at the output (as in inverse multiplexing) to these time slots so that the time slots of the frame period at the input can be realigned back into the same frame period at the output. In addition, within the same frame period, the order of the time slots may not be preserved and a time-slot interchanger is needed to reorder the time slots (although the controller can integrate this consideration easily into the call setup procedure).

The second method for multirate switching is to design the switch such that one can specify not only the desired output of a connection, but also the internal path to be taken. By insisting that the switch sets up n subchannels over the same internal path, there will be no need for time-slot realignment at the output. This is illustrated in Fig. 7.2(b). Although there is no need for time-slot realignment at the output, some new complexities are introduced as a result.

First, the call setup is more complicated because the overall switch must recognize the logical relationship of all the time slots and must then attempt to switch them over the same path, maintaining the sequence. In addition, there is also the performance penalty in terms of the increased blocking probability as compared to the first method. This is due to the phenomenon of bandwidth fragmentation: one may not find n idle time slots in any particular path (especially when n is large) although there are more than n idle time slots in total over different paths, in which case the call is not blocked in the first method, but is blocked in the second method. For the same blocking performance, the second method generally requires the number of alternative paths

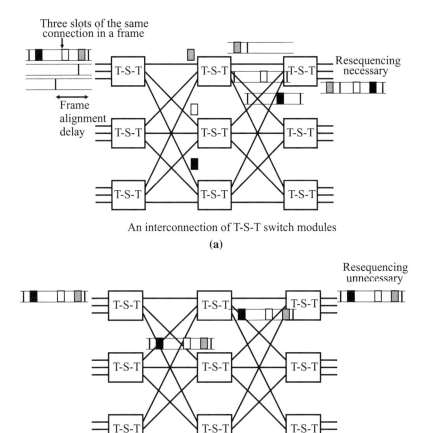

FIGURE 7.2 (a) Using basic rate switch to switch multirate connections; (b) using multirate switch to switch multirate connections.

to be increased. For the switching network in Fig. 7.2, this means that the number of middle-stage switch modules must be increased.

To summarize, the first limitation of STM is that the present-day network can only offer the basic rate transport because most network equipments are not designed to handle multirate switching. The second limitation is that although multirate switching is possible by modifying the switches, additional complexity will be incurred at both the software (managing the connections) and hardware levels. Even if multirate switching were available, there would still be the third limitation that the channel rate must be an integral multiple of the basic rate of the switch. For example, a DS-1 switch modified to handle multirate switching will not be able to accommodate channels at 100 kbps, 1 Mbps, or 2 Mbps.

To deal with these limitations, the proponents of ATM argued that the concept of frames should be removed so that time slots can be assigned independently to channels: time slots are not assigned in a periodic fashion to channels and they are given to the channels only when there is information to be transported on that channel. In other words, time slots are not dedicated to channels in a synchronous fashion. Without the frames, the output (destinations) of a time slot is not implicitly derivable from its position anymore. Therefore, a header that specifies the output must be attached to the time slot so that the switch can use it to route the time slot to the correct output. With this, what we have is basically the framework for fixed-length packet switching where each time slot carries a fixed-length packet! With the additional introduction of virtual-circuit routing (to be discussed in Section 7.4) so that information from a source is routed through a single path over the network, there is no need for inverse multiplexing. Also, without the frame structure, there is no need to limit the bandwidth assignment to an integral multiple of the basic rate. This new framework also gives rise to new issues and considerations, and they will be discussed in the next few sections.

In addition to regular user data, network equipment also needs to coordinate with each other to achieve certain functions. For instance, to set up an end-to-end connection from source to destination, several switches will be involved. The traffic status at different parts of the network may also need to be disseminated. In STM, control channels can be defined for the purpose of transporting control information. However, the bandwidth of the control channels will typically need to be overengineered so as to accommodate control functions needed in future services. Also, control information, unlike continuous traffic like voice and video, may arrive in a bursty fashion. For efficiency, one can send the control information in packet form in the control channel.

One advantage of ATM is that the transport mechanism is already packet-based, and only as much bandwidth as is required by the control channels needs to be assigned. Among the five-byte headers, only certain bits need to be predefined for the indication of whether the cell is a control or a regular data cell. If 5 bytes are not enough to distinguish the different levels or shades of control functionalities needed, one may further define different types of control cells using some of the bytes of the 48-byte payload. To draw an analogy, control information is like instructions in a computer in that they both achieve certain functions by invoking operations of the underlying systems. ATM cells are a very simple mechanism that allows a rich set of "instructions" to be defined: each 53-byte cell is like a 53-byte instruction.

7.2 DELAYS IN ATM NETWORK

Let us consider the delays incurred by packets in an ATM network. With the aid of Fig. 7.3, we shall trace the delays of a packet as it travels from its source to the receiver. At the source, the information must be packetized into fixed-length packets called cells. In the ATM standard, each cell is 53 bytes of which 48 bytes are dedicated

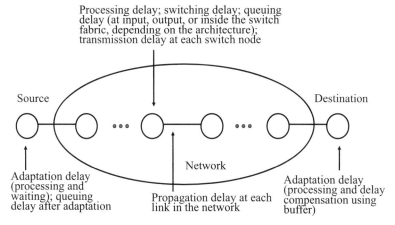

FIGURE 7.3 Delay components of cells from source to destination.

to carrying the source traffic[1] and 5 bytes are the packet header. The packetization of the information to be transported is part of the so-called adaptation process. The delay incurred during the process is called the adaptation delay.

There are two ways in which adaptation delay could be incurred. If the information to be sent is a long message that arrives in its entirety as a data block, then the only delay that is incurred is the delay incurred by the hardware or software that "segments" the message into cells and then computes and inserts the cell headers. This is the *adaptation processing delay*. Instead of block messages, information can also arrive as a data stream. For example, a data stream can be a byte stream in which the bytes arrive one after another with a certain time gap between successive bytes. In this case, some delay is incurred during the adaptation process to wait for enough number of bytes (i.e., 48 bytes) to arrive to form a cell. This is the *adaptation waiting delay*. For such stream information, there are therefore two delay components: the waiting delay followed by the processing delay. The processing delay is generally a fixed, constant delay. The waiting delay can be fixed or varying depending on how the source generates the bytes. The waiting delay, for example, is fixed for a voice source that generates data at a rate of 64 kbps. Specifically, the waiting delay will then be about $48 \times 8/64 = 6$ ms.

When a cell travels from node to node, there is also the propagation delay due to the limit on how fast signals can travel, the speed of which must be lower than the speed of light. In other words, the propagation delay of a link is the time required for a bit of information to travel from the transmitter end to the receiver end of a transmission link. The end-to-end propagation delay is simply the sum of the propagation delays incurred on all links in the end-to-end path. The propagation delay does not vary widely over time for a fixed path and can be considered as fixed. The temperature may

---

[1] In actuality, less than 48 bytes are used to carry the source traffic (see the appendix) because of additional overhead bytes within the 48 bytes.

have an effect on the propagation delay of, say, a fiber link, because the temperature expansion and contraction of the glass may affect the fiber length, but the effect is generally insignificant.

As a cell travels from node to node, it may also incur processing delay and waiting delay at each node. First, consider the source node again. After adaptation, the cell will be forwarded to the transmitter end of the transmission link. Because there could be a number of cells already waiting to be transmitted, the cell concerned will incur a *queueing delay* while waiting for its turn to be transmitted. For example, after a long message has been adapted, the resulting cells may be put into a buffer waiting to be transmitted. The cells at the later part of the message would wait for a longer time than cells at the earlier part of the message.

As the cell travels through nodes in the network, within each node, the cell will be processed (e.g., VCI translation, switching, etc.) upon arriving at each node. The switching part of the processing delay can be varying or fixed, depending on the switch implementation. Recall that the arrivals of packets to a switch are not scheduled in an ATM network, and several cells may be targeted for the same output at the same time. This leads to the contention problem that we have already covered in Chapters 3 and 4. As has been seen, some solutions to the contention problem (e.g., using input buffers to store packets that have lost contention) may cause random and variable delays. In short, some *input delay* is incurred at the input. Part of the input delay is the processing delay for VCI translation and so on. Depending on the switch architecture, part of the delay can be random and variable delay due to waiting at an input queue for output access.

In addition, cells may also incur delay at an output after they have arrived at the outputs. There are two possible delay components at an output, the queueing delay and the transmission delay. The queueing delay is incurred when more cells arrive at the output than can be transmitted immediately. Some of the cells must then be buffered for transmission some time in the future. This can happen in switches that are capable of forwarding more than one cell to an output in each time slot. The transmission delay is simply the amount of time needed to transmit one cell. For an ideal output-buffered switch, the output queueing delay is the only delay component that is random and varying.

The random and varying queueing delays at the nodes in turn cause the end-to-end delay from source to destination to be random and time varying. The variation in the end-to-end delay from cell to cell is often referred to as the delay jitter. For many applications, it is desirable to remove this delay jitter so that data can be forwarded to the user (or end application) in a continuous and periodic fashion. For example, for voice conversation, data are injected into the network in a continuous and periodic fashion. While traveling through the network, the cells may incur varying delays and the time gap between adjacent cells may be variable by the time they arrive at the receiver because of network delay jitter. It is necessary to remove the jitter before the audio is played out.

A *smoothing buffer* is generally used to remove the delay jitter. The principle of the smoothing buffer works like that of a water dam. Water may arrive to the dam in a bursty manner that depends on the rainfall. However, the water flows out of the dam at a constant rate. At the receiver, instead of immediately presenting the arrived data to the user, the receiver buffers it in the smoothing buffer first. By introducing a sufficiently large delay at the buffer, it is possible to deliver the data to the user in a continuous fashion. This additional delay is the *adaptation delay* incurred at the receiver end. The idea is that cells that have suffered a large delay within the network will incur a small receiver adaptation delay, and vice versa. Specifically, let T_n be all other delays and T_a be the adaptation delay introduced by the smoothing buffer. The idea is to make $T_n + T_a$ a constant: when T_n is large and T_a is small, and vice versa.

7.3 CELL SIZE CONSIDERATION

We have already covered the motivations for fixed-length cell switching from the viewpoint of traditional circuit switching. These were the considerations that gave rise to the ATM proposal originally. In that context, cells are basically time slots of 53 bytes. Instead of motivating cell switching from the shortcomings of circuit switching, let us examine fixed-length cell switching in the context of packet switching. Specifically, we now consider fixed-length versus variable-length packet switching and the issue of setting packet size.

Transport using fixed-length cells may incur more overhead than using variable-length packet because of the segmentation process. Consider the sending of a message. As shown in Fig. 7.4, if the underlying network transports information in the form of fixed-length cells, it is then necessary to partition the packet into many cells before transmission. The header of each cell must contain enough information for the nodes in the network to be able route it to the desired destination, and this information must be repeated for successive cells. This incurs much overhead as compared to transporting the long message as a long packet. In a variable-length packet, the packet can be as long as the message to be transmitted.

Let L_m be the length of the message, and let L_o be the header size. For the sake of argument, suppose that L_o in the cell network and the variable-length packet network are the same. Then, the number of bytes needed to transmit the message as a packet

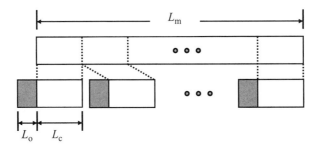

FIGURE 7.4 Segmentation of a message into cells during the adaptation process.

in the latter is

$$B_p = L_m + L_o. \tag{7.1}$$

The fractional overhead is $(B_p - L_m)/L_m$. The number of bytes needed to transport the message in the cell network is

$$B_c = \lceil L_m/L_c \rceil \times (L_c + L_o), \tag{7.2}$$

where L_c is the number of bytes in a cell excluding the header. The fractional overhead is $(B_c - L_m)/L_m$. When L_m is not exactly a multiple of L_c, the last cell will only be partially packed and some bytes are not transporting useful information. If the message is long, the round-off is insignificant, and we have

$$B_c \approx L_m + L_m L_o/L_c \gg B_p. \tag{7.3}$$

The penalty is primarily in the form of higher header overhead.

If the message is short, the header overhead penalty is reduced. However, the round-off penalty due to the message size L_m not being a multiple of cell size L_c will be more significant. This overhead tends to be more severe the shorter the message.

The above discussion shows that cell transport is less efficient than variable-length packet transport in terms of bandwidth usage. Cell transport has two advantages over variable-length packet transport. The first is that it is easier to design switches to have high throughput when the incoming traffic is fixed-size cells. The switch designs discussed in the previous chapters were based on the assumption of cells. It can be shown that the throughputs of the switches will decrease when the input packets are variable length and the switches are not operated in a time-slotted manner. Furthermore, since the operation of the switch is not time-slotted, it is necessary for the switch to keep track of the durations of the transport of the packets through the switch. This leads to additional complexity in switch design.

The second advantage of cell transport can be explained with the aid of Fig. 7.5. In the figure, a multiplexer is used to multiplex traffic from two inputs onto one output. Suppose on one input there is a long packet and on the other input there is a short packet that arrives slightly later. That is, the transmission of the long packet on the output has begun when the shorter packet arrives. Then, the short packet must wait for a long time before it can be transmitted. On the other hand, if the long packet has been segmented into smaller cells before transmission, then the short packet can be transmitted before the other cells of the long packet are transmitted, as shown in the figure. Thus, cell transport coupled with the use of an appropriate scheduling scheme can prevent a long packet from hogging the use of the output transmission capacity. This issue of unfairness, however, is perhaps less of a concern when the output link has very high transmission rate: even if the short packet has to wait for the transmission of the long packet to finish, the waiting time is not too long anyway. In any case, this unfairness issue was *not* the motivating factor for cell network as far as the ATM proposal is concerned.

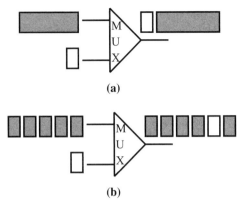

FIGURE 7.5 (a) Multiplexing variable-length packets; (b) multiplexing cells.

Given that we want to adopt cell transport, there is then the consideration of the cell size. If the cell size is small, the header overhead will then be large. If the cell size is too large, on the other hand, the adaptation waiting delay of low-rate real-time services, such as voice, may be too large. Recall that ATM was originally initiated by people in the telephone network community. The decision to set the cell size to 53 bytes is mainly motivated from the consideration of the transport of voice. Smaller cell size may be inefficient. Cell size much larger than this (say 100 bytes) may result in echo problems. The voice transmitted by the source, through the headset of the receiver, may be fed back to the source through the reverse path of a duplex voice connection. Recall that the bit rate of digital voice is 64 kbps. Each sample is one byte in length. Therefore, the sampling rate is 8000 samples per second. The adaptation delay both ways, assuming 100-byte cell, is already $2 \times 100/8 = 25$ ms. This itself results in noticeable echoes even in the absence of all other delays.

7.4 CELL NETWORKING, VIRTUAL CHANNELS, AND VIRTUAL PATHS

There are many options and alternatives for the operation and architecture design of a cell network. This book by and large assumes those options adopted by the ATM standard. This section explains the principles and rationale behind them.

7.4.1 No Data Link Layer

In traditional packet-switched networks, the links between nodes are not reliable and the bit error rate is high. Therefore, the *data link layer* (more specifically, automatic repeat request (ARQ) in the data link layer) is introduced to coordinate and automate the retransmission of data in case of data corruption. The advent of fiber-optic technology brings along very reliable transmission. An important assumption in the ATM standard is the use of reliable links. With reliable links, the ATM proponents argue that there is no need of a data link layer, which only makes networking more complicated and less efficient. Thus, for ATM networks, there is no data link layer. The recent development of wireless ATM may challenge the assumption of reliable links, since the transmission medium may not be as reliable as is assumed. Nevertheless, the ATM standard has been defined, and other ways must be sought for reliable communication. A possibility is to introduce ARQ at a layer higher than the ATM layer.

7.4.2 Cell Sequence Preservation

To simplify the implementation of network services and applications over ATM networks, the standard specifies that ATM networks must deliver cells in the correct sequence from source to destination. In other words, if you write an application program that sends data from a source workstation to a destination workstation over an ATM network, the communication "pipe" is expected to be FIFO so that the order in which cells are sent at the source is also the order in which they are received at the destination. This makes it easier for the users to develop network applications and services over the ATM network.

7.4.3 Virtual-Circuit Hop-by-Hop Routing

An implication of the sequence-preservation requirement is that the ATM network must be a virtual-circuit (or connection-oriented) network. A connection must be set up *a priori* before data can be sent over the network. If the network were to be a datagram (connectionless) network, cells might then travel over different physical routes to their destinations, and the different delays of these routes might introduce out-of-sequence arrivals at the receiver.

There are two possible routing mechanisms in a virtual-circuit network: source routing and hop-by-hop routing. In source routing, the sequence of nodes to be traversed by a packet is encoded in the packet header. In a cell network, the cells have a fixed-length header, and therefore there is an upper bound on the number of bytes that can be used to encode the sequence of nodes. In other words, there is a limit on the maximum number hops if source routing were to be adopted.

The ATM standard adopted hop-by-hop routing. The nodes traversed are not explicitly encoded in the header. Rather, there is a fixed-length virtual-circuit identifier in the header that is used to identify cells belonging to a particular virtual circuit. Upon the arrival of a cell to a node, its virtual-circuit identifier is used to retrieve from a lookup table the output link of the node to which the cell must be forwarded. The cell is then switched and transported on this output link, that connects to the node next in sequence in the path. In this way, the sequence of nodes to be traveled is derived on a hop-by-hop basis, and there is no limit on the maximum number of nodes that can be in the route. The virtual-circuit identifiers of different virtual circuits on the same physical link must be distinct; otherwise, there will be confusion during the table lookup.

7.4.4 Virtual Channels and Virtual Paths

There are two types of virtual circuits in the ATM virtual-circuit hierarchy: the virtual channel (VC) and the virtual path (VP). The VC is at a lower level and the VP is at the higher level. To draw an analogy with the telephone network multiplexing hierarchy, the VC is analogous to the 64 kbps DS-0 voice channel and the VP is analogous to the 1.5 Mbps DS-1 channel that consists of 24 voice channels. A difference is that the bandwidths of VCs or VPs may be varying rather than fixed.

The reader is referred to Fig. 7.6(a) in the following explanation of VCs and VPs. A VC is an end-to-end entity associated with a source and a destination. Thus, it may pass through several nodes within the ATM network. A VC has two termination points: the head end and the tail end. The source injects cells into the ATM network through the head end and the destination removes cells from the ATM network through the tail end. A VC is specified in terms of a sequence of VPs rather than physical links from source to destination. The cells of the VC pass through these VPs while traveling from source to destination. To a VC, the VPs are like logical links. A VP, in turn, consists of a succession of physical links. Each physical link connects two physically adjacent switch nodes.

As in a VC, a VP also has two termination points. The head end is located at either a source or an output of a switch node within the ATM network. The tail end is located at either a destination or an input to a switch node. Several VCs can be multiplexed onto a VP at the VP's head end, and the VCs contained in a VP are demultiplexed from the VP at the VP's tail end. Thus, as the cells of a VC travel through the sequence of its associated VPs, they can be multiplexed and demultiplexed several times, depending on the number of VPs. The situation is analogous to that in a telephone network in which DS-0 channels are multiplexed onto and demultiplexed from the higher level DS-1 channels.

With reference to Fig. 7.6(a), we see that in a way it is correct to say that a VP consists of a number of VCs or that a VC consists of number of VPs. This is a point that often causes confusion. The former statement is correct if we interpret it to mean that several VCs may be multiplexed onto a VP. A VC, however, does not necessarily terminate at the tail end of a VP. In this case, the VC will be demultiplexed from the VP and remultiplexed onto yet another VP. Thus, we see that the latter statement is correct in that a VC may traverse several VPs from source to destination.

The network in Fig. 7.6(a) can be broken down into two levels for the understanding of its routing mechanism. At the higher level, which is shown in Fig. 7.6(b), we have a VP network with all the physical nodes and links as its underlying building blocks. At the lower level, which is shown in Fig. 7.6(c), we have a VC network with a subset of physical nodes and VPs as its building blocks.

Between the two terminal points of a VP, the cells of the VP may pass through several physical nodes. At the intermediate (nonterminating) nodes, all arriving cells of the same VP are routed to the same physical output link. In other words, although the cells of a VP may be from different VCs within the VP, no attempt is made to distinguish these cells as far as routing at an intermediate node is concerned. Cells of different VPs may, however, be routed to different physical outputs of an

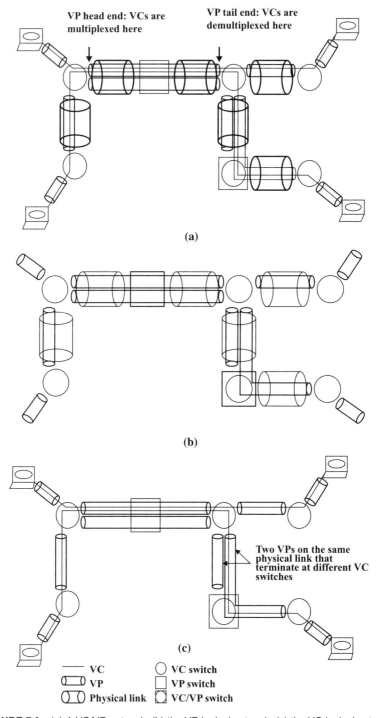

FIGURE 7.6 (a) A VC/VP network; (b) the VP logical network; (c) the VC logical network.

intermediate node when the VPs do not traverse the same sequence of physical links. These intermediate nodes are sometimes called VP switches, since they distinguish cells based on their VPs rather than VCs in the routing process.

The nodes to which the termination points of a VP are attached are called VC switches, and they form the nodes of the lower level VC network shown in Fig. 7.6(c). The network in the figure is a logical network rather than a physical network. The VP switches are transparent in the logical network, since VCs do not terminate there. To a VC, a VP is like a link between two nodes. At the input of a VC switch, the cells from a terminating VP are distinguished based on their VCs, and cells from different VCs may be routed to different output VPs at the VC switch. Note that we say *output VPs* rather than *output physical links* because each output physical link may contain several VPs that, to the VCs, are different logical entities. Different output VPs at the VC switch, even if they were on the same physical output links, could lead to different VC switches at their tail ends, as shown by the example in Fig. 7.6(c).

In general, a switch can be a VC, VP, or VC/VP switch. As explained above, a VP switch is *not* a terminating node for VPs or VCs, and a VC switch is a terminating node for VPs but not VCs. (*Note*: VCs terminate at their destinations outside the network.) A VC/VP switch is simply one in which some VPs terminate and some VPs do not (see Fig. 7.6(a) for an example). In this switch, the cells of nonterminating VPs are not distinguished based on their VCs, and the cells of terminating VPs must be distinguished based on their VCs for further routing.

We can draw an analogy between VP/VC networking and transportation by bus. Cells of a VC are like people who are from a common origin wanting to go to a common destination. If there is no direct bus from the origin to the destination, then these people must take several bus routes in succession to reach the destination. The different bus routes are like VPs. Each bus route goes through several streets or roads, which are analogous to physical links. At the beginning of each bus route, people (possibly from different origins and to different destinations) board the bus, and at the end of the bus route, they either reach their destinations or continue on to another bus route. This is analogous to multiplexing VCs onto and demultiplexing VCs from a VP. A VP switch is like a nonterminal road junction where nobody is allowed to board or leave the bus. A VC switch, on the other hand, is like a bus station where many bus routes end and begin.

7.4.5 Routing Using VCI and VPI

With hop-by-hop routing, virtual-circuit identifiers must be encoded in the cell headers so that a switch may use them to decide how to route cells. The identifiers are used as an index to retrieve routing information from a routing table. Each input of a switch typically has a routing table. The routing information associated with a virtual circuit is added to the table during the setting up of the virtual circuit, and it is removed from the table when the virtual circuit is torn down.

With the two-level hierarchy in which there are two types of virtual circuits, VCs and VPs, each cell has two identifiers in its header: the virtual-channel identifier (VCI) and the virtual-path identifier (VPI). Figure 7.7 illustrates how routing is performed

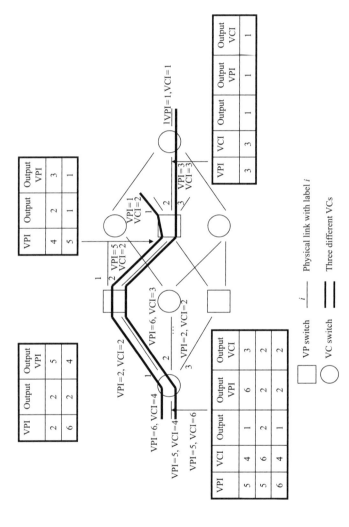

FIGURE 7.7 VP/VC routing; illustration of the routes of three VCs and their relationships to the routing tables.

at the VC and VP switches and the corresponding routing tables. Each input link (physical link) of a VP switch has a routing table that is used to determine how to route cells arriving on the input. The VPI of a cell is used as the index to look up an entry in the table, and the entry contains two elements (*output link* and *output VPI*), which indicates the output link to which the cell must be switched and the new VPI value it should adopt at the output link.

The cells of a VP may change its VPI value as they travel from a physical link to another physical link within the VP. The reasons for allowing the VPI to change from link to link, even though the VP has not terminated, are to simplify network operation and to reduce blocking probability due to insufficient VPIs. These are also the reasons for allowing the VCI of a VC to change, the details of which will be elaborated shortly. A point also to be noted is that the cells with the same VPI value but on different physical links may actually belong to different VPs.

While a VP switch uses only the VPI for routing, a VC switch uses both VPI and VCI. As shown in Fig. 7.7, the VPI–VCI combination is the index for reference to an entry in the routing table. This is because the cells of a VC can only be identified by the combination of VPI and VCI: cells with different VPIs but the same VCIs, or with different VCIs but the same VPIs, are cells of different VCs. An entry of the routing tables at a VC switch consists of three elements: *output link*, *output VPI*, and *output VCI*).

The VCI of a VC is allowed to change as its cells travel from source to destination. While the VPI of the cells can change from physical link to physical link, the VCI can change only when the cells pass through a VC switch; the VCI does not change when cells pass through a VP switch. The changing of VPI and VCI does not cause confusion as long as the VPI–VCI combinations of different VCs on a link are different. An analogy is to compare the situation to that of a person wearing shirts of different colors or adopting different names on different occasions: as long as the color or name is unique on each occasion (i.e., no other person wears the same color or uses the same name), we can use it to identify the person.

There are two reasons why the VCI of a VC and the VPI of a VP are allowed to change: simpler operation and lower blocking probability during call setup. Consider the alternative of requiring the VPI and VCI to remain the same. Let us focus on the setting up of a new VC over the lower level VC network. Suppose that a sequence of VPs for the VC has been identified. Each VP in the sequence may have some VCI values already used by existing VCs within the VP. To set up the new VC, the sequence of VC switches at the termination points of the VPs must coordinate with each other to come up with a commonly available VCI not already used in any of the VPs. For instance, if VCI = 11 is already used in the first VP, and VPI = 10 is already used in the second VP, then neither VCI = 10 nor VPI = 11 can be assigned to the new VC.

The coordination of the VC switches to come up with a unique VCI assignment throughout the path required a complicated distributed control algorithm and a lot of control information to be sent around among the VC switches (see Problem 7.7). The alternative of allowing the VCI to change from VP to VP is much more desirable. Each VC switch simply keeps track of the unused VCIs on each of its output virtual paths. During call setup, the source chooses one unused VCI to be used for the new VC and informs the next VC switch in sequence that this VCI will be used. The next VC switch will then use this VCI to identify the arriving cells of the VC in the future. The next VC switch in turn maps this input VCI to yet a different VCI that is available on the output VP next in the sequence. In this way, only adjacent nodes coordinate with each other and only one short message needs to be passed between two adjacent nodes.

Insisting on using the same VCI from end to end also increases the likelihood of blocking probability (see Problem 7.7). It could be that each VP has at least one unused VCI but that there is no common VCI that is not used in all the VPs. Allowing VCI to change from VP to VP ensures that the call will not be blocked in this way.

We have assumed a node is either a VC or a VP switch in the above discussion. Recall that we could also have a node that performs both VC and VP routing. In other words, at this node, some VPs are nonterminating and routed directly, while other VPs are terminating and their VCs need to be taken apart and routed individually. Logically, the node (see Fig. 7.8) appears to consist of three switches, two VP switches and one VC switch. The first VP switch routes all the terminating VPs to the VC switch; the nonterminating VPs are forwarded directly to the inputs of the second VP switch. The VC switch demultiplexes the VCs of the terminating VPs and remultiplexed them onto new VPs. The second VP switch multiplexes the new and nonterminating VPs onto output links.

Figure 7.8 depicts the logical functions performed at a VP/VC switch. In actual implementation, any VC switch is also capable of performing VP routing and can

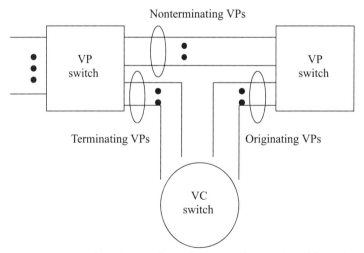

In actual implementation, using appropriate routing tables, only one physical switch is needed to realize the VP/VC switching function

FIGURE 7.8 The logical structure of a VP–VC switch.

therefore be used as a VP/VC switch. That is, one physical switch can simultaneously perform both VP and VC routing. In the routing table, all the VCs of the nonterminating VPs will simply be mapped to the same VPs without their VCIs being changed, as illustrated in the following example in which input VCI = 1 is nonterminating.

Input VPI	Input VCI	Output	Output VPI	Output VCI
1	0	2	3	0
1	1	2	3	1
1	2	2	3	2
1	3	2	3	3
⋮	⋮	⋮	⋮	⋮

A more efficient (in terms of routing table size) method is to divide the table into two subtables. The first subtable operates at the VP level. The VPI of the cells of a nonterminating VP will simply be mapped to an output VPI without their VCIs being changed. Thus, only one entry is needed for each nonterminating VP. For a terminating VP, this entry consists of a pointer to a location in the second subtable rather than the output VPI value. At this location of the second subtable, there are entries that indicate how the VPI and the VCIs of the cells of the terminating VP should be changed. In this way, the routing table memory requirements of the VP/VC switch can be reduced.

7.4.6 Motivations for VP/VC Two-Tier Hierarchy

Having discussed the operation of VP/VC networks, let us now examine the motivations for this two-tier hierarchy. First, it simplifies network management and operation in several ways:

1. Call setup is simplified. Setting up a call involves finding a route from source to destination, allocating bandwidth over the path, and updating the routing tables of the switches in the route. A call is usually established by setting up a VC to carry its traffic. A VC, in turn, is set up over a sequence of VPs. Typically, these VPs are existing VPs that have previously been set up rather than new VPs: new VPs are needed only if the existing VPs are not suitable for the VC (e.g., when the VPs do not have sufficient bandwidth to accommodate the VC or when the VPs run out of VCIs), which should occur rarely in a well-operated network. Thus, VP switches are usually not involved in the setting up of a VC and no actions are required of them even though the VC may pass through them. Since only the VC switches are involved, calls can be established much quicker than when all the switches are involved.
2. A VP switch is less complex than a VC switch. A VP switch does not need to examine the VCIs of cells, and therefore, its routing tables can be much

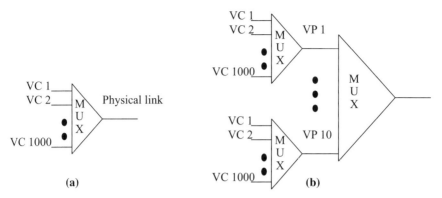

FIGURE 7.9 (a) Directly multiplexing a large number of VCs onto a physical link; (b) two-tier multiplexing hierarchy.

smaller. As the call is ongoing, a VP switch does not need to monitor the status and guarantee the performance experienced by individual VCs. It only monitors and guarantees the performance of VPs.

3. Multiplexing is simplified. Let us focus on the local problem of multiplexing traffic from VCs, with and without the two-tier hierarchy. Figure 7.9(a) shows a multiplexer that statistically multiplexes the traffic from, say, 1000 VCs onto a physical link directly. The traffic from the VCs may not arrive at the multiplexer in a predictable fashion. As shall be detailed in the next chapter, a scheduling policy is needed to determine the order in which the input traffic from different VCs will be transmitted on the output physical link. Among other requirements, the scheduling policy must guarantee the performance (say, in terms of cell loss or delay) experienced by individual VCs. This becomes increasingly difficult as the number of VCs being multiplexed increases.

Figure 7.9(b) shows the alternative of using the two-tier hierarchy. The VCs are divided into groups (say, 100 VCs to each group) and each group is first multiplexed onto a VP. The VPs are then multiplexed onto the physical link. The design of a complex multiplexer has been decomposed to that of a number of less complex multiplexers. Now, the simplification does not come without cost. In terms of bandwidth usage, it is better to multiplex (see Problem 7.8) a large number of VCs directly onto the physical link. This allows all VCs to share a larger pool of transmission bandwidth. For the two-tier system, if each VP is allocated a fraction of the capacity of the physical link, there could be situations in which some VPs are underutilized while other VPs are congested. Nonetheless, to the extent that there are already many VCs being multiplexed onto each VP, this phenomenon is less likely to occur (see Problem 7.8).

From the overall networking viewpoint, it is generally worthwhile to set up a VP between two physical locations when there is sufficient traffic between them. This way, the bandwidth usage penalty of the two-tier hierarchy will be small and the management and control of traffic between the two end points can be simpler, since there is no need to take apart the traffic of individual VCs at the intermediate VP switch nodes. It is simpler to deal with fewer channels of high bandwidths than many channels of small bandwidths.

VPs can also be used to segregate the traffic of different services in an integrated service network. Some services, say real-time video or voice, have very stringent delay requirements but may tolerate some data loss. The traffic is more or less continuous. Other services, such as computer data transport, may tolerate some delay but not loss, and the traffic can be bursty. Different services and applications can have very diverse requirements and traffic characteristics, and it is generally difficult to intermix their traffic and to guarantee their respective performance requirements at the same time.

One simple way to deal with the problem is to set up different logical networks on the same physical network (see Fig. 7.10). The "links" in the logical networks are made up of VPs rather than physical links. A VP belongs exclusively to one logical network only and all the traffic on it belongs to one type (or similar type) of service. On each logical network, one then has to worry only about multiplexing services of the same type, and the objectives and goals can be more clearly defined. The bandwidth on a physical link is partitioned and allocated to the VPs on it, and there is no interference among traffic of different VPs.

The bandwidth allocated to each VP is quasi-static in the sense that it does not change rapidly over time. The bandwidth changes are slow, say, on the timescale of hours. For instance, one VP could be carrying voice traffic, and during business hours, more bandwidth can be assigned to it. During the evenings, perhaps the VP does not need as much bandwidth, and some of the excess bandwidth can be assigned to VPs carrying entertainment video programs. In this way, efficient bandwidth usage and simple network management and control can be achieved simultaneously.

Segregation of virtual private networks (VPNs) is another way in which VP networking can be useful. Several virtual private networks, each belonging to a customer of the network provider, can be set up over the same physical network. A customer could be, for example, a private company. The bandwidth on each VP is dedicated and not shared to prevent the interference of traffic of different VPNs. Different customers can use their own virtual networks in whichever way they want.

7.5 ATM LAYER, ADAPTATION LAYER, AND SERVICE CLASS

We have discussed cell networking with respect to the ATM standard. From the end user's viewpoint, in many situations it would be nice if the end user does not have to worry about the intricacies of cell networking when using the network. Also, enhanced capabilities can be added to the basic service provided by a cell network to simplify and facilitate the development of network services and applications. The above two functions are provided by the adaptation layer. The adaptation layer exists only at the two ends of a VC and it is *not* an entity within the ATM network, although its design is closely tied to the principles of the ATM network.

296 BASIC CONCEPTS OF BROADBAND COMMUNICATION NETWORKS

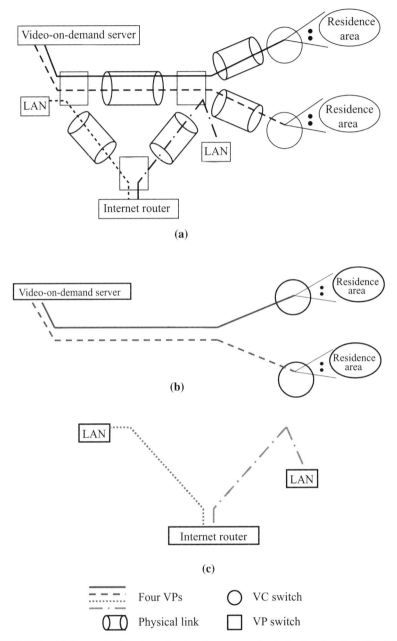

FIGURE 7.10 (a) Overlaying two logical networks on one physical network using VPs; (b) first logical network in (a), video-on-demand network; (c) second logical network in (a), computer data network.

One task performed by the *adaptation* layer is to convert the data to be sent at the source into cells before transmission and to reassemble the cells reaching the receiver back into the original form. In addition, certain services may have particular requirements. For example, for services that require the end-to-end delay to be constant, the adaptation layer may handle cell delay variation introduced by the network via the use of an elastic buffer (see Section 7.2). For services that deliver data to the adaptation layer in stream mode (see Section 7.2) but require the delay to be not too large, the adaptation layer may limit the maximum adaptation waiting delay at the source and send out partially filled cells if necessary. Given that there is no data link layer in the ATM network, end-to-end error protection will be required for services that demand high integrity in data delivery if the network links are unreliable (e.g., wireless data links). This protection can be applied at the adaptation layer through forward error control and ARQ. In addition to the above functions, the adaptation layer at the receiver may also be responsible for the recovery of source clock. In short, the purpose of the adaptation layer is to make the underlying network easier to use from the user's perspective. What functions are implemented at the adaptation layer depends on the services to be supported.

In the ATM standard, the adaptation layer is divided into two sublayers. Each of the sublayers may add overhead information to the data to be transmitted in order to perform its job. The lower sublayer is the segmentation and reassembly (SAR) sublayer. It is responsible for segmenting source data into cells and reassembling received data from cells at the receiver. The higher sublayer is the convergence sublayer (CS). It is responsible for managing the flow of data to and from the SAR sublayer. Depending on the services, the CS performs functions like synchronizing the receiver and sender's clock, handling delay variations, and so on, as described in the previous paragraph.

The ATM standard divides the services to be supported by the ATM network into four classes, and different ATM adaptation layers (AALs) are defined accordingly to facilitate their support. As shown in the following table, each class is associated with certain characteristics. The appendix goes into the details of AAL headers adopted by the standard committee. Here, we confine ourselves to a brief discussion of the basic principles.

Class A	Class B	Class C	Class D
Constant bit-rate traffic	Variable bit-rate traffic		
Connection-oriented			Connectionless
Fixed delay required		Fixed delay not required	

With reference to the above table, there are two important aspects to a service. The first is the characteristics of the traffic generated by the source and the second is the required support and performance to be guaranteed by the network. For the first consideration, the traffic can be generated by the source either in a continuous or a bursty fashion. The former presents a constant-rate data stream to the network and the latter a variable-rate data stream. The sources are commonly referred to as constant bit-rate (CBR) and variable bit-rate (VBR) services, respectively. An example of a continuous bit-rate traffic source is the telephone, in which traffic is generated continuously at a rate of 64 kbps. An example of a variable bit-rate traffic source is the computer, in which traffic is generated only when the user has something to send.

For the second consideration, there are two aspects. First, a user can demand either connection-oriented services or connectionless services. The ATM network itself is connection-oriented and therefore offering connection-oriented services is relatively straightforward. However, it is sometimes desirable to overlay a connectionless network on top of the ATM network. One can define a set of VPs within the ATM network for the exclusive use of the traffic generated connectionless services, as illustrated by the Internet example in Fig. 7.10(c). The Internet Protocol (IP) is connectionless and the Internet router in the figure is an example of a connectionless server. Basically, this server performs connectionless routing. In the later section, we shall discuss the support of IP services over ATM in detail. The two ends of a VP of the connectionless network are connected to connectionless servers, end users of the connectionless communication service, or LANs connected to the end users. The interface of the ATM network with the overlaid connectionless network is through the adaptation layer between the VP termination points and the connectionless servers, users, or LANs. In other words, the offering of connectionless services is made possible by the adaptation layer and the connectionless servers outside the ATM network. Consider the delivery of an IP packet. At the source, the adaptation layer divides the packet into many cells and routes them through the ATM network to a connectionless server. The adaptation layer at the connection server reassembles the IP packet from the cells. The server examines the IP address in the packet and decides how to further route the packet. The packet is then forwarded to the VP associated with the route to be taken. Of course, before the data enter the VP, they must go through the segmentation process at the adaptation layer.

A service may also have certain performance requirements. Some services, such as telephone voice, may require a fixed delay from end to end. As already discussed in Section 7.2, the ATM network may introduce time-varying delay. The adaptation layer is responsible for handling this delay variations by introducing additional compensating delay at the receiver. In addition, for services that have very stringent requirements on the integrity of data delivered, the adaptation layer is responsible for providing error protection at the receiving end.

To deal with the diverse traffic characteristics and the requirements of services, several types adaptation layers can be defined. Exactly how many different types are required is a controversial issue and some people argue that one universal adaptation will suffice. Currently, there are five different AALs in the ATM standard.

AAL1 is for the adaptation of constant bit-rate, connection-oriented services that require fixed end-to-end delay. That is, the end-to-end delay must be constant and the introduction of delay compensation at the receiver adaptation layer is necessary. Examples of these services are voice, constant bit-rate video, and circuit emulation. Circuit emulation basically emulates the circuit-switched environment in which a user can lease a circuit from the telephone company. For these services, within the ATM network, an amount of bandwidth equal to the bit rate of the service must be dedicated exclusively for the associated connection.

AAL2 is for variable bit-rate, connection-oriented services with fixed delay requirement. An example is the delivery of video data that have been compressed with a variable-rate compression algorithm: compression reduces the data amount by removing redundant information in the data, and depending on the scene contents, the compression ratio may vary over time, giving rise to variable bit-rate compressed output. Bandwidth allocation is especially difficult (relative to other service classes) within the ATM network for this class of services. If no exclusive bandwidth is allocated, then it is difficult to bound end-to-end delay tightly. If bandwidth equivalent to the peak rate is allocated, then bandwidth usage will be very inefficient and it somewhat defeats the original purpose of data compression. If a bandwidth below the peak rate is allocated, then the delay may be large when the source generates data at close to the peak rate because the network is not ready to deal with it. Many issues remain to be investigated further, and perhaps for this reason, AAL2 is the least well-standardized among all the AALs.

AAL3 and AAL4 are for connection-oriented and connectionless variable bit-rate services, respectively. There is no strict delay requirement and the end-to-end delay can be varying. The standard committee has found the specifications of AAL3 and AAL4 to be so similar that there is no strong reason to distinguish between the connection-oriented and connectionless services at the adaptation layer. Consequently, AAL3 and AAL4 have been combined into one adaptation layer type, AAL3/4. AAL3/4 does not perform all functions required by connectionless services. For instance, functions like routing and network addressing are handled at a higher layer (through a connectionless server, for example). Thus, for a source with multiple ongoing sessions simultaneously, the data of sessions destined for different destinations must be separated by the higher layer before they enter the adaptation layer. The data of separate sessions are handled by separate adaptation-layer processes, and each adaptation-layer process is associated with a VC or VP. For multiple sessions with a common destination, AAL3/4 allows their data to be multiplexed at the adaptation layer so that they can be handled by one adaptation-layer process and they will use only one VC or VP to reach the destination.

AAL3/4 is primarily for datadelivery services. After the standardization of AAL3/4, it was found that it is very inefficient (the header overhead in the AAL3/4 SAR sublayer is rather high) for such services and does not provide sufficient data protection. Subsequently, AAL5 was developed to replace it. The reader can find more details in the appendix.

For services with no stringent delay requirements, no exclusive bandwidth needs to be dedicated in the ATM network. The traffic of this service class, for instance, may use the leftover bandwidth on a link. The kind of network service provided to the users that does not have hard bandwidth guarantee is called the available bit-rate (ABR) service. The ABR service matches well with the AAL5 protocol.

The above has briefly discussed the different AALs in the ATM standard. A more detailed discussion requires us to look into the AAL cell headers and explains the motivations for each field in the headers and how they are used. This discussion can be found in the appendix.

7.6 TRANSMISSION INTERFACE

ATM switches and equipment are connected by physical links. Bits of the cells are coded and transmitted on these physical links according to some specifications, which are sometimes referred to as the transmission interfaces, and they belong to the physical layer of the network protocol stack. Examples of the common transmission interfaces are the 100 Mbps 4B/5B[2] TAXI interface, the 45 Mbps DS-3 interface, the SONET/OC-3c, and so on. The specifications include the way bits are encoded, overhead bytes used for synchronizing the transmitter and receiver, error protection, and so on. Each of these interfaces can be used to transmit data other than ATM cells, and additional specifications on how cells are packed and transported are generally needed when used for ATM purposes.

It is possible to have an ATM switch that has input/output ports with different transmission interfaces. However, the transmitter end and the receiver end of a physical link must be of the same interface for compatibility. Some of the standards (e.g., TAXI) are purely for transmission purposes only (i.e., they transmit bits or bytes with no regard to the applications) and they cause no confusion. Other standards, such as DS-3 and SONET/OC-3, are also commonly used to transport STM data, and they were originally designed by the telephone community to have a fixed-size frame for such a purpose. The use of these interfaces on ATM equipment may be wrongly perceived as combining the ATM and STM multiplexing techniques. It turns out that DS-3 or OC-3 frames can also be used to transport cells with the understanding that the data in them are not TDM data. If the two pieces of ATM equipment connected to the two ends of a DS-3 or OC-3 line conform to the same cell-packing standard, cells can be extracted from the frames at the receiving end without any problem. If the equipment is an ATM switch, the extracted cells from the physical/transmission layer are then forwarded to inputs of the switch for further routing.

7.7 APPROACHES TOWARD IP OVER ATM

Although ATM is suitable for wide-area networks (WANs) as well as for local-area networks (LANs), the deployment of an end-to-end ATM infrastructure may be expensive compared to the other competing technologies such as Fast Ethernet and Gigabit Ethernet in LANs. Also, there is a lack of applications deployed directly on an ATM network infrastructure. Typically application developers concentrated

---

[2] A standard block code that transforms 4 bits into 5 bits for error protection and synchronization purposes at the physical layer.

on IP stack for end-to-end data communications, and ATM is therefore mainly used as a lower layer technology for WANs, which in most cases carries IP-related traffic.

IP is a connectionless network layer protocol capable of supporting point-to-point and multicast communication. Each IP datagram is routed from its source to its destination in a hop-by-hop manner via a number of IP routers. The delivery service is called *best-effort* as no quality-of-service (QoS) guarantee is provided. However, with a surge of business activities on the Internet, users are demanding more bandwidth and QoS support for multimedia applications. These applications generally require the underlying network to provide predictable and bounded packet delay, loss rate, and minimum bandwidth. The best-effort service of Internet can no longer satisfy these requirements. Owing to its QoS support as well as the ability to provide high bit rate, ATM seems to be the most promising candidate as an underlying link-layer mechanism for carrying IP packets.

There are many challenges and problems in transmitting IP traffic over ATM networks. Recall that ATM is connection-oriented. It is fast, but could be expensive and ineffective for short-lived applications. On the other hand, IP traffic is connectionless, and there are no overhead and delay associated with connection set-up in a connection-oriented network. However, it is in general more difficult to provide QoS guarantee on a differentiated basis (i.e., different QoS for different applications) in a connectionless network. ATM technology provides an evolutionary path for IP for QoS support.

7.7.1 Classical IP over ATM

Recently, various methodologies have been proposed for supporting IP traffic on top of ATM backbone infrastructure. Classical IP over ATM (RFC1577) is one of the early attempts. The term "classical" here grew out of a need to make no changes in the internetworking paradigm when deploying ATM. The prime concern of this model is not on how to take full advantage of ATM for efficient data forwarding. In contrast, it largely negates the potential benefits of ATM in exchange for preserving the connectionless nature of IP for facilitating easy migration to ATM. For example, in classical IP over ATM, an ATM network is separated into logical IP subnets (LISs) interconnected by IP routers. In this environment, direct ATM connections between IP hosts in separate LISs are prohibited even if the underlying ATM topology is capable of supporting them. An ATM virtual connection (VC) originating within a given LIS can only extend as far as a router at the LIS boundary where the contents of the received ATM cells will be reassembled into IP packets, each of which is then subjected to an IP forwarding decision. These IP packets will be resegmented into ATM cells at the next intra-LIS router along the journey and sent along a default VC within this LIS (see Fig. 7.11).

In actuality, the reassembly and resegmentation of IP packets at each router along the path severely restrict the potential benefits of ATM. Also, the necessity for an address resolution mechanism for this approach to map a next hop's IP address to its

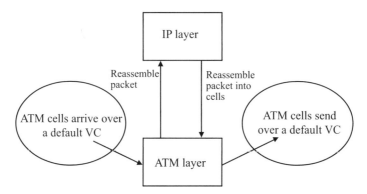

FIGURE 7.11 Overhead of IP layer forwarding with classical models.

ATM address greatly limits the overall size of the ATM network. In addition, because the end-to-end path is forced to traverse certain LIS boundary routers, a convoluted path may result, which depends on the physical positioning of those boundary routers. Furthermore, to accommodate certain IP protocols, each LIS must provide intra-LIS broadcast, which is typically implemented using a point-to-point VC from every node in the LIS to a multicast server and a single point-to-multipoint VC from the multicast server to every node in the LIS it serves. This implementation imposes a limit on the number of nodes in each LIS, which depends on the number of VCs the multicast server can support.

7.7.2 Next Hop Resolution Protocol

The Next Hop Resolution Protocol (NHRP) was developed to utilize the potential benefits of ATM, which are lost with the classical IP over ATM. NHRP is an inter-LIS address resolution mechanism that maps a destination's IP address to the destination's ATM address if the destination resides within the same ATM cloud as the sources. In cases where the destination resides outside the ATM cloud containing the source, NHRP returns the ATM address of the source ATM cloud's egress router that is closest to the destination. Once the source receives the NHRP response, it can then set up a direct cut-through VC to the destination or to the closest egress router to the destination in cases where the destination is outside the ATM cloud, using standard ATM signaling/routing protocols (Fig. 7.12).

Although NHRP overcomes some of the weaknesses of the classical IP over ATM model, it has its own restrictions. First, setting up a separate VC for every single data flow is unlikely to yield optimal results, especially in large ATM clouds within the Internet. This is because in such an environment the number of IP flows traversing the cloud may be very large. Using a separate VC for each flow in this case may result in an unmanageable number of VCs at switches within the cloud. Second, providing such a cut-through VC may be unnecessary and even undesirable for certain short-lived flows. In these cases, it would be hard to justify the associated

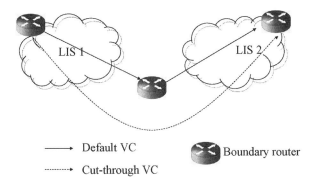

FIGURE 7.12 Classical service path and cut-through service path.

overhead of the end-to-end connection and its setup, especially for flows that make no assumptions about QoS anyway. Also, NHRP cannot directly support multicast, although certain elements of NHRP may be used to facilitate shortcuts within certain multicast scenarios. Moreover, an NHRP solution necessitates routing/signaling functionality in both the ATM and IP layers, which further increases the overall complexity.

7.7.3 IP Switch and Cell Switch Router

There have been a number of schemes proposed based on similar hybrid ATM switch/IP router designs, which allow coexistence of hop-by-hop IP forwarding with direct VC cut-through mode of service in order to provide each data flow with the most suitable mode of service while maintaining desirable network conditions such as a manageable number of VCs at each switch. Amongst these schemes, IP switch and cell switch router (CSR) are the most well known and established, and thus they are the focus of our discussion in this subsection.

The IP switch and CSR hybrid switch/routers contain all the usual functionality of conventional IP routers and thus can support connectionless IP forwarding services. However, their usage depends on the topological configurations. The IP switch does not support the ATM user–network interface (UNI) standards, so it is incapable of interfacing with conventional ATM switch in existing ATM networks. For example, it can work for the topology shown in Fig. 7.13 but not for that in Fig. 7.14. The CSR, on the other hand, is UNI compatible and therefore capable of interconnecting ATM subnets in a similar way to the LIS border routers in the classical IP over ATM model. The difference is that unlike the classical model, the CSR is also capable of providing direct VC cut-through between the adjacent subnets for selected data flows. Consequently, the valid configurations for it include that of Figs. 7.13 and 7.14.

In IP switch and CSR, a dedicated cut-through VC for a data flow can be set up by associating an incoming VC with an outgoing VC for switching cells of that flow directly in hardware without IP forwarding (see Fig. 7.15). This cut-through

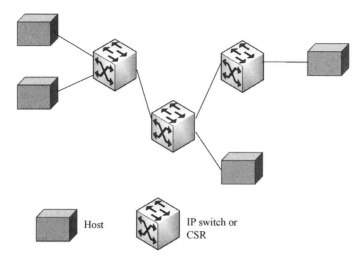

FIGURE 7.13 Topology for IP switch or CSR.

service differs from that offered by NHRP and traditional ATM signaling in that the switching table associations are no longer made on an end-to-end basis. Instead, each hybrid switch router makes a decision independently on whether to implement local cut-through. The rules for making such decisions can be configured by network management and will typically result in cut-through for flows of any suitable higher layer protocol. For example, TCP FTP flows are suitable since they are of sufficient duration to justify the overhead associated with cut-through setup. User Datagram

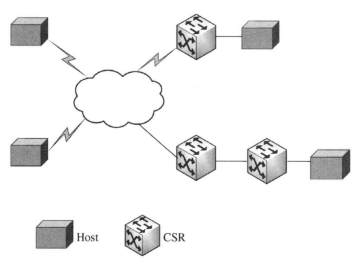

FIGURE 7.14 Topology for CSR.

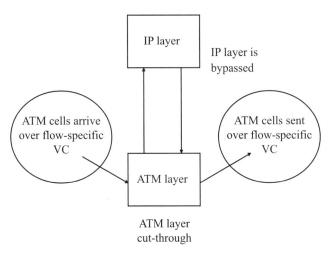

FIGURE 7.15 Cut-through service of a hybrid switch/router.

Protocol (UDP) flows carrying Network Time Protocol (NTP) traffic, on the other hand, are not suitable since each UDP flow typically consists of a single packet. The higher layer protocol can be determined by inspecting the packet headers during IP processing of the first packet of a flow. Once the cut-through decision has been made and the switching table associations installed, all further packets of the flow will receive local cut-through service at each hybrid switch/router along the path. The end-to-end service received by the flow is essentially the same as that obtained using the end-to-end signaling approach of NHRP.

Using a concatenation of local cut-throughs is advantageous in a number of aspects. One of the advantages is that with the hybrid switch/routers, the end-to-end route is determined entirely, using the underlying IP routing protocols. This means that the ATM routing/signaling and address resolution protocols are no longer required, which leads to a reduction in complexity. Note that CSRs must still support ATM UNI signaling in order to connect to adjacent CSRs that are reachable across ATM subnets. Also, hybrid switch/routers are well suited for use with Resource Reservation Protocol (RSVP), which is the protocol of choice for setting up QoS over IP networks and will be discussed in the next chapter. RSVP control messages would travel over the default VCs and would receive full IP processing at each hybrid switch/router, where they could initiate setup of flow-specific VC cut-throughs according to the QoS information contained within the RSVP messages. In addition, any VC associations set up by the hybrid switch/routers are soft-state, which means that they need to be continually refreshed in order to avoid timeout. The use of soft-state rather than hard-state helps to maintain much of the connectionless nature of IP. Another key advantage of hybrid switch/routers over NHRP is that they offer full support for cut-through multicast trees by accommodating branch points at the ATM layer. In spite of its point-to-point nature, NHRP could be used to emulate multicast through a number of point-to-point VCs, although this would be bandwidth-inefficient since many of the multicast VCs may share common links that consequently carry the same data more than once.

7.7.4 ARIS and Tag Switching

In the hybrid switch/router approaches, setup of VCs is either topology-driven for default VCs in ATM subnets or traffic-driven for flow-specific VCs. The traffic-driven approach requires cooperation between edge devices and hybrid switch/routers in order to make a decision on whether to establish a "cut-through" path across the hybrid switch/router. The topology-driven approach, on the other hand, requires a protocol to distribute the routing information, as well as a way to label the cells with short headers used by the ATM switches for forwarding. The fundamental differences in the two approaches are that the amount of control traffic in a traffic-driven scheme is equal to the number of individual flows (which may grow very high), while the amount of control traffic in a topology-driven approach is constant, based on the number of IP destinations. IBM's Aggregated Route-Based IP Switching (ARIS) and Cisco's Tag Switching architecture are approaches to IP over ATM in which VC association is completely topology-driven. Both ARIS and Tag Switching use VC cut-through for all traffic, including best-effort. They can do this without causing "VC explosion" since they are able to offer a choice of granularities according to the network environment. In this way, the cut-through VCs of both ARIS and Tag Switching can have a coarser granularity than the per-flow cut-through VCs of the hybrid switch/routers.

ARIS introduces the concept of "egress identifier" type to define granularity. For each value of "egress identifier," the ARIS protocol establishes a multipoint-to-point tree that originates at routing domain ingress integrated switch routers (ISRs) and terminates at the router domain egress ISR for that particular egress identifier. The multipoint-to-point tree will also be a multipoint-to-point VC if VC merging is used, as described later. Here, ISR is the name used to refer to an ARIS-compatible switch. The identification of the egress ISR for a particular egress identifier is obtained from the routing protocols. Thus, if the egress identifier represents IP destination prefixes, a separate multipoint-to-point tree is set up per IP destination prefix. This is illustrated in Fig. 7.16, which shows a multipoint-to-point tree that is set up from ingress ISRs A, B, C, and D to egress ISR E. When a packet arrives at one of these ingress ISRs, the forwarding table is consulted to determine which outgoing interface as well as virtual-path identifier/VC identifier (VPI/VCI) label to be used. Cells from the packet are then switched along the tree completely at the ATM layer until they reach egress ISR E, where they are again reassembled.

The ARIS protocol mechanisms for setting up the tree vary depending on whether VC merging is used. VC merging is when cells arriving on separate incoming links of an ISR are routed onto the same VC of an outgoing link of the ISR. With AAL5, which has no intra-VC multiplexing identifier, VC merging is only possible provided no interleaving of cells from different AAL5 frames occurs. Otherwise, it is not possible to reconstruct each AAL5 frame at the destination since it is difficult to determine which cells belong to which frames. To support VC merging, switches need to buffer

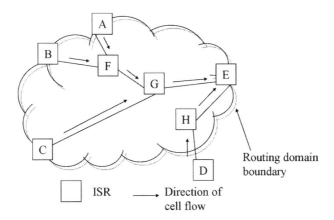

FIGURE 7.16 ARIS multipoint-to-point tree.

cells of each incoming AAL5 frame until the full frame has arrived, store the full frame for a period of time determined by the scheduler, and then transmit the frame so that it occupies a contiguous sequence of cells on the output link, as shown in Fig. 7.17. VC merging reduces the number of consumed VCs but introduces latency due to buffering of AAL5 frames. However, this increase in latency will still be less than that for the case of IP forwarding, while the switching speed will be close to that attainable without VC merging. If VC merging is not used by the ISRs, buffering of AAL5 frames is unnecessary since mapping each input VC to a separate output VC allows us to reconstruct each AAL5 frame at the destination while cells of frames

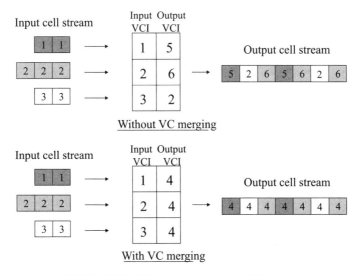

FIGURE 7.17 VC merging versus non-VC merging.

from different input VCs are interleaved on the output link. The interleaving process is illustrated in Fig. 7.17.

Tag Switching uses a tag information base (TIB) in each tag switch router (TSR) in order to provide the mapping between an incoming interface and tag (VPI/VCI value) of an incoming cell to the outgoing interface and the outgoing tag of the cell. The TIB entries can be generated either explicitly or using the Tag Distribution Protocol. In the latter case, a separate TIB entry is created for each route in the forwarding information base (FIB). The FIB is the information base in IP routers that is used to forward IP packets. In addition, the FIB is extended to include a tag entry for each route. That is, when a packet first arrives at the ingress TSR for the tag switching network, the FIB forwards the packet to the next hop while labeling the outgoing cells with the indicated tag value. In this case, each TSR will switch the cells directly at the ATM layer using the TIBs of each subsequent TSR traversed.

There are a lot of similarities between ARIS and Tag Switching mechanisms. First, both mechanisms can provide support for multicast and explicit routes. Second, both use default VCs between the hybrid switch/routers in order to implement hop-by-hop forwarding for their control protocols as well as for the IP routing protocols. Third, the ARIS and Tag Switching architectures include protocol mechanisms to avoid setup of switched path loops. Also, they are both able to correctly implement time-to-live (TTL) decrement for cut-throughs. That is, when a packet is reassembled at the egress router following VC cut-through, its TTL value will be the same as if it had undergone hop-by-hop IP forwarding instead. Furthermore, apart from the throughput improvement obtained by ARIS and Tag Switching through bypassing the IP layer, the use of underlying ATM technology also makes them very suitable for offering QoS support.

7.7.5 Multiprotocol Label Switching

The works of ARIS and Tag Switching are currently receiving much attention within the networking community, and this has resulted in the Internet Engineering Task Force (IETF) setting up the Multiprotocol Label Switching (MPLS) Working Group in order to standardize these schemes. The MPLS Working Group of the IETF is concerned with the label switching concept of ARIS and Tag Switching in general and not just with the special case of label switching in an ATM environment.

MPLS is designed to incorporate many elements of ARIS and Tag Switching, including label distribution protocols and mechanisms, topology-based assignment, support for VC merging, as well as multicast, QoS, and traffic engineering. It has a wide range of usage; some of the examples include IP-over-ATM, high-performance forwarding, QoS, and traffic differentiation.

In MPLS, labels are assigned to packets for transport across packet- or cell-based networks, which is just similar to ATM networks and frame relay networks. Like ARIS, the forwarding mechanism throughout the network is label swapping, in which units of data carry a short, fixed-length label that tells switching nodes along the packets path how to process and forward the data. The label may correspond to an ATM VPI/VCI.

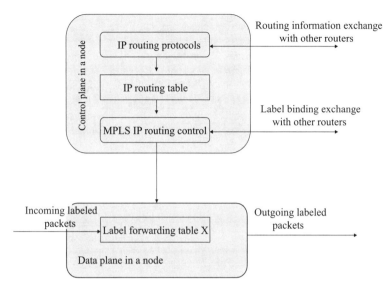

FIGURE 7.18 Basic architecture of an MPLS node.

The MPLS architecture is divided into two separate components: the control component and the forwarding component. The control component is responsible for intercepting the information from the IP routing protocols and creating and maintaining label-forwarding information, which is called bindings, among a group of interconnected label switches. In addition, it is responsible for signaling and topology discovery. The forwarding component is responsible for maintaining the label-forwarding database and using the label switch to perform the forwarding of data packets based on labels carried by packets. Figure 7.18 shows the basic architecture of an MPLS node that performs IP routing.

MPLS requires the IP routing protocols to operate together. Every MPLS node must rely on the IP routing protocols to exchange IP routing information with other MPLS nodes. Similar to traditional routers, the IP routing protocols populate the IP routing table, but it is not directly used. Instead, it is now used for determining the label binding exchange, where adjacent MPLS nodes exchange labels for individual subnets that are contained within the IP routing table.

The concept of MPLS architecture network is very similar to the differentiated services (discussed in the next chapter). Any router or switch that is capable of implementing label distribution procedures and forwarding packets based on labels is called label switch router (LSR), which is similar to that in ARSI. A group of LSR forms a domain called MPLS domain. When a packet enters a particular MPLS domain from the non-MPLS domain, a label is being added for the incoming packets, which is called label imposition, at the ingress point of the MPLS domain, and the label is removed when it leaves the MPLS domain, which is called label disposition, at the egress point of MPLS domain. Both of the actions are done by the edge-LSR, which is located at the boundary of the MPLS domain. The edge-LSR keeps both the

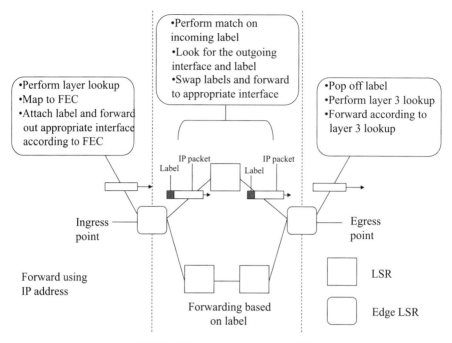

FIGURE 7.19 Basic operation of MPLS.

IP routing table and label routing table. When the non-boundary LSR encounters a labeled packet, it will only use the label-forwarding table to determine the next hop address for the packet.

To perform label imposition, the edge-LSR needs to understand where the packet is headed and which label should be assigned to the packet. This actually involves two steps. The first step is to divide the entire set of possible packets into a set of IP destination prefixes. In MPLS architecture, the result of the first step is called forwarding equivalence classes (FEC), which describes a group of IP packets that are forwarded in the same manner. The criteria for FEC can be fields in IP packet, such as destination IP subnet and the IP precedence value. The second step is to map the IP destination prefix to an IP next hop address. In MPLS, it is possible based on other criteria, rather than the shortest path, to determine the next hop. This feature is very desirable as the shortest path may not necessarily be the best path with QoS. In addition, this feature is a must for the implementation of per-hop QoS. The relationship between different components of MPLS is shown in Fig. 7.19.

Each LSR keeps two tables for holding the relevant information for MPLS forwarding component. The first one, known as label information base (LIB), holds all labels assigned by the LSR and the mappings of the labels from other neighbors in different MPLS domains. This information is distributed by using label distribution protocol (LDP). The second table, known as label-forwarding information base (LFIB), stores only the labels that are currently used by the forwarding component of MPLS.

As previously mentioned, each packet enters an MPLS network at an ingress LSR and exits the MPLS network at an egress LSR. Here, LDP is used to create the unidirectional path from ingress to egress point for each FEC in the MPLS domain, which is referred to as label switched path (LSP). It is necessary because the forwarding tables at each LSR must be populated with the mappings from the incoming interface/label value to outgoing interface/label value. The creation of LSP is connection-oriented as the path needs to be set up before any traffic flow.

The MPLS architecture document does not assign a single protocol for the distribution of labels between LSRs. Several approaches to label distribution can be used depending on the requirements of the hardware and the administrative policies used on the network. Only the approaches related to traffic engineering are covered.

From the previously mentioned architecture of MPLS, if the LSP can be controlled in such a way that it will only be set up as long as certain QoS requirements can be satisfied for that path, then MPLS can be easily used for traffic engineering, which is referred to as the process where data are routed through the network according to a management view of the availability of resources and the current and expected traffic. The MPLS domain uses LSPs to provide tunnel-like topological isolation, and temporal isolation if the LSPs have associated QoS guarantees. Hence, one of the main uses of MPLS is for building virtual private network (VPN).

Traffic-engineered and/or QoS-enabled LSPs are conventionally referred to as constraint-routed LSPs (CR-LSPs). Two solutions exist for the explicit signaling of CR-LSPs. One solution are borrows from existing RSVP called M-RSVP, and the other solution requires adding new functionality to the base LDP, referred to as CR-LDP. At abstract level, both the protocols have a lot of similarity between their functions. Both of them can form strict or loose specification of the route to be taken by the LSP that is initialized by edge-LSR and specify QoS parameters to be associated with the LSP. The details of the protocol will not be covered here.

APPENDIX 7.A ATM CELL FORMAT

This appendix describes the ATM cell formats and the function of each field in the header. The overhead at the AAL is also described.

7.A.1 ATM Layer

An ATM cell is 53 bytes in length. Out of the 53 bytes, 5 bytes are overhead and 48 bytes are payload for carrying information from and to the layer above the ATM layer. There are two formats for the 5 bytes of overhead, as shown in Fig. 7.20(a) and (b).

A function of any standard is to allow equipment of different vendors to interoperate. As such, conformation to the cell formats is important at the "interface" between equipment. The header format in Fig. 7.20(a) applies to the UNI (user–network interface), and the header format in Fig. 7.20(b) applies to the NNI (network–node interface). As the names suggest, the former refers to the interface between the

312 BASIC CONCEPTS OF BROADBAND COMMUNICATION NETWORKS

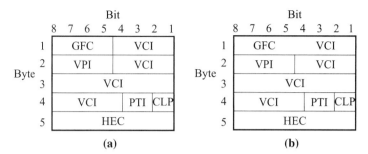

FIGURE 7.20 (a) The cell header format for the UNI; (b) the cell header format for the NNI.

user's equipment and the network, and the latter refers to the interface between two ATM nodes. The only difference between these two formats is that the UNI format has a 4-bit GFC (generic flow control) field whereas the NNI format does not.

The GFC is not carried from end to end and has local significance only for flow control on the customer site. The user's terminal equipment may not be directly connected to an input port of a public ATM switch. In fact, a number of users may be connected to a multiaccess local-area or metropolitan-area network (such as, e.g., a DQDB (distributed queue dual bus) network), which is in turn connected to the ATM network. The GFC field provides a mechanism for the flow control of traffic of multiple user terminals on the multiaccess network. Alternatively, traffic from a number of users may be multiplexed onto a shared trunk that is connected to the ATM network. The GFC field also allows the multiplexer to regulate the transmission of the users' traffic to the multiplexer.

The GFC field is absent in the NNI header. The 4 bits of GFC and 8 bits of VPI in the UNI header become the 12 bits of VPI in the NNI header. This means that there can be 16 times more VPs on a physical link in the ATM network of the network provider, which is reasonable since VPs are expected to be more widely used in the higher network hierarchy. The rest of the header formats are the same for the UNI and the NNI.

There are two bytes for VCI. After that, three bits are reserved for PTI (payload type). In the ATM network, both the traffic of the user and the control traffic pertaining to operations, administration, and management (OAM) are carried in cells. The PTI field is used to distinguish among different traffic types. The values of the PTI are defined as follows:

PTI	Cell Type
000 and 001	User cell, no congestion
010 and 011	User cell, congestion
100	Segment OAM cell for a VC
101	End-to-end OAM cell for a VC
110	Resource management cell
111	Reserved for future functions

The first bit is used to differentiate between user and OAM cells. If it is 0 then the cell is a user cell, and if it is 1 then the cell is an OAM cell. For the user cell, the second bit is set if the cell encounters congestion in the network. The third bit is available to the users for end-to-end indication purposes. The ATM network does not examine this bit and will not modify it. For instance, in AAL 5, the bit is used to indicate whether the cell is the last cell of a packet, that has been segmented into several cells during adaptation.

For the OAM cell, PTI = 100 or 101 means that the cell is related to the OAM function of a VC. The VPI and VCI of the cell will be set to be same as that of the user cells of the same VC. If PTI = 100, then the cell is related to the OAM function of the segment of a VC across the UNI. Segment OAM cells will be removed and examined at the end of the segment (i.e., at the first public network node through which the VC traverses). If PTI = 101, then the cell is related to the end-to-end OAM function of a VC. End-to-end OAM cells are passed unmodified by all intermediate nodes, although the contents may be monitored by any node in the path. These cells are removed and examined at the end point of the VC. Finally, if PTI = 110, then the cell is a resource management cell, and at the moment PTI = 111 is undefined and reserved for future use.

After PTI, there is one CLP (cell loss priority) bit for indicating the priority of the cell when it is necessary to discard cells in the network during congestion. Cells with lower priority are dropped first before the cells with higher priority.

One byte of HEC (header error control) is used to store the CRC (cyclic redundancy code) computed over the five-byte header using the CRC polynomial $x^8 + x^2 + x + 1$. Since the header of a cell may change from node to node (consider, for instance, that the VPI and VCI may change), the CRC needs to be checked and recomputed at every node. With the CRC polynomial, the header has a minimum distance of 4, meaning two headers must differ in at least four bits. A point to emphasize is that this HEC is for the protection of the header only, and the 48-byte payload is not protected in the ATM network.

The CRC is used for both error detection and correction. The normal or default mode is the single-bit error-correction mode. In this mode, if a single-bit error is detected, it is corrected and the cell is forwarded, and the receiver goes into the detection mode. Note that with a minimum distance of 4, it is possible to treat a three-bit error as a single-bit error and "correct" the header the wrong way. More generally, odd number of errors will be treated as single-bit errors and even number of errors will be treated as multiple-bit errors. If multiple-bit errors are detected during the normal mode, the cell will be discarded and the receiver goes into the detection mode. In the detection mode, no attempt is made to correct errors, and all cells with errors detected will be discarded: this, for instance, prevents odd bit errors from being treated as single-bit errors, which occurs during the correction mode. As soon as a header with no error is encountered, the receiver goes back to the correction mode. This operation is devised with very high-quality links (e.g., fiber-optic link) in mind. Random bit errors in the link and the detector are rare and independent, and encountering more than a bit error in the header due to noise in the physical channel is even rarer. The detection mode is used to guard against burst errors, that may occur when there is an

CSI	SN	SNP	SAR-SDU
1 bit	3 bits	4 bits	47 bytes

FIGURE 7.21 AAL1 SAR protocol data unit.

equipment problem (e.g., an equipment fault on the transmitter side that causes it to output random bit patterns).

7.A.2 Adaptation Layer

The overhead for the adaptation layer is included as part of the 48 bytes of the payload of the ATM layer. Recall that the adaptation layer is divided into two sublayers: the CS and the SAR sublayers, with the former above the latter. Figure 7.21 shows the SAR protocol data unit (PDU)[3] for AAL1. One byte is the overhead and the other 47 bytes are the payload from the CS sublayer (i.e., each CS PDU is 47 bytes).

The three-bit SN (sequence number) in Fig. 7.21 is incremented once for every cell. It is used to detect missing cells due to loss (e.g., discarded by the network during congestion) or errors. With the three-bit SN, however, it is not possible to detect multiples of eight missing cells.

The SNP (sequence number protection) consists of two parts: a three-bit CRC and one parity bit. The CRC is computed over the SN using the polynomial $x^3 + x + 1$. The parity bit is computed over the other seven bits. The SNP is capable of single-bit error correction and multiple-bit error detection.

The one-bit CSI (convergence sublayer indicator) is used for signaling purposes as well as for the indication of the absence or the presence of a CS function. The use of this field is optional. Strictly speaking, this bit belongs to the CS layer. In the absence of a CS function, the CSI is set to 0, in which case the whole 47 bytes of payload at the SAR layer is used to transport user data. There is no additional predefined overhead at the CS layer. Basically, this means that the user can define the usage of the 47 bytes in whatever way that is suitable for the application. Note that $CSI = 0$ does not mean that the application has no need for CS functions such as clock recovery, error protection, and so on. It simply means that it is up to the user to decide how the various CS functions are to be implemented within the 47 bytes.

The CSI bit may be used in a number of ways. For instance, the CSI bits of successive ATM cells may be used to convey timing information from the source to the receiver so as to synchronize the clock of the receiver to that of the source. For circuit emulation of DS1 and DS3 over ATM, the timing information is conveyed over the CSI bits of cells with odd SN values. AAL1 also provides the so-called *structured data transfer*, in which case the CSI may be set 1 to indicate that 1 byte

---

[3]In conformance to normal usage, we refer to the packet or message unit generated by layer *A* as layer-*A* PDU. The PDU includes the overhead. The term layer-*A* service data unit (SDU) excludes the overhead and refers only to the payload.

ATM CELL FORMAT

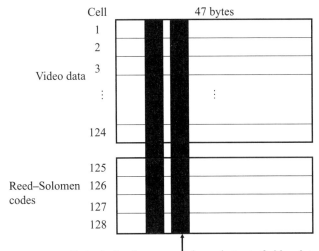

FIGURE 7.22 AAL1 CS forward error correction for transporting video data.

in the SAR payload is reserved for CS usage and only 46 bytes are for the user data.

Another function that can be performed by the CS is forward error correction. The example in Fig. 7.22 shows the CS for the transport of video data with forward error correction. The video data is arranged in 124 rows of 47 bytes each, and each row is packed into one cell. Additional four cells are used for error-protection purposes. An error-correction code (e.g., the Reed–Solomen code) is computed over the ith bytes of all rows and it forms the four ith bytes in the four error-protection cells. After the codes are computed, they are put into cells separated from the data cell. Up to four lost cells can be recovered at the receiver.

Figure 7.23 shows the SAR PDU for AAL2. The SN field contains the sequence number of the cell. The source generates traffic at variable rate and some cells could be partially filled only. A message may be contained in one or more cells. The IT (information type) field is two bits in length, and it indicates whether the cell contains the BOM (begin of message), COM (continuation of message), or EOM (end of message): 10 for BOM, 00 for COM, 01 for EOM, and 11 for a message equal to or shorter than one cell in length (i.e., the cell contains the whole message). The LI (length indicator) at the trailer of the cell indicates the number of bytes used in partially filled cells. At present, AAL2 is not well defined and the length of each field requires further study.

SN	IT	SAR-SDU	LI	CRC
	2 bits			

FIGURE 7.23 AAL2 SAR protocol data unit.

SN	IT	RES/MID	SAR-SDU	LI	CRC
2 bits	4 bits	10 bits	44 bytes	6 bits	10 bits

FIGURE 7.24 AAL3/4 SAR protocol data unit.

There are two modes for AAL3/4:

- *Message mode:* In this mode, the layer above the AAL passes data to the AAL as messages that could be fixed or variable in length. Each message is passed in its entirety at one time. At the receiver end, data are also passed from the AAL to the layer above in discrete messages.
- *Streaming Mode:* This service provides the transport of long variable-length data units. In this mode, the layer above the AAL may pass data to the AAL in separate chunks. Each pass may contain only part of the whole data unit, and several passes may occur at separate times. A reason for not passing a data unit as a whole could be that the data unit has not been created by the application entirely. The AAL may send out cells associated with a data unit before all the data have arrived. In this way, the adaptation delay could be reduced.

Figure 7.24 shows the SAR-PDU format of AAL3/4. Four bytes in each cell are used as the overhead. The two-bit ST (segment type) field is the same as the IT field in AAL2, and it is used to indicate BOM, COM, EOM, or SSM (single segment message). The four-bit SN is used for detection of lost and inserted cells. The LI field is six bits in length: an EOM or SSM cell may be only partially filled (i.e., less than 44 bytes) and the LI indicates the number of bytes used. The 10 CRC bits are computed over the whole SAR PDU using the generating polynomial $x^{10} + x^9 + x^5 + x^4 + x + 1$.

The 10-bit MID (multiplexing identifier) field is used in AAL4 (i.e., connectionless service) only; the bits are reserved for future usage in AAL3. In AAL4, this means that up to 2^{10} connectionless packets (messages) can be transmitted on the same ATM connection simultaneously. For instance, there could be two connectionless servers (e.g., IP routers) connected to the two ends of the ATM connection sitting on top of the adaptation layer. At the transmitter end, the connectionless server aggregates all the connectionless traffic to be routed to the other connectionless server and forwards it via the ATM connection. Each IP packet can be assigned an MID, and the cells of the packet will have the same MID. At the receiving end, the connectionless server in turn examines the MID field of each incoming cell and reassembles the cells with the same MID back to the packet before deciding how to further route it.

The multiplexing/demultiplexing mechanism is needed only if the cells of different messages are interleaved during the transmission over one ATM connection. In this case, successive cells arriving at the receiving end may belong to different messages and the MID field is needed to distinguish them apart. For illustration, a situation where this may arise is shown in Fig. 7.25. There are several connectionless servers connecting to the same ATM switch via different inputs of the switch. The connectionless servers may have traffic to be routed to a remote connection server,

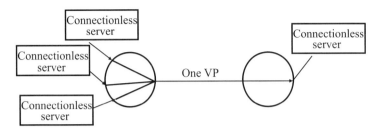

FIGURE 7.25 Multiplexing the cells from several connectionless servers onto one VP.

that is connected to a remote ATM switch. To improve efficiency, the traffic from all the transmitting connection servers may be statistically multiplexed onto the same ATM connection (say, a VC) for transport to the remote ATM switch.

In situations where the cells of different messages are not interleaved over an ATM connection, there is no need for the MID, since any particular message will be transmitted as successive cells. The ST field will be sufficient for reassembling the message at the receiver end. In the above example, instead of one VC, separate VCs can be used to transport the cells from different connectionless servers to the remote connectionless server. There will then be no need for the use of MID. Of course, there will be several AAL processes, one for each VC, interfacing with the receiving connectionless server.

In the situation where cells of different messages are interleaved, instead of the MID, one could in principle encode the routing information in all cells of a message. Each message forwarded by the connectionless server to the AAL must contain the explicit routing information (e.g., IP address for IP routing) for the receiving connectionless server to further route the message. The problem is that the routing information may be much lengthier than the MID (e.g., the IP address is 32 bits) and will therefore create larger overhead if it is incorporated in all cells of the message.

The overhead in the CS of AAL3/4 consists of four bytes of header and four bytes of trailer, as illustrated in Fig. 7.26. The one-byte CPI (common part indicator) signals how the rest of the overhead fields are to be interpreted. Until further studies by the standard committee, this field is always set to 0. The zero value indicates that the counting units in the two-byte BASize (buffer allocation size) and the one-byte Length fields are bytes.

The BASize indicates the maximum number of bytes needed to store the CS PDU at the receiver. It is used to reserve buffer at the receiver for the reception of the PDU. It is possible for the cells associated with the CS PDU to arrive at the peer convergence

CPI	BTag	BASize	CS-SDU	PAD	AL	ETag	Length
1 byte	1 byte	2 bytes	$1-2^{16}$ bytes	0–3 bytes	1 byte	1 byte	2 bytes

FIGURE 7.26 AAL3/4 CS protocol data unit.

sublayer at the receiver separated in time. This could be the case, for instance, when the interface between the CS and AAL3/4 at the sender is in the streaming mode. With 16 bits, the maximum size of the CS PDU is 2^{16} bytes. When the message mode is used, the BASize corresponds to the payload of the CS PDU (or SDU), and the contents of the BASize and Length fields are the same. The BASize and Length fields may be different when the streaming mode is used. In this case, the BASize is set to be equal to or greater than the actual CS SDU length. Typically, this is set to be the maximum possible CS SDU length, since in the streaming mode the sender itself may not know the exact length of the data unit when it sends out the cell containing the BASize field (because the data unit may not have arrived in its entirety). The exact length of the SDU is indicated in the Length field, that is contained in the last cell of the data unit.

The one-byte BTag (beginning tag) and the one-byte ETag (end tag) fields are set to the same value, which changes for each successive CS PDU. They are for error detection purposes. If the BTag and ETag of a received CS PDU do not match, then some cells must have been lost. Although the four-bit SN of the AAL3/4 SAR sublayer is also meant for detecting missing cells, the mechanism at the SAR sublayer may not be strong enough and in many situations, lost cells may not be detected. Thus, the BTag and ETag can be viewed as an enhancement to the cell loss detection capability.

The one-byte AL (alignment) field serves no other purpose than to make the trailer 32-bit long. The reason is that many computers are 32-bit machines that work on 32 bits in each cycle. Between the CS PDU payload and the trailer, there will be 0–3 unused bytes for padding purposes. The reason is again to make sure that the whole CS PDU is a multiple of 32 bits in length.

The overhead in the AAL3/4 SAR layer is rather high: 4 bytes out of every 44 bytes. Also, the 10-bit CRC may not offer enough protection at the cell level, and there is no CRC at the CS PDU level; generally, the longer the block length of the data, the better the efficiency achieved in error protection (i.e., for a given data corruption probability, the longer the block length, the smaller the ratio of CRC bits to the data bits), and this would have argued for putting the CRC in the CS. For these and other reasons, the computing industry initiated the AAL5.

There is no overhead in the AAL5 SAR sublayer. All 48 bytes of the payload of the ATM layer are used to carry the CS PDU. Figure 7.27 shows the format of the AAL5 CS PDU and how it is being segmented into cells at the SAR sublayer. One bit of the PTI (payload type) field of the ATM-layer header is used to indicate the end of the PDU. That is, if the cell is user cell (first bit of the PTI is set to 0) and if the last bit of the PTI is set to 1, then the cell contains the last segment of the PDU. If this is the case, then the last eight bytes of the payload of the cell are the overhead associated with the convergence sublayer of AAL5. The one-byte UU (user-to-user indication) field is intended for signaling between two end users and it is carried transparently by AAL5. The usage of the one-byte CPI is currently not defined. The two-byte Length field is the number of bytes in the CS SDU, not including the padding in the last cell. The 32-bit CRC is computed over the entire CS PDU, including the padding and the trailer. Notice that strictly speaking, AAL5 has violated the layering principle in that a bit of the overhead of the ATM layer has been "stolen" for use in the AAL layer.

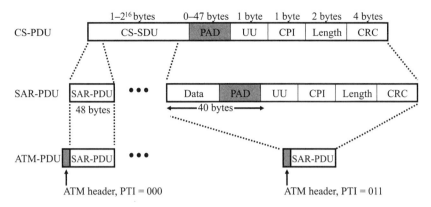

FIGURE 7.27 AAL5 CS protocol data unit and its segmentation.

This, however, does not cause any confusion as long as the designers of the AAL5 and ATM hardware or software have taken it into account.

PROBLEMS

7.1 Consider the second method of multirate switching depicted in Fig. 7.2(b). The main idea is that all the n channels of the same input connection must be routed over the same internal path. Each input/output link is DS-1, which consists of 24 DS-0s. Suppose that each connection consists of either one or two DS-0 channels (i.e., n is either 1 or 2). How many middle-stage modules are needed for the overall switch to be strictly nonblocking, assuming the switch modules in all stages are nonblocking?

7.2 Does ATM switching solve the problem of bandwidth fragmentation in a switch consisting of multiple stages of switch modules? Has the problem of increased blocking probability during call setup of a connection due to bandwidth fragmentation been solved?

7.3 Consider sending the cells from a voice source over an ATM network. Suppose that the delay suffered by each cell in the network ranges from 2 to 10 ms. How much time after the arrival of the first cell can the receiver start to play out the audio? How should the buffer be dimensioned to avoid data overflow?

7.4 What is the worst message length for a cell network in terms of the overhead?

7.5 Suppose that each cell has a fixed header size of 5 bytes. The packet to be fragmented into cells has a fixed size of 100 bytes. What is the overhead if the cell size is fixed to be 32 bytes (excluding the 5-byte header)? What is the overhead if the cell size is 48 bytes instead? Which cell size is more efficient for these fixed-length packets?

320 BASIC CONCEPTS OF BROADBAND COMMUNICATION NETWORKS

7.6 The network shown in Fig. 7.7 consists of a number of VC and VP switches.
  (a) Identify the end-to-end path of the VC from user A to user B with VPI = 2 and VCI = 3 at the source.
  (b) Write down the VPI/VCI entry in the VPI/VCI table needed for routing the VC traffic at the last node just before the traffic reaches user B. (*Note:* the answer is not unique.)
  (c) Indicate the VP termination points (if any) in the path.
  (d) Indicate the VC termination points (if any) in the path.

7.7 Consider the problem of setting up a new VC over a sequence of n VPs. Suppose that the VCI of the VC is not allowed to change from VP to VP and that we want to use the same VCI throughout. Suppose that on each of the VPs, a fraction p of the VCI values has already been used, and a fraction $1 - p$ has not been used. Assume that there are altogether m possible VCI values on each VP.
  (a) The source node first randomly chooses a VCI not already used in the first VP, it then asks the next VC switch whether this VCI has been used in the next VP. If no, the next VC switch asks the VC switch further downstream whether the VCI has been used; if yes, the source is asked to choose another VCI and repeat the process. This is repeated until a unique VCI throughout the path can be found. For small p, how many attempts by the source will be needed before a unique VCI can be found? Can you think of a better way to reduce the number of attempts?
  (b) Estimate the probability of not having a unique VCI throughout in terms of p, n, and m, assuming the VCIs already used in different VPs are uncorrelated? For $p = 0.75, n = 4$, and $m = 256$, what is the blocking probability?

7.8 This problem examines the shortcoming of the VP/VC two-tier multiplexing hierarchy depicted in Fig. 7.8 and the extent to which the shortcoming is important. Suppose that all the VCs behave the same way. For each VC, half of the time, data are arriving at the rate of r bits per second and half of the time there is no data arriving. Suppose that there are n VCs being multiplexed directly onto a physical link of $0.6rn$ bits per second. Define the probability of congestion as the probability that the arrival rate of the data into the multiplexer is higher than the departure rate, $0.6rn$. Suppose we use the two-tier hierarchy with 10 VPs, each of which have a bandwidth of $0.06rn$ bits per second, and there are $0.1n$ VCs being multiplexed onto each VP. What is probability of congestion for each of the first-level multiplexers? How do the two cases differ when n is small, say, $n = 20$, and when n is large, say, $n = 1000$.

7.9 The text discussed the mapping of input VPI–VCI pair to an output VPI–VCI pair at an input of a VC switch. This works for point-to-point connections. For a multicast connection in a multicast switch, there are several outputs, and therefore, several output VPI–VCI pairs must be mapped to the input VPI–VCI pair. Consider a 256 × 256 switch in the following.

(a) Suppose the switch supports only point-to-point connections. Assume that the VPI is 12 bits and VCI is 16 bits. Estimate the routing table size at an input.

(b) For a switch that supports multicast connections, suggest a structure for the routing table at an input, assuming that the number of outputs for a general connection (point-to-point or multicast) can range from 1 to 256. Based on this routing table, estimate the memory required at an input.

(c) For the multicast switch that is made up of a cascade of a copy network and a point-to-point switch, what is the problem (other than the large memory requirement) of doing the mapping at the input? (*Hint*: Before packet replication, the master packet represents many packets that will be replicated, what output VPI–VCI should it have?)

(d) Consider the cascade multicast network again. Suppose that for point-to-point connections, the mapping will be the same as before. But for multicast connections, the BCN mechanism will be used at the outputs of the copy network to retrieve both the output and the output VPI–VCI pair of a packet. Estimate the memory requirement of the lookup table at an output of the copy network.

(e) Consider a bus-based multicast switch such as the knockout switch. In this case, the outputs of a multicast cell pick up the cell from the bus. Discuss the design of the routing tables and their locations in such a switch.

7.10 With respect to the discussion of the use of MID in AAL4 in the appendix, can AAL5 be used to transport messages from different connectionless servers over the same ATM connection? How?

7.11 Suppose we want to send a short message of 1000 bytes, is AAL3/4 more efficient or AAL5? Is there any situation under which AAL3/4 is more efficient than AAL5? Do you think AAL3/4 will be used in practice to transport connectionless data?

8

NETWORK TRAFFIC CONTROL AND BANDWIDTH ALLOCATION

A broadband integrated network carries a large volume of traffic, and the traffic may be from many different types of sources and have diverse characteristics. It is necessary to allocate network resources and control this traffic to achieve two fundamental goals:

- Satisfy the QoS (quality of service) requirements of network users.
- Maximize network usage.

Controls at different timescales are exercised to achieve these two goals. Recall that many logical networks can be set up on the same physical network by VP networking. Control can be exercised on the VPs in logical networks. Each logical network can be used by many end users who establish VCs over it, so we also have control at the VC connection level. Finally, the delivery of cells over a VC must also be controlled to meet certain QoS objectives, such as delay and cell loss probability.

In a virtual-path (VP) network, the virtual paths and the bandwidths on them may be reconfigured dynamically over time. For instance, if VPs are used to segregate services, different services may demand different amount of network resources at different times of the day. To maximize network usage and service quality, bandwidths can be reallocated once in a while (say, every half hour or so) among the different logical networks according to their usage. In addition, new VPs may be set up and old VPs may be torn down in a logical network dynamically to reflect the change in traffic demands within it. The control timescale at the VP level is large compared to that in cell-level control.

At the VC connection level, we have the control functions of call admission, routing, and bandwidth allocation. Since each VP may contain many VCs, the amount

---

Principles of Broadband Switching and Networking, by Tony T. Lee and Soung C. Liew
Copyright © 2010 John Wiley & Sons, Inc.

of control actions is larger at the VC level than at the VP level: VC-level control occurs more frequently in the network. Depending on the service, a VC may last for a long or short time. A telephone call will probably last a few minutes, and the control timescale of the associated VC will be smaller than that at the VP-level control. A movie, on the other hand, may last up to 2 h, and the control timescale is not necessarily smaller than that at the VP level.

For an overview, let us briefly walk through the process of call admission control at the VC connection level. The purpose of call admission is basically to determine whether to admit or reject a connection request based on a projection of the consequences of admitting the call. If accepting the call will degrade the QoS (say, the delay or the cell loss probability) of the existing connections to an unsatisfactory level, then the call should be rejected. At the same time, if the QoS demanded by the call cannot be satisfied, then the call should either be rejected or negotiation should be initiated to see if a lower QoS is acceptable to the end user requesting the call.

Closely tied to call admission is the issue of routing. There may be several alternative routes in the network from the source to the destination of a new connection request. Whether the call is admitted depends on whether an acceptable route can be found. Some paths may not satisfy the QOS criteria and some may. Among the latter, the "optimal" route should be chosen such that the network resources are used in the most efficient way. The definition of the optimal route is not as trivial as it might first appear. For instance, if by optimality, we mean the route that can maximize the number of accepted calls in the future (or minimize the call blocking probability), then some statistics of the future call arrivals must be available. In addition to routing, bandwidth allocation (i.e., how much bandwidth is to be allocated to the connection) should also be considered as part of the call admission process. In short, the issues of call admission, routing, and bandwidth allocation are interrelated.

Once a call is accepted, control at the cell level will be exercised throughout the call's duration. Since each VC may have many cells, cell-level control is very much more frequent than connection-level control. Also, each cell lasts for a very short time, and therefore the control timescale is very small. For this reason, cell-level control is often performed by specialized hardware rather than software. For example, the switches we have discussed in the previous chapters are specialized hardware for routing cells. This is in contrast to VC- and VP- level control that is often implemented in software for flexibility.

Cell-level control can occur at several places in the network. Control may be applied at the boundary of the network to make sure that the connection is not pumping cells into the network at a rate higher than some pre-agreed rate. This is sometimes called the policing function. The end user may also perform "traffic shaping" (e.g., intentionally delay some cells at the edge or space out their transmission in time) at the source so that the traffic presented to the network conforms to the pre-agreed rate.

Once the cells enter the network, their flow may also be monitored and controlled. Consider a connection traversing several nodes in a given direction. A node is said to be upstream (downstream) of another node if cells of the connection traverse the node before (after) passing through the other node. In feedback flow control (also called closed-loop control), the downstream traffic conditions are fed back upstream to regulate the traffic flow. In feedforward flow control (also called open-loop control), there is no feedback signal.

There are several possibilities for closed-loop control: among them are link-to-link flow control and end-to-end flow control. In link-to-link flow control, the traffic condition at a node is fed back to the immediate upstream node. If the downstream link is congested, then the upstream node will slow down the transmission of cells to the downstream node. This in turn may cause cells to back up at the upstream node, and through feedback, the node further upstream will be informed to cut down the rate of its traffic flow. Through a chain of this "back-pressure" mechanism, eventually the source will be told to reduce its transmission of traffic.

In end-to-end feedback flow control, the feedback to the source comes from the destination rather than through the back-pressure mechanism in link-to-link control. There are many possibilities for end-to-end feedback. In ATM network, recall that there is a bit in the PTI (see the appendix of Chapter 7) field in the ATM cell header that is used for congestion indication. In ABR service, the source sets this bit to zero. As a cell travels through the nodes in the network, this bit will be set to one at congestion points. The destination, upon receiving this cell, can then inform the source of the occurrence of congestion (or the absence of it) through a path in the reverse direction.

In open-loop control, since there is no feedback on the downstream traffic conditions, the rate at which the source pumps traffic into the network must be tightly controlled so that it conforms to some pre-agreed rate. Otherwise, if many sources transmit large amounts of traffic simultaneously, congestion in the network may occur, giving rise to large cell loss rates when switch buffers start to overflow. If all traffic sources conform to their pre-agreed rates, the transport of their cells in the network can be scheduled such that their desired QoS can be satisfied. Feedforward flow control eliminates a problem with feedback flow control in a high-speed network in which the propagation delay is large compared to how fast the statistics of cell traffic can change: by the time the feedback signal arrives at the source, the traffic conditions at the downstream nodes may have already changed; it is also possible that by the time a traffic congestion signal reaches the source, the source may have already pumped out a large number of cells, making the congestion even worse.

Feedforward flow control, on the other hand, may not be able to achieve a statistical multiplexing gain as high as that in feedback flow control. In case of light traffic at the downstream node, the source may not be able to take advantage of that by pumping in more traffic, since there is no feedback telling it so.

The current ATM standard reflects the following as far as feedback and feedforward flow control is concerned. For services that have stringent delay requirements, feedforward flow control can better meet the requirement. This will be the case for AAL1 and AAL2 traffic, and the targeted network services are CBR and VBR. For data services such as those supported by AAL3/4 and AAL5, delay requirement is less stringent. The targeted network services are either ABR (available bit rate) or UBR (unspecified bit rate). In the ABR service, for instance, the available capacity for the transmission of the traffic from a source may change over time in a way that depends on network usage by other users. The network capacity available at any time is reflected in the feedback signal. Having overviewed the general issues and methods of traffic control, the remaining sections of this chapter will treat them at a more fundamental level. As a start, we will first introduce a traffic model for cell traffic that will be useful for the study of flow control.

8.1 FLUID-FLOW MODEL: DETERMINISTIC DISCUSSION

A *fluid-flow model* can be used to describe the flow of cell traffic in the network. This model treats traffic as fluid flow. The fluid-flow model is a powerful and intuitive tool for visualizing the flow traffic in the network without the need for complicated queueing analysis. This is especially true when we have a deterministic description of the "worst-case" input traffic pattern that a source can pump into the network. This is the case, for example, when traffic policing is applied at the edge of the network. When the input traffic is stochastic and not known deterministically, the model can also be subjected to rigorous mathematical analysis, as shall be seen in Section 8.2. By adopting the fluid-flow model, closed-form analytical solutions can often be obtained where they are not so readily available otherwise. In other words, the analysis can often be carried out further with the fluid-flow model than by using the cell-traffic model.

To explain the fluid-flow model, let us focus on the traffic of a simple "on–off source." Figure 8.1(a) shows the cell-traffic model. During *on* time a cell is generated every T seconds, and during *off* time no cell is generated. Packetized voice traffic exhibits this kind of behavior. The on time is the duration of a talkspurt when the

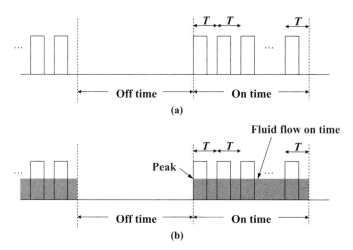

FIGURE 8.1 (a) On–off cell-traffic source; (b) approximating on–off cell traffic with fluid-flow traffic.

speaker speaks, and the off time is simply when the speaker idles. The average data rate during on time is therefore $1/T$ cells/s or

$$r_{\text{on-time}} = \frac{53 \times 8}{T} \text{ bits/s.} \quad (8.1)$$

The on and off times are in general not fixed and may vary over time. Let the average off time be denoted by $1/\lambda$ and the average on time be denoted by $1/\mu$. Then the average rate of the traffic is

$$r_{\text{ave}} = \frac{\frac{1}{\mu} r_{\text{on-time}}}{\frac{1}{\lambda} + \frac{1}{\mu}} = \frac{\lambda r_{\text{on-time}}}{\lambda + \mu} \text{ bits/s.} \quad (8.2)$$

Figure 8.1(b) shows the approximation of the cell-traffic model with the fluid-flow model. During on time, instead of cells being separated by intervals of T, we make the approximation that the arrivals are continuous in time at rate r_{on}. Thus, instead of cell arrivals, we have "fluid" arrivals. The peak rate of the on–off fluid-flow source is $r_{\text{peak}} = r_{\text{on}}$. Both models have the same total number of bits arrived in each on period, but the bit arrivals are distributed evenly over time in the fluid-flow model.

Consider the statistical multiplexing of traffic from several on–off sources, as illustrated in Fig. 8.2. Suppose that the buffer capacity is B bits and the transmission bandwidth is C bits/s. Then, as far as the buffer occupancy is concerned, the multiplexer is analogous to a leaky bucket of size B with a hole at the bottom. Fluid is poured into the bucket at a rate corresponding to the traffic generated by the sources, and fluid leaks out of the bucket at rate C.

Figure 8.3 shows the simple case in which there is only one traffic source and $r_{\text{peak}} > C$. Both the rate-versus-time and cumulative traffic-versus-time curves are drawn. By drawing the cumulative arrived and departed traffic, we can find out the delay of a bit i (or a drop of fluid) and the buffer occupancy at time t. As shown in

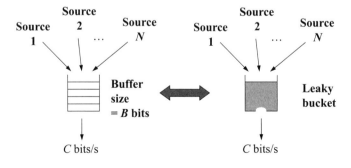

FIGURE 8.2 An analogy between a transmission buffer of output rate C and a leaky bucket.

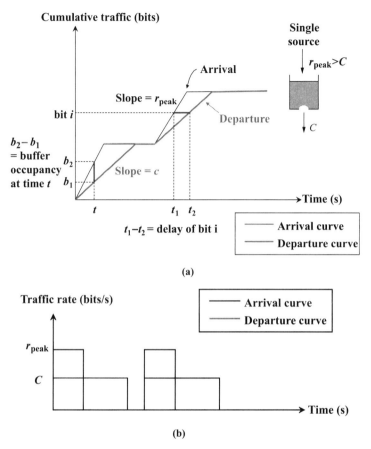

FIGURE 8.3 A leaky bucket with a single source; (a) cumulative traffic-versus-time arrival and departure curves; (b) traffic rate-versus-time arrival and departure curves.

Fig. 8.3, the delay of a bit i corresponds to the horizontal difference between the arrival and departure curves at bit position i. The buffer occupancy at time t corresponds to the vertical difference between the two curves at time t.

Figure 8.4 shows the case in which there are multiple sources. The cumulative arrived traffic is a piecewise linear curve whose slope changes according to how many sources are on: specifically, when m sources are on, the slope is mr_{peak}. In the figure, we assume that the traffic is normalized so that $r_{\text{peak}} = 1$. The departure curve has a maximum slope of C. When the departure curve is below the arrival curve, the backlog in the buffer starts to build up. If we assume that the system works as hard as possible to clear up the backlog, the slope of the departure curve is C whenever the backlog is nonzero. That is, with the leaky-bucket analogy, fluid will continue to leak out at the maximum rate C so long as there is fluid in the bucket.

When there is no backlog and the total arrival rate from all sources is less than or equal to C, then the fluid that goes into the bucket immediately flows out. In this

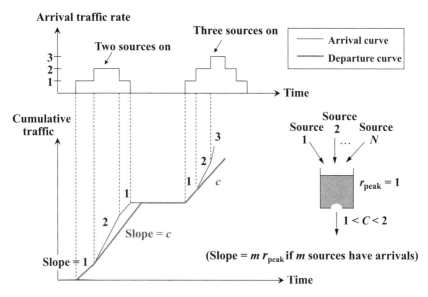

FIGURE 8.4 The arrival and departure curves of a fluid-flow multiplexer.

situation, the cumulative departed traffic equals the cumulative arrived traffic, and the departure rate equals the arrival rate. When there is no backlog and the arrival rate is larger than C, then backlog immediately builds up, and the departure rate will be C.

The curves in Fig. 8.4 only show the "global" view of a multiplexer. The QoS as perceived by individual traffic streams cannot be seen from the curves. The QoS perceived by a traffic stream depends to a large extent on the scheduling algorithm used by the multiplexer. When several streams have backlog in the multiplexer, the multiplexer has a choice of whose data to send out first, and the order in which the data are sent out is determined by a scheduling algorithm. Different scheduling schemes may give rise to different QoS.

Let us assume that the traffic of each stream must be served in a first-come-first-serve (FCFS) manner. That is, for a particular stream, the bit that arrives before another bit from the same stream must also depart before this other bit. Interpreting each stream as one VC in the ATM network, the FCFS requirement makes sure that the data will be delivered to the end destination in the correct order. For sequence preservation of the data of a VC, however, it is not necessary that the bits from different streams to be served in any particular order: for example, sending out a bit of a stream before a bit of another stream that arrives earlier would not compromise the sequence order of either stream.

Let us first return to the cell-traffic model. The round-robin scheduling scheme treats all streams as equal and serves them in a round-robin fashion. Conceptually, there is one buffer for each traffic stream. The multiplexer selects cells from the nonempty buffers in a round-robin fashion. That is, the head-of-line cells of the nonempty buffers are served in a cyclic fashion.

We would like to approximate the round-robin service discipline with a fluid-flow *service* model that is compatible with the fluid-flow source traffic model. That is, instead of serving data in a cell-by-cell manner, we assume that data are infinitely divisible (i.e., like fluid) and can be served in infinitesimal amount by infinitesimal amount fashion. A more detailed discussion of the relationship between the round-robin service discipline and the fluid-flow service discipline will be presented in Section 8.4.

Roughly, the round-robin service discipline with cell traffic can be approximated by the fluid-flow service discipline with fluid traffic as follows: instead of scanning through the streams with backlog, all streams with backlog will be served simultaneously in the fluid-flow service discipline. Specifically, the output capacity C will be divided evenly among the traffic streams with backlog: if there are m backlog streams, then the departure rate of each stream will be C/m.

Figure 8.5 gives an example in which there are two traffic streams: for traffic stream i, $A_i(t)$ and $D_i(t)$ are the cumulative arrived traffic and cumulative departed traffic;

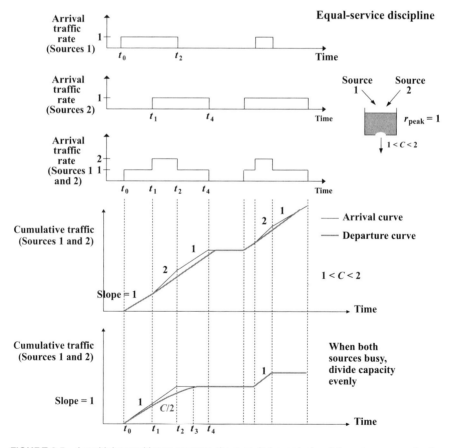

FIGURE 8.5 A multiplexer with two sources: the cumulative arrival and departure curves for the total traffic and the traffic of stream 1.

the r_{peak} of each source is 1 and $1 < C < 2$. From time t_0 to t_1, there are arrivals from stream 1 but not from stream 2, so the arrived traffic departs immediately, since the whole capacity C is dedicated to stream 1 and $C > 1 = r_{\text{peak}}$. From time t_1 to t_2, there are arrivals from both streams 1 and 2, the total arrival rate, 2, is more than C; C is divided evenly among the two streams and each of them is served at the rate of $C/2$; backlog starts to build up for both streams. From time t_2 to t_4, only stream 2 has arrivals. But at time t_2, there is still backlog for stream 1 and therefore the service rates for both streams remain at $C/2$. At time $t_3 (t_2 < t_3 < t_4)$, the backlog of stream 1 is cleared, and therefore from time t_3 to t_4, stream 2 is served at rate C.

Strictly speaking, the preceding discussion was overly simplified and it neglected a subtle difference between the cell-traffic and the fluid-flow models. In the cell-traffic model, as far as the service discipline is concerned, each stream is in one of the two possible states: when its buffer is empty, it is in the *idle* state. An idle stream will not receive any service, and a busy stream will wait for its turn to receive service. At any one time, only one of the busy streams is served, and during its service the whole capacity C is dedicated to it to transmit a cell. In the fluid-flow service model, we assume that the capacity C can be divided and allocated to the busy streams so that they can be served simultaneously, each at a lower rate than C. In the discussion in the preceding paragraph, we attempted to divide the capacity evenly among the busy streams. But such *even distribution* of capacity cannot always be achieved in the fluid-flow model, as described below!

Consider an example. Suppose that there are $m = 2$ streams with backlog. However, there is another stream with no backlog, which is in the on state and therefore has traffic arriving at the rate of 1. Suppose that $C = 4$. Then, how should C be divided among the three streams? Dividing it evenly means each stream should be served at rate $C/3 = 4/3$, but this is more than the arrival rate of the third stream, and we are certainly not making full use of the capacity C. If we divide C only among the first two streams, then backlog immediately builds up in the third stream, making it also a busy stream and therefore C must be divided among three streams again. In other words, the service rate of the third stream alternates between 0 and $C/3$ within an infinitesimal amount of time.

A problem is with the previous definition of "busy streams." As far as the service discipline is concerned, for the fluid-flow model, it is more appropriate to treat the streams as having three possible states: (1) *idle state*: the buffer is empty and the arrival rate is zero; (2) *busy state*: the buffer is nonempty; and (3) *partially busy state*: the buffer is empty but the arrival rate is nonzero. In the above example, the third stream, which is partially busy, will be served at its input rate 1, and the two busy streams will be served at rate $(C - 1)/2 = 3/2$. Of course, in situations where $r_{\text{peak}} > C$, it is not possible to have the partially busy state.

In general, the fluid-flow service discipline that attempts to approximate the round-robin discipline will first see if it is possible to serve each non-idle stream at rate $C/(m + n)$, where m is number of busy streams and n is the number of partially busy streams. Among the partially busy streams, those with arrival rate $r_i < C/(m + n)$ will be served at rate r_i. If none of the partially busy streams has arrival rate less than

$C/(m+n)$, then all non-idle streams will be served at rate $C/(m+n)$. Furthermore, those partially busy streams with $r_i > C/(m+n)$ quickly become busy.

Let us suppose that some partially busy streams do in fact have $r_i < C/(m+n)$. Then, these partially busy streams remain partially busy and there is no backlog buildup for them. The multiplexer will attempt to allocate the remaining transmission capacity (after deducting the rates already allocated to the first group of partially busy streams) among the rest of the streams. Let us say, there are C' capacity remaining and n' partially busy streams remaining. Then, equal division of C' yields $C'/(m+n')$. However, the arrival rate of the n' partially busy streams may be less than $C'/(m+n')$, in which case the rate allocated to them will also be equal to their input rates, and there is no buildup of backlog and these partially bust streams remain partially busy. This process is repeated until all remaining non-idle streams have arrival rate greater than the equipartition of the remaining capacity.

Given a particular service discipline, if we know the arrival patterns of the traffic streams, then we can predict the delay and backlog experienced by each of them. We shall see in Section 8.4 that if the traffic streams are shaped at the source using a leaky-bucket mechanism before they enter the network, then analyzing piecewise linear curves as above in a deterministic way can give us the worst-case delay and backlog within the network. In general, the arrival patterns could be stochastic rather than deterministic. In the stochastic situation, drawing arrival and departure curves gives us some sense of the performance to be expected, but does not readily yield statistical performance data such as average delay and the variance of delay. In the next section, we shall focus on a stochastic analysis of the system and derive performance data such as average delay and buffer-overflow probability.

8.2 FLUID-FLOW ON–OFF SOURCE MODEL: STOCHASTIC TREATMENT

We have discussed the on–off source model but have not mentioned the statistical distributions of the on and off periods, which are needed in a stochastic analysis. In the following analysis, we assume that the on and off periods, T_{on} and T_{off}, are independent and exponentially distributed. Furthermore, we assume that the unit of time measurement has been normalized so that the average on time is 1. The on time and off time have the following probability densities:

$$p_{T_{on}}(t) = e^{-t},$$
$$p_{T_{off}}(t) = \lambda e^{-\lambda t}, \qquad (8.3)$$

where $1/\lambda$ is the average off time. Figure 8.6(a) shows the input traffic to which a multiplexing system with two such on–off sources is subjected. The arrivals from the two sources are assumed to be independent. The *state* $S(t)$ of the system is defined to be the number of on sources, so it is either 0, 1, or 2. The associated continuous-time Markov chain is shown at the bottom of the figure. At state 0, the rate of transition to state 1 is 2λ. That is, in an infinitesimal amount of time Δt, the probability of

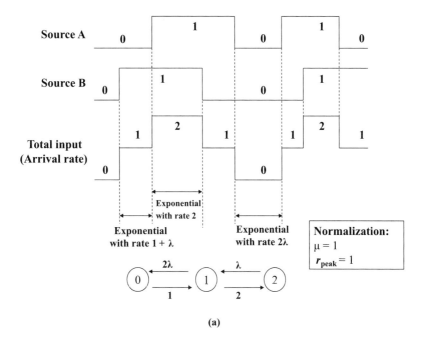

FIGURE 8.6 (a) Input traffic of a multiplexing system with two on–off sources and its associated Markov chain; (b) the Markov chain for the input traffic of a multiplexing system with N sources.

transition, $\Pr[S(t + \Delta t) = 1 | S(t) = 0] \approx 2\lambda \Delta t$. More exactly, the rate of transition has the following interpretation:

$$\begin{aligned}
\Pr[S(t + \Delta t) = 1 | S(t) = 0] &= \Pr[S(\Delta t) = 1 | S(0) = 0] \\
&= \Pr[\text{source 1 or source 2 is on at time } t + \Delta t] \\
&= \Pr[\text{source 1 turns on in time interval } \Delta t] \\
&\quad + \Pr[\text{source 2 turns on in time interval } \Delta t] + O(\Delta^2 t) \\
&= 2 \int_0^{\Delta t} P_{T_{\text{off}}}(\tau) d\tau \\
&= 2(1 - e^{-\lambda \Delta t}) + O(\Delta^2 t) \\
&= 2\lambda \Delta t + O(\Delta^2 t),
\end{aligned} \qquad (8.4)$$

where $O(\Delta^2 t)$ represents all terms of order $\Delta^2 t$ or higher: this includes, for example, the probability of multiple transitions, such as both sources turn on in time interval Δt. By the same argument, we can show that

$$\Pr[S(t+\Delta t) = 0 | S(t) = 0] = 1 - 2\lambda \Delta t + O(\Delta^2 t),$$
$$\Pr[S(t+\Delta t) = 2 | S(t) = 0] = O(\Delta^2 t),$$
$$\Pr[S(t+\Delta t) = 0 | S(t) = 1] = \Delta t + O(\Delta^2 t),$$
$$\Pr[S(t+\Delta t) = 1 | S(t) = 1] = 1 - (1+\lambda)\Delta t + O(\Delta^2 t),$$
$$\Pr[S(t+\Delta t) = 2 | S(t) = 1] = \lambda \Delta t + O(\Delta^2 t),$$
$$\Pr[S(t+\Delta t) = 0 | S(t) = 2] = O(\Delta^2 t),$$
$$\Pr[S(t+\Delta t) = 1 | S(t) = 2] = 2\Delta t + O(\Delta^2 t),$$
$$\Pr[S(t+\Delta t) = 2 | S(t) = 2] = 1 - 2\Delta t + O(\Delta^2 t). \quad (8.5)$$

Let $P_i(t)$ be the probability of state i at time t. We can write

$$P_i(t+\Delta t) = \sum_j P_j(t) \Pr[S(t+\Delta t) = i | S(t) = j], \quad i = 0, 1, 2. \quad (8.6)$$

From (8.5) and (8.6), we can get

$$\frac{dP_0}{dt} = \lim_{\Delta t \to 0} \frac{P_0(t+\Delta t) - P_0(t)}{\Delta t} = -2\lambda P_0(t) + P_1(t). \quad (8.7)$$

Note that the terms $O(\Delta^2 t)$ get canceled out in the limit $\Delta t \to 0$. Also, note that the coefficients in (8.7) are associated with the transition rates: 2λ is the transition rate out of state 0 and 1 is the transition rate into state 0 from state 1. In a similar fashion, the following can be derived from (8.5) and (8.6), yielding the same interpretation of the transition rates:

$$\frac{dP_1}{dt} = -(1+\lambda)P_1(t) + 2\lambda P_0(t) + 2P_2(t),$$
$$\frac{dP_2}{dt} = -2P_2(t) + \lambda P_1(t). \quad (8.8)$$

At equilibrium (as $t \to \infty$), $dP_i/dt \to 0$. Substituting this into (8.7) and (8.8), we get

$$P_1 = 2\lambda P_0,$$
$$P_2 = \frac{\lambda}{2} P_1 = \lambda^2 P_0, \quad (8.9)$$

where P_i denotes the equilibrium probability $P_i(t \to \infty)$. Using the fact that $P_0 + P_1 + P_2 = 1$ in (8.9), we can get

$$P_0 = 1/(1 + 2\lambda + \lambda^2) = 1/(1+\lambda)^2. \qquad (8.10)$$

In general, we note that

$$P_i = \binom{2}{i} \left(\frac{1}{1+1/\lambda}\right)^i \left(\frac{1/\lambda}{1+1/\lambda}\right)^{2-i}, \quad i = 0, 1, 2. \qquad (8.11)$$

This could have been obtained using the following more straightforward argument than the Markov chain analysis. At any arbitrary time, the probability of a source being on, denoted by q, is the ratio of the average on time divided by the sum of the average on and off times (i.e., a "cycle" of on and off times). This is $q = 1/(1+1/\lambda)$. Given that the sources are independent, the number of on sources is then given by the binomial distribution with parameter q, which yields (8.11).

Equation (8.11) does not tell us much about the multiplexing system. It only describes the source behavior. To study a multiplexer in more detail, we have to define the state to include the buffer occupancy in addition to the number of on sources. In this case, straightforward argument as in the previous paragraph does not work. This is the motivation for the more roundabout Markov chain approach: we can modify the argument to study the buffer occupancy.

Before moving onto the study of buffer occupancy, Fig. 8.6(b) shows the Markov chain of the general case with N sources. From state i, $1 < i < N$, the system can only evolve to state $i-1$ or $i+1$ in the next transition. At the boundary states, 0 and N, the transition can go only one way. The equilibrium probability is given by

$$\begin{aligned}P_i &= \binom{N}{i} \left(\frac{1}{1+1/\lambda}\right)^i \left(\frac{1/\lambda}{1+1/\lambda}\right)^{N-i} \\ &= \binom{N}{i} \lambda^i \left(\frac{1}{1+\lambda}\right)^N, \quad i = 0, \ldots, N.\end{aligned} \qquad (8.12)$$

We now consider a multiplexer system with infinite buffering space. Let $X(t)$ be the random variable for the buffer occupancy level and $S(t)$, as discussed before, be the random number denoting the number of on sources at time t. For a stable system in which $X(t)$ does not grow to infinity, the output capacity C must be larger than the average arrival rate from the N sources,

$$\bar{r} = \frac{N}{1+1/\lambda}. \qquad (8.13)$$

In the fluid-flow model, $X(t)$ is a continuous variable. Let the probability density that $S(t) = i$ and $X(t) = x$ be denoted by $p_i(t, x)$. We want to find the dynamic equation governing the evolution of $p_i(t, x)$. For $i = 1, 2, \ldots, N-1$, $p_i(t + \Delta t, x)$ can be expressed in terms of the probability densities at time t as the sum of three terms, corresponding to the events of no transition and transitions from two neighboring states $i-1$ and $i+1$. Let E_{none}, E_{on}, and E_{off}, be the events of no transition, an off source turns on, and an on source turns off in time Δt, respectively. The events of multiple transitions have probability densities of order $\Delta^2 t$. We have

$$p_i(t+\Delta t, x) = A_i + A_{i-1} + A_{i+1} + O(\Delta^2 t),$$

where

$$A_i = p(S(t)=i, E_{\text{none}}, X(t)=x-(i-C)\Delta t),$$
$$A_{i-1} = p(S(t)=i-1, E_{\text{on}}, X(t)=f(x)),$$
$$A_{i+1} = p(S(t)=i+1, E_{\text{off}}, X(t)=g(x)), \tag{8.14}$$

where $p(\cdot)$ are probability densities, and we leave $f(x)$ and $g(x)$ undefined for the time being.

The term A_i is associated with no transition in the time interval Δt. Given that there is no transition, away from the boundary where $X(t)=0$, the departure rate from the buffer is C and the arrival rate to the buffer is i. This means that the net increase in buffer occupancy from time t to time Δt is $(i-c)$. Therefore, in order that $X(t+\Delta t)=x$, we must have $X(t)=x-(i-C)\Delta t$. We can write

$$A_i = \Pr[E_{\text{none}}|S(t)=i, X(t)=x-(i-C)\Delta t] \; p_i(t, x-(i-C)\Delta t)$$
$$= [1 - \{(N-i)\lambda + i\}\Delta t] \; p_i(t, x-(i-C)\Delta t) + O(\Delta^2 t)$$
$$= [1 - \{(N-i)\lambda + i\}\Delta t] \left\{ p_i(t,x) - (i-C)\Delta t \frac{\partial p_i}{\partial x} + O(\Delta^2 t) \right\} + O(\Delta^2 t)$$
$$= [1 - \{(N-i)\lambda + i\}\Delta t] p_i(t,x) - (i-C)\Delta t \frac{\partial p_i}{\partial x} + O(\Delta^2 t), \tag{8.15}$$

where we have used the Taylor's series expansion in the second last line.

The term A_{i-1} in (8.14) is associated with an off source turning on in time interval Δt. The question is what should $f(t)$ be. Depending on when the source turns on within the time interval Δt, $f(t)$ may differ. Let $t+\Delta t'$ ($0 < \Delta t' \leq \Delta t$) be the time the source turns on. We can write

$$A_{i-1} = \int_0^{\Delta t} d(\Delta t') \; p(S(t)=i-1, E_{\text{on}}, \text{on transition at } t$$
$$+ \Delta t', X(t) = x - (i-C)\Delta t + \Delta t'), \tag{8.16}$$

where we have made use of the fact that the on transition is at time $t+\Delta t'$ to derive that $X(t) = x - (i-C)\Delta t + \Delta t'$. Note that (8.16) makes use of the marginal probability-density formula: $p(A=a) = \int_{\text{all } b} db \; p(A=a, B=b)$. Continuing from (8.16),

$$A_{i-1} = \int_0^{\Delta t} d(\Delta t') \; p(S(t) = i - 1, E_{\text{on}}, X(t) = x - (i - C)\Delta t + \Delta t')$$
$$\times p(\text{on transition at } t + \Delta t' | S(t) = i - 1, E_{\text{on}}, X(t)$$
$$= x - (i - C)\Delta t + \Delta t'). \tag{8.17}$$

Now, the conditional probability density $p(\text{on transition at } t + \Delta t' | S(t) = i - 1, E_{\text{on}}, X(t) = x - (i - C)\Delta t + \Delta t')$ does not depend on $X(t)$ given E_{on} and $S(t)$. We have

$$p(\text{on transition at } t + \Delta t' | S(t) = i - 1, E_{\text{on}}, X(t) = x - (i - C)\Delta t + \Delta t')$$
$$= p(\text{on transition at } t + \Delta t' | E_{\text{on}}, S(t) = i - 1)$$
$$= \frac{\{N - (i-1)\}\lambda e^{-\{N-(i-1)\}\lambda \Delta t'}}{1 - e^{-\{N-(i-1)\}\lambda \Delta t}}, \tag{8.18}$$

where, in the last line, the numerator is the unconditional probability density that one of the $\{N - (i-1)\}$ off sources turns on at time $\Delta t'$, and the denominator is the probability that one of the off sources turns on within the time interval Δt. Now,

$$p(S(t) = i - 1, E_{\text{on}}, X(t) = x - (i - C)\Delta t + \Delta t')$$
$$= \Pr[E_{\text{on}} | S(t) = i - 1, X(t) = x - (i - C)\Delta t + \Delta t']$$
$$p_{i-1}(t, x - (i - C)\Delta t + \Delta t')$$
$$= \{N - (i-1)\}\lambda \Delta t \times p_{i-1}(t, x - (i - C)\Delta t + \Delta t') + O(\Delta^2 t)$$
$$= \{N - (i-1)\}\lambda \Delta t \times \left[p_{i-1}(t, x) - \{(i - C)\Delta t + \Delta t'\}\frac{\partial p_{i-1}}{\partial x} + O(\Delta^2 t) \right]$$
$$+ O(\Delta^2 t). \tag{8.19}$$

Substituting (8.18) and (8.19) into (8.17) and integrating, we can get

$$A_{i-1} = \{N - (i-1)\}\lambda \Delta t p_{i-1}(t, x) + O(\Delta^2 t). \tag{8.20}$$

Note that the term associated with $\partial p_{i-1}/\partial x$ after integration has a coefficient of order $O(\Delta^2 t)$. It will disappear when we take the limit $\Delta t \to 0$ later. So, although we have taken the exact time of transition $t + \Delta t'$ into consideration and the term $\partial p_{i-1}/\partial x$ arises from this consideration, the term turns out to be not important after all. We could have ignored at the outset the $\partial p_{i-1}/\partial x$ term and just consider the transition from state $i - 1$ to state i without taking into account the change in x. The above derivation, however, shows exactly how this can be justified.

338 NETWORK TRAFFIC CONTROL AND BANDWIDTH ALLOCATION

Using a similar argument, it is routine to show that

$$A_{i+1} = (i+1)\Delta t p_{i+1}(t, x) + O(\Delta^2 t). \tag{8.21}$$

From (8.14), (8.15), (8.20), and (8.21), we can get

$$\begin{aligned}\frac{\partial p_i}{\partial t} &= \lim_{\Delta t \to 0} \frac{p_i(t + \Delta t, x) - p_i(t, x)}{\Delta t} \\ &= \lim_{\Delta t \to 0} \frac{A_i + A_{i-1} + A_{i+1} - p_i(t, x)}{\Delta t} \\ &= -\{(N-i)\lambda + i\} p_i(t, x) - (i - C)\frac{\partial p_i}{\partial x} \\ &\quad + \{N - (i-1)\}\lambda p_{i-1}(t, x) + (i+1) p_{i+1}(t, x), \end{aligned} \tag{8.22}$$

which is the dynamic equation governing the evolution of the probability densities $p_i(t, x)$. Solving this equation completely will give us the transient as well as equilibrium probability densities. For a stable system $\partial p_i/\partial x \to 0$ as $t \to \infty$, and the resulting $p_i(t = \infty, x)$ is the equilibrium probability density. Let us now study the equilibrium probability density. Define

$$f_i(x) = \lim_{t \to \infty} p_i(t, x).$$

From (8.22) we have

$$(i - C)\frac{df_i}{dx} = -\{(N-i)\lambda + i\} f_i(x) + \{N - (i-1)\}\lambda f_{i-1}(x) + (i+1) f_{i+1}(x).$$

Notice that instead of $\partial f_i/\partial x$, we wrote df_i/dx to signify the fact that there is no dependence on t. Instead of probability density, we can focus on the probability distribution that turns out to be more convenient when matching the boundary conditions later on. Define

$$F_i(x) = \Pr[X(t) \le x] = \int_0^x f_i(x')dx'.$$

In the above, we need to know $f_i(x)$ at $x = 0$. Equation (8.23), however, is valid only for $x \ge 0_+$: in the derivation of A_i in (8.15), we have assumed that $x > 0$ so that $x - (i - C)\Delta t$ is not negative. Note also that there may be a finite probability that the buffer size is 0 that translates to $f_i(0)$ being an impulse. Integrating (8.23) from $x = 0_+$, we have

$$\begin{aligned}&(i - C)\{f_i(x) - f_i(0_+)\} \\ &= -\{(N-i)\lambda + i\}(F_i(x) - F_i(0)) + \{N - (i-1)\}\lambda(F_{i-1}(x) - F_{i-1}(0)) \\ &\quad + (i+1)(F_{i+1}(x) - F_{i+1}(0)). \end{aligned} \tag{8.23}$$

It turns out that the above can be further reduced to

$$(i - C)\frac{dF_i(x)}{dx} = -\{(N - i)\lambda + i\}F_i(x) + \{N - (i - 1)\}\lambda F_{i-1}(x)$$
$$+ (i + 1)F_{i+1}(x). \quad (8.24)$$

which is of the same form as (8.23) with f_i replaced by F_i, because

$$(i - C)\frac{dF_i(x)}{dx}\bigg|_{x=0^+} = -\{(N - i)\lambda + i\}F_i(0) + \{N - (i - 1)\}\lambda F_{i-1}(0)$$
$$+ (i + 1)F_{i+1}(0). \quad (8.25)$$

Equation (8.25) can be derived using a similar argument as the derivation (8.23), but instead of probability density, probability distribution is examined. Briefly, as in (8.14) we start out by observing

$$\Pr[S(t + \Delta t) = i, X(t + \Delta t) = 0] = A_i + A_{i-1} + A_{i+1} + O(\Delta^2 t),$$

where

$$A_i = \Pr[S(t) = i, E_{\text{none}}, X(t) \leq -(i - c)\Delta t],$$
$$A_{i-1} = \Pr[S(t) = i - 1, E_{\text{on}}, X(t) \leq f],$$
$$A_{i+1} = \Pr[S(t) = i + 1, E_{\text{off}}, X(t) \leq g], \quad (8.26)$$

where, as mentioned before, f and g depend on the analysis of the transition time but turn out to be not important when the limit $\Delta t \to 0$ is taken: that is, both f and g can be treated as 0. Using the same argument as before, it can be shown that

$$A_i = [1 - \{(N - i)\lambda + i\}\Delta t]\Pr\{S(t) = i, X(t) = 0\}$$
$$- (i - c)\Delta t p_i(t, 0_+) + O(\Delta^2 t),$$
$$A_{i-1} = \{N - (i - 1)\}\lambda \Delta t \Pr\{S(t) = i - 1, X(t) = 0\} + O(\Delta^2 t),$$
$$A_{i+1} = (i + 1)\Delta t \Pr\{S(t) = i + 1, X(t) = 0\} + O(\Delta^2 t). \quad (8.27)$$

From the above, (8.25) can be derived by considering the equilibrium probability distribution in exactly the same way as discussed before. Note that in (8.27), $A_i = 0$ for all $i > C$ because $\Pr\{S(t) = i, X(t) = 0\} = 0$ for all $i > C$ due to the fact that any empty buffer fills up in no time at all in the fluid-flow model when the input rate is greater than the output rate. Equation (8.25) remains valid, however, so long as we have this understanding: indeed, this fact will be used later to match the boundary conditions.

We now move to the second part of the derivation: solving the dynamic equation (8.24). Define an $(N+1)$-dimensional vector

$$\mathbf{F}(x) = [F_0(x)\ F_1(x)\ F_2(x)\ \cdots\ F_N(x)]^T. \tag{8.28}$$

Then (8.24) can be written in vector form as

$$\mathbf{D}\frac{d}{dx}\mathbf{F}(x) = \mathbf{M}\mathbf{F}(x),\ x \geq 0, \tag{8.29}$$

where $\mathbf{D}$ is an $(N+1) \times (N+1)$ diagonal matrix:

$$\mathbf{D} = \text{diag}\{-c,\ 1-c,\ 2-c,\ \ldots,\ N-c\} \tag{8.30}$$

and $\mathbf{M}$ is an $(N+1) \times (N+1)$ tridiagonal matrix:

$$\mathbf{M} = \begin{bmatrix} -N\lambda & 1 & 0 & \cdots & 0 & 0 \\ N\lambda & -((N-1)\lambda+1) & 2 & \cdots & 0 & 0 \\ 0 & (N-1)\lambda & -((N-2)\lambda+2) & \cdots & 0 & 0 \\ \vdots & \vdots & \vdots & \vdots & \vdots & \vdots \\ 0 & 0 & 0 & \cdots & -(\lambda+(N-1)) & N \\ 0 & 0 & 0 & \cdots & \lambda & -N \end{bmatrix}.$$

$$\tag{8.31}$$

Equation (8.29) is a system of linear differential equations. The standard way of solving it is as follows. Let us assume that $\mathbf{F}$ has the form

$$\mathbf{F} = e^{sx}\boldsymbol{\Phi}, \tag{8.32}$$

where

$$\boldsymbol{\Phi} = [\phi_0\ \phi_1\ \cdots\ \phi_N]^T.$$

We want to find s and $\boldsymbol{\Phi}$, which are referred to as the eigenvalue and eigenvector, respectively. Substituting (8.32) into (8.29), we get

$$\mathbf{D}se^{sx}\boldsymbol{\Phi} = \mathbf{M}e^{sx}\boldsymbol{\Phi}, \tag{8.33}$$

which gives

$$(\mathbf{D}^{-1}\mathbf{M} - s\mathbf{I})\boldsymbol{\Phi} = 0, \tag{8.34}$$

where $\mathbf{D}^{-1}$ is the inverse of $\mathbf{D}$ and $\mathbf{I}$ is the $(N+1) \times (N+1)$ identity matrix. In order that the solution to $\boldsymbol{\Phi}$ is not the trivial solution (i.e., $\boldsymbol{\Phi} = 0$), the matrix $(\mathbf{D}^{-1}\mathbf{M} - s\mathbf{I})$

must not have an inverse: otherwise, multiplying the LHS and RHS of (8.34) by the inverse will yield the trivial solution. To have no inverse,

$$\det(\mathbf{D}^{-1}\mathbf{M} - s\mathbf{I}) = 0. \tag{8.35}$$

This translates to solving for the roots of an $(N+1)$-order polynomial equation in s. The $(N+1)$ roots, $s_0, s_1, \ldots, s_N$ are the eigenvalues. Corresponding to each root s_i is an eigenvector $\mathbf{\Phi}_i$ that can be solved by substituting s_i into (8.34).

After solving for all the eigenvalues and eigenvectors, the general solution of (8.29) is given by

$$\mathbf{F}(x) = \sum_i a_i e^{s_i x} \mathbf{\Phi}_i, \tag{8.36}$$

where the coefficients $\{a_i\}$ are determined by boundary conditions. Although in principle the above general method can be used, numerical instability may result if we rely on numerical computation by a computer. Also, the special structures of the matrices $\mathbf{D}$ and $\mathbf{M}$ allow us to carry the analysis further to yield further insights.

Because of the diagonal structure of $\mathbf{D}$ and tridiagonal structure of $\mathbf{M}$, we can write from (8.29) and (8.32) the following:

$$s(i - C)\phi_i = \lambda(N + 1 - i)\phi_{i-1} - \{(N - i)\lambda + i\}\phi_i + (i + 1)\phi_{i+1},$$
$$0 \leq i \leq N. \tag{8.37}$$

Define the generating function of ϕ_i as

$$\tilde{\Phi}(z) = \sum_{i=0}^{N} \phi_i z^i. \tag{8.38}$$

It is straightforward to verify that

$$\frac{\tilde{\Phi}'(z)}{\tilde{\Phi}(z)} = \frac{sC - N\lambda + N\lambda z}{\lambda z^2 + (s + 1 - \lambda)z - 1}, \tag{8.39}$$

where $\tilde{\Phi}'(z) = d\tilde{\Phi}/dz$. Let z_1 and z_2 denote the roots of the denominator $\lambda z^2 + (s + 1 - \lambda)z - 1$:

$$z_1, z_2 = \frac{-(s + 1 - \lambda) \pm \sqrt{(s + 1 - \lambda)^2 + 4\lambda}}{2\lambda}. \tag{8.40}$$

We can write

$$\frac{\tilde{\Phi}'(z)}{\tilde{\Phi}(z)} = \frac{sC - N\lambda + N\lambda z}{\lambda(z - z_1)(z - z_2)} = \frac{C_1}{z - z_1} + \frac{C_2}{z - z_2}, \tag{8.41}$$

where

$$C_1 = \frac{sC - N\lambda + N\lambda z_1}{\lambda(z_1 - z_2)}, \tag{8.42}$$

$$C_2 = N - C_1. \tag{8.43}$$

From (8.41), we have the following indefinite integration:

$$\int \frac{d\tilde{\Phi}}{\tilde{\Phi}} = \int dz \left(\frac{C_1}{z - z_1} + \frac{C_2}{z - z_2} \right), \tag{8.44}$$

the solution of which is $\ln \tilde{\Phi} = C_1 \ln(z - z_1) + C_2 \ln(z - z_2) + \text{constant}$, or

$$\tilde{\Phi}(z) = K(z - z_1)_1^C (z - z_2)_2^C = K(z - z_1)_1^C (z - z_2)^{N-C_1}, \tag{8.45}$$

where K is a constant. The eigenvector of an eigenvalue is unique modulus, a constant. That is, if Φ_i is an eigenvector of s_i, so is $K\Phi_i$, where K is a constant. At this point, only the relative magnitudes of the components of an eigenvector are important. When all the eigenvectors and eigenvalues are found, boundary conditions will be matched using (8.36), at which point, the constant a_i will be adjusted. And so we may ignore K in (8.45) and write

$$\tilde{\Phi}(z) = (z - z_1)^{C_1}(z - z_2)^{N-C_1}. \tag{8.46}$$

In (8.42), C_1 is expressed in terms of the eigenvalue s, but s is not derived yet. We may turn the problem around and take a different standpoint: if C_1 is known, then the associated eigenvalue s can be found. There are two important observations on $\tilde{\Phi}(z)$ that narrow down the possibilities for C_1:

1. By definition in (8.38), $\Phi(z)$ is a polynomial of z of degree N.
2. From (8.40), it can be seen that $z_1 \neq z_2$.

The two observations, applied on (8.46), imply that C_1 must be an integer in $[0, N]$. That is,

$$\tilde{\Phi}(z) = (z - z_1)^k (z - z_2)^{N-k}, \quad k = 0, 1, \ldots, N. \tag{8.47}$$

As shown below, each k corresponds to two eigenvalues. Substituting $C_1 = k$ as well as z_1 and z_2 from (8.40) into (8.42), we can get the following second-order polynomial of s:

$$A(k)s^2 + B(k)s + C(k) = 0, \quad k = 0, 1, \ldots, N, \tag{8.48}$$

where

$$A(k) = (N/2 - k)^2 - (N/2 - C)^2, \qquad (8.49)$$

$$B(k) = 2(1 - \lambda)(N/2 - k)^2 - N(1 + \lambda)(N/2 - C), \qquad (8.50)$$

$$C(k) = -(1 + \lambda)^2 \{(N/2)^2 - (N/2 - k)^2\}. \qquad (8.51)$$

Note from the above that $A(k) = A(N - k)$, $B(k) = B(N - k)$, and $C(k) = C(N - k)$. Therefore, substituting k and $(N - k)$ in (8.48) yields the same eigenvalues. It is easily seen that the number of eigenvalues for $k = 0, \ldots, \lfloor N \rfloor$ is $(N + 1)$ for N odd. For N even, $k = N/2$ yields a repeated real root in (8.48), and so the number of distinct eigenvalues is also $(N + 1)$. It can also be shown that among the $(N + 1)$ eigenvalues, $N - \lfloor C \rfloor$ are negative, one is 0, and $\lfloor C \rfloor$ are positive. Furthermore, they are all distinct. Label the negative eigenvalues $s_0, s_1, \ldots, s_{N-\lfloor C \rfloor - 1}$, the 0 eigenvalue s_N, and the positive eigenvalue $s_{N-\lfloor C \rfloor}, \ldots, s_{N-1}$. The positive eigenvalues correspond to unstable solutions in (8.36) because $e^{s_i x} \to \infty$ as $x \to \infty$, and therefore they can be eliminated from consideration (i.e., the associated $a_i = 0$ in (8.36)), since they do not make physical sense. The negative and zero eigenvalues correspond to stable solutions.

Associated with $k = 0$ is the zero eigenvalue and the largest negative eigenvalue. Label this largest negative eigenvalue by s_0. It is

$$s_0 = \frac{-(1 + \lambda - N\lambda/C)}{(1 - c/N)} = -\frac{(1 + \lambda)(1 - \rho)}{(1 - c/N)}, \qquad (8.52)$$

where $\rho = N\lambda/C(1 + \lambda)$ is the traffic load. Now, as $x \to \infty$, $e^{s_i x} \to 0$ if $s_i < 0$. Therefore, only the term corresponding to the zero eigenvalue is left in (8.36). That is,

$$\mathbf{F}(\infty) = a_N \mathbf{\Phi}_N. \qquad (8.53)$$

Thus, we can write

$$\mathbf{F}(x) = \mathbf{F}(\infty) + \sum_{i=0}^{N-\lfloor C \rfloor - 1} a_i e^{s_i x} \mathbf{\Phi}_i. \qquad (8.54)$$

The value of $\mathbf{F}(\infty)$ can be derived either by substituting $s = 0$ into (8.34) and solving for the eigenvector or by the direct observation that

$$F_j(\infty) = \Pr[i \text{ sources are on}] = \binom{N}{j} \left(\frac{1}{1 + 1/\lambda}\right)^j \left(\frac{1/\lambda}{1 + 1/\lambda}\right)^{N-j}$$

$$= \binom{N}{j} \lambda^j \left(\frac{1}{1 + \lambda}\right)^N. \qquad (8.55)$$

Now that we know how to solve for the eigenvalues $s_0, \ldots, s_{N-\lfloor C \rfloor -1}$, the next step is to solve for the corresponding eigenvectors. Given a k and an associated eigenvalue s_i, we can use (8.40) to solve for z_1 and z_2. All these are then substituted into (8.47) and the coefficients of the resulting polynomial are the eigenvector components. Specifically, the ith component of the eigenvector is

$$\phi_i = (-1)^{N-i} \sum_{i=0}^{k} \binom{k}{j} \binom{N-k}{i-j} z_1^{k-j} z_2^{N-k-i+j}, \quad 0 \le i \le N. \quad (8.56)$$

The rest is to solve for the coefficients a_i in (8.52) by matching boundary conditions. Consider the boundary at $x = 0$ (i.e., buffer is empty). If the number of on sources $j > C$, the capacity of the output channel, then $\Pr[x = 0] = 0$, since the buffer occupancy is always increasing. Using this fact on (8.54), we have

$$F_j(0) = 0 = F_j(\infty) + \sum_{i=0}^{N-\lfloor C \rfloor -1} a_i(\mathbf{\Phi}_i)_j, \quad \lfloor C \rfloor + 1 \le j \le N, \quad (8.57)$$

where $(\mathbf{\Phi}_i)_j$ is the jth component of the eigenvector $\mathbf{\Phi}_i$. The above equation gives us $N - \lfloor C \rfloor$ equations that can be used to determine the $N - \lfloor C \rfloor$ coefficients a_i. However, it is difficult to get a closed-form solution for large N. A closed-form solution can be obtained by a different method that is beyond the scope of this book.[1] It is given by

$$a_i = -\left(\frac{\lambda}{1+\lambda}\right)^N \prod_{\substack{j=0 \\ j \ne i}}^{N-\lfloor C \rfloor - 1} \frac{s_j}{s_j - s_i}, \quad 0 \le i \le N - \lfloor C \rfloor - 1. \quad (8.58)$$

Several performance measures of interest can be derived. Let us first examine the probability that the buffer content exceeds $\bar{x}$, denoted by $G(\bar{x})$. In the above analysis, the buffer size is assumed to be infinite. In a real system, the buffer size is limited by the physical constraint. The probability $G(\bar{x})$ gives us a rough idea on how likely that a real system with a buffer size of $\bar{x}$ will overflow. In general, $G(\bar{x})$ is an upper bound on the overflow probability (see Problem 8.2), which is the proportion of time the finite buffer will overflow in a long stretch of time. The derivation of $G(\bar{x})$ is as follows:

$$G(\bar{x}) = \Pr\{\text{Queue length} > \bar{x}\}$$
$$= 1 - \sum_{i=0}^{N} F_i(\bar{x}) = 1 - (1 1 \cdots 1)\mathbf{F}(\bar{x})$$

---
[1] See Ref. [AMS82] for details.

$$= 1 - 1 - (1 1 \cdots 1)\mathbf{F}(\infty) - (1 1 \cdots 1) \sum_{i=0}^{N-\lfloor C \rfloor - 1} a_i e^{s_i \bar{x}} \mathbf{\Phi}_i$$

$$= - \sum_{i=0}^{N-\lfloor C \rfloor - 1} a_i e^{s_i \bar{x}} (1 1 \cdots 1) \mathbf{\Phi}_i. \tag{8.59}$$

To approximate a finite-buffer system with a large buffer size, we are interested in large $\bar{x}$. As $\bar{x} \to \infty$, the asymptotic behavior of $G(\bar{x})$ will be dominated by the term with the largest eigenvalue s_0, which is given by (8.52). Therefore, for large $\bar{x}$

$$G(\bar{x}) \sim -a_0 e^{s_0 \bar{x}} (1 1 \cdots 1) \mathbf{\Phi}_0. \tag{8.60}$$

For eigenvalue s_0, we have

$$z_1 = 1 - \frac{N}{C}, \quad z_2 = \frac{1}{\lambda} \frac{1}{N/C - 1}, \quad k = 0, \tag{8.61}$$

and

$$\tilde{\Phi}_0(z) = \{z + (N/C - 1)\}^N. \tag{8.62}$$

Now,

$$(1 1 \cdots 1)\mathbf{\Phi}_0 = \tilde{\Phi}_0(z = 0) = (N/C)^N. \tag{8.63}$$

Substituting the above and (8.58) into (8.60), we get

$$G(\bar{x}) \sim \rho^N \left\{ \prod_{j=0}^{N-\lfloor C \rfloor - 1} \frac{s_j}{s_j - s_0} \right\} e^{s_0 \bar{x}}. \tag{8.64}$$

Recall that the traffic intensity $\rho = N\lambda/C(1 + \lambda)$. Thus, the probability $G(\bar{x})$ varies with three parameters: the average off period $1/\lambda$ for a source, the output capacity C, and the total number of sources N. Figure 8.7 plots $G(\bar{x})$ versus buffer size $\bar{x}$. We have the following observations from (8.64) and the graph:

1. $G(\bar{x})$ is asymptotically linear in log scale with $\bar{x}$, and the slope is s_0. The asymptotical approximation is less accurate for small $\bar{x}$. This can be seen directly from (8.64).
2. For fixed C and λ, $G(\bar{x})$ becomes larger as the number of sources N increases (see Fig. 8.7(a)). This is expected because the system is then subjected to higher traffic load.
3. For fixed traffic intensity ρ and C, $G(\bar{x})$ increases as N increases. The explanation is that although the input traffic intensity ρ remains the same, the traffic is more bursty than before with a larger number of sources.

4. For fixed ρ and λ, $G(\bar{x})$ becomes smaller as N increases (see Fig. 8.7(b)). To explain this, observe that C varies in a way that is proportional to N when ρ and λ are fixed. That is, doubling N also doubles the output capacity C. Resource sharing is achieved at a higher level because more sources are sharing a larger pool of bandwidth with bandwidth per source being constant.

Instead of using $G(\bar{x})$ to approximate the overflow probability in a finite-buffer system, one may also derive the solution for the finite-buffer system directly. The eigenvalues and eigenvectors derived remain valid, and the difference is in matching the boundary conditions to find out a_i's. This is considered in Problem 8.3. Also, the overflow probability is not the probability of data loss due to overflow. The former is related to the proportion of time the system overflows and the latter is related to the proportion of input traffic that is discarded due to buffer overflow. This is considered in more detail in Problem 8.4.

Continuing on the infinite-buffer system, we now consider the delay. The average delay can be derived easily using the Little's law. Let $f(x) = \sum_{j=0}^{N} f_j(x)$ and $F(x) = \sum_{j=0}^{N} F_j(x)$. The average buffer occupancy is given by

$$E[X] = \int_0^\infty x f(x) dx = \int_0^\infty 1 - F(x) dx. \quad (8.65)$$

The second equality is a standard probability result and is derived in Problem 8.5. The average delay is given by $E[X]/\bar{r}$, where $\bar{r}$ is the average traffic arrival rate, $N\lambda/(1+\lambda)$.

To derive the delay distribution, a subtle point to note is that $F_j(x)$ is the probability distribution observed at an arbitrary point in time, which is not the same as the probability distribution observed by an arbitrary arrival. This is because the arrival rate during larger j is higher than during lower j, and therefore more arrivals see larger j than indicated by $F_j(x)$. Let

$$H_j(x) = \Pr[\text{an arrival sees state } j \text{ and } X \leq x] \quad (8.66)$$

Consider an infinitely long stretch of time T, the fraction of time the system is in state j is $TF_j(x)$. Therefore, the total amount of traffic arrivals seeing state j and $X \leq x$ is $jTF_j(x)$. The total amount of arrived traffic is $T\bar{r}$. Therefore,

$$H_j(x) = \frac{jTF_j(x)}{T\bar{r}} = \frac{jF_j(x)}{N\lambda/(1+\lambda)}. \quad (8.67)$$

We have

$$\Pr[\text{an arbitrary arrival sees } X \leq x] = \sum_{j=0}^{N} H_j(x). \quad (8.68)$$

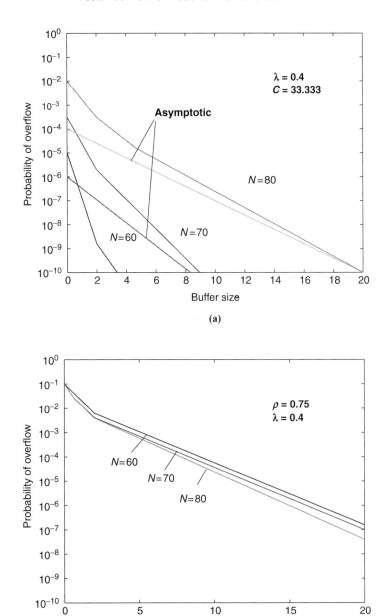

FIGURE 8.7 Probability of overflow versus buffer size for (a) fixed λ and C, and (b) fixed ρ and λ.

An arbitrary arrival seeing $X \leq x$ has a waiting time or delay of $W \leq x/C$. Thus, the delay distribution

$$\Pr[\text{delay} \leq t] = \sum_{j=0}^{N} H_j(Ct). \qquad (8.69)$$

8.3 TRAFFIC SHAPING AND POLICING

Traffic shaping and policing are functions performed at the boundary of the network at the user–network interface (UNI). The idea is that there is some sort of a contract between the user and the network that specifies the traffic that the user can inject into the network. From the user's viewpoint, the network is required to provide certain performance guarantee: for instance, the network may guarantee the user that the maximum delay experienced by a packet in the network will not exceed a certain bound. However, without any constraint on the user traffic, the network will not be able to offer performance guarantee to individual users, especially when many users start to pump excessive amount of data into the network. In such a situation, well-behaved users may experience bad performance because of greedy users. Thus, it is essential that each user be constrained somewhat. As part of the user–network contract, the allowable traffic from each user is specified in terms of a set of traffic parameters.

Traffic policing is performed by the network to ensure that the parameters specified by the user are being complied. Traffic in excess of the specification may be discarded at the UNI or transmitted as low-priority traffic that may be discarded in the network when congestion arises.

The source traffic (e.g., the traffic output from a video encoder) can have characteristics that do not conform to the specification in the contract. For instance, the source traffic may be very bursty in nature while the contract specifies that the input rate of the traffic into the network must be constant. Traffic shaping can be performed by the user on the source traffic in order to conform to the specification. Instead of injecting the source traffic into the network immediately when a burst occurs, the user may buffer the excess data outside the network so that they can be sent out later. In this way, traffic injected into the network is smoother than the traffic generated by the source.

The simplest form of a traffic-shaping device is perhaps the *simple leaky bucket* shown in Fig. 8.8(a). Here, the contract between the user and the network is specified in terms of a parameter: the channel capacity ρ. The user cannot inject bursty traffic into the network and only CBR traffic is allowed. Any data in excess of those allowed are temporarily stored in a buffer of size β.

We now use the fluid-flow model to discuss several traffic-shaping mechanisms. First, in the simple leaky bucket of Fig. 8.8(a), the buffer is analogous to a buckct with a hole at the bottom through which fluid can flow out at the rate of ρ. Figure 8.8(b)

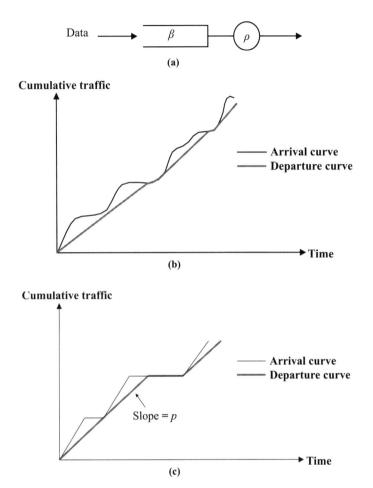

FIGURE 8.8 (a) A simple leaky bucket; (b) an example of arrival and departure curves; (c) another example of arrival and departure curves.

shows the departure pattern (traffic allowed into the network) that corresponds to a particular arrival pattern. Note that the slope of the departure curve can never be larger than ρ. When there is backlog in the buffer, the departure rate is ρ. When there is no backlog, the departure rate equals the arrival rate. Figure 8.8(c) shows the departure pattern that corresponds to another arrival pattern. The arrivals could correspond to that of an on–off source in which the traffic arrives at a rate greater than ρ during on time. In general, it is easy to see that any piecewise linear arrival curve will give rise to a piecewise linear departure curve.

In the literature, the term "leaky bucket" is more commonly used to refer to the dual-leaky-bucket system shown in Fig. 8.9(a). There are two buckets, one corresponding to the buffer for the data and the other corresponding to the buffer for a pool of tokens. Before a unit of data can enter the network, it must acquire and remove a corresponding unit of token from the token bucket. Tokens are generated at the rate of ρ and the bucket size for the tokens is σ. When the token bucket is full, the excess tokens being generated are simply discarded.

In the model, as long as there are enough tokens in the token bucket, any arriving data can depart immediately: that is, there is no restriction on the instantaneous departure rate. Over the long term, however, the average rate at which the data enter the network cannot be higher than ρ, the token generation rate. For the same arrival patterns in Fig. 8.8(b) and (c), Fig. 8.9(b) and (c) shows the departure patterns in the dual-bucket system. In addition to the arrival and departure patterns, the total amount

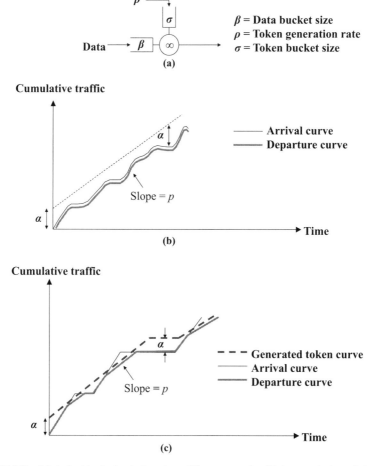

FIGURE 8.9 (a) A dual-leaky-bucket system; (b) an example of token, arrival, and departure curves; (c) another example of token, arrival, and departure curves.

of tokens that have been generated is also shown, assuming initially the token bucket is full.

Note that the data that arrive when the token bucket is nonempty can depart immediately. Comparing Figs. 8.8(b) and 8.9(b), we see that in Fig. 8.9(b), all arrivals depart immediately because the token bucket is never empty. Also note that it is not possible for both the token bucket and the data bucket to be nonempty at the same time: the data backlog would have used the tokens to depart the system. Because of this, with the fluid-flow model, the departure curve tracks either the token curve (when the data bucket is nonempty) or the arrival curve (when the token bucket is nonempty). Problem 8.6 makes use of this fact to perform a stochastic analysis of the system.

Compared with the simple leaky bucket, the dual leaky bucket allows a certain degree of burstiness in the traffic that enters the network. In the simple leaky-bucket system, the maximum amount of data allowed into the network in a duration of T is ρT. Assuming a full token bucket initially, the dual-bucket system can allow a maximum amount of data equal to $\rho T + \sigma$. For small T, we see that the data can enter the network as a large burst. But over a long period (i.e., $\rho T \gg \sigma$), the data departure rate averages out to be approximately ρ in the worst case.

In actual implementation, the tokens and token bucket need not be implemented as actual physical entities. We only need a counter to keep track of the amount of tokens in the token bucket. For discrete cell traffic (as opposed to the fluid-flow traffic), each time a cell departs, the counter is decremented, and each time a token is generated (assuming the tokens are also generated in a discrete manner), the counter is incremented.

Figure 8.10(a) shows a more sophisticated three-bucket system. In addition to the two buckets in the dual-bucket system, an additional data bucket is added to the output of the dual-bucket system. This additional bucket has a bucket size of σ and a service rate of C. There are two reasons for this model. The first is that it models the physical situation more realistically. For instance, the link between the UNI and the network may have only a finite capacity C rather than ∞, which is assumed in the dual-bucket system. The second applies to the situation where the physical link has capacity more than C: the additional constraint is introduced so as to reduce the burstiness of traffic entering the network. The maximum amount of traffic that can enter the network in this system in a period of time T is $\min(\rho T + \sigma, CT)$.

Note that it does not make sense in the three-bucket system to have a $\rho > C$. Otherwise, the traffic entering the third bucket may saturate the buffer when the arrivals are bursty, resulting in lost data; meanwhile, the traffic cannot depart at a rate higher than C anyway: that is, increasing ρ beyond C does not relax the burstiness constraint on the data entering the network. If $\rho = C$, then the system is equivalent to a simple leaky-bucket system (see Problem 8.8). Thus, only when $C > \rho$ does it make sense to have a three-leaky-bucket system. When $C > \rho$, given that the token bucket size is σ, the backlog in the third bucket can never be more than σ (why?), and therefore as shown in the figure, the size of the third bucket is also σ. Figure 8.10(b) and (c) shows the departure curves for the same arrival patterns as discussed before. In general, with $C > \rho$, the departure curve of the

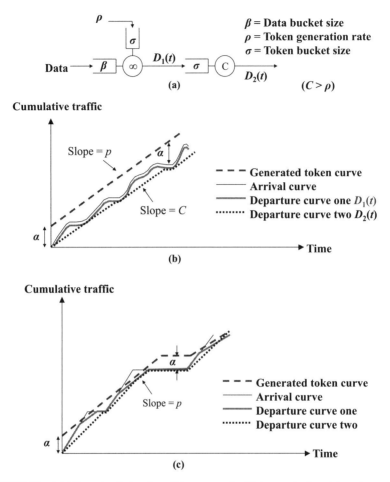

FIGURE 8.10 (a) A three-bucket system; (b) an example of token, arrival, and departure curves; (c) another example of token, arrival, and departure curves.

three-leaky-bucket system is between those of the simple and dual-leaky-bucket systems.

Once the traffic enters the network, it may traverse several nodes and links before reaching its destination. The fluid-flow model can again be used to study the delay and backlog within the network. As an example, consider Fig. 8.11(a), in which a dual-leaky-bucket traffic shaper is used to control the traffic entering the network. In the network, the traffic passes through a switch and exits on one of its outputs.

Recall that a physical link in the network can generally be shared among many VCs. In this example, the switch assigned a bandwidth of ρ' to the VC concerned. Figure 8.11(b) shows the token generation curves, and the arrival and departure curves

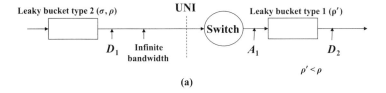
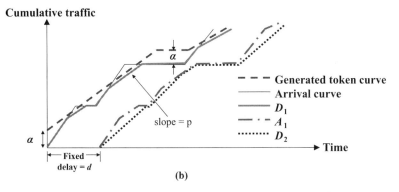
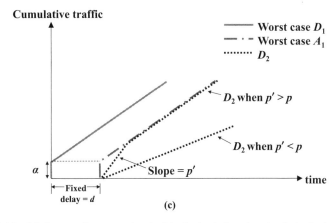

FIGURE 8.11 (a) A cascade curve of a dual leaky bucket and a simple leaky bucket; (b) an example of cumulative curves; (c) worst-case arrival curve and its associated departure curves.

at various locations. Note that the traffic leaving the leaky bucket will suffer a fixed amount of propagation delay from the output of the leaky bucket to the switch and then a fixed amount of processing delay in the switch before arriving at the targeted switch output (we assume that the switch is an output-buffered switch so that the only queueing point, where delay can be variable, is at the switch output).

With an amount of bandwidth ρ' assigned to it, it is as if there is a simple leaky bucket at the switch output controlling the traffic of the source that leaves the switch. So, we can again use the fluid-flow diagram to study the delay at the output. First, suppose that $\rho' > \rho$. A question is what is the worst-case delay. The corresponding fluid-flow diagram is depicted in Fig. 8.11(c). We shall not provide a rigorous proof for the following results, as they are pretty obvious intuitively.

The worst-case delay is incurred whenever the source pumps in as much traffic as is allowed by the dual leaky bucket, and this corresponds to the linear curve shown in the diagram. Note that at time 0, the traffic entering the network is an impulse of σ, and since we assume the link between the source and the switch has infinite bandwidth, this impulse arrives at the switch output after a fixed delay d. The more realistic situation in which the link has a limited capacity C is explored in Problem 8.9.

As shown in Fig. 8.11(c), since $\rho' > \rho$, the maximum backlog occurs at time d, where d is the fixed delay between the output of the dual leaky bucket and the output of the switch. The maximum delay is suffered by the first bit of the data that enters the switch output, and this is σ/ρ'. Also illustrated in Fig. 8.11(c) is the fact that the delay and backlog will be unfounded if $\rho' < \rho$, since the slope of the departure curve will then be smaller than the slope of the arrival curve.

The above discussion concerns static assignment of bandwidth ρ' to the VC concerned. More generally, to use the output capacity more efficiently, ρ' could be a function varying with time t. For instance, when there is little traffic arriving from the other VCs sharing the same output link, we could increase $\rho'(t)$ if there is a lot of traffic from the VC concerned. This is related to the flow control issues that will be discussed in the next Section.

8.4 OPEN-LOOP FLOW CONTROL AND SCHEDULING

Open-loop control, or feedforward flow control, deals with the issue of regulating the flow of the traffic of a session through the network without feedback. The issues in open-loop flow control are listed as follows:

1. *Service guarantee*: During call setup, a session may have been guaranteed a certain amount of end-to-end transport capacity. An issue is how to ensure each session receives its proper amount of transmission capacity as its packets pass from node to node. For instance, a *greedy* session that pumps a lot of data into the network should not jeopardize the service of other sessions.

2. *Efficient use of link capacity*: Another issue is how to make use of the link capacity in an efficient manner. Ideally, if there are packets waiting to use the link, a packet should be transmitted. Also, it could turn out that the sum total of the guaranteed capacities to all the sessions passing through the link is less than the link transmission capacity. This extra capacity, if not used, will be wasted. An issue is how to allocate the extra capacity to the sessions in a fair manner so that their packets can be delivered as soon as possible.

The two issues above are addressed by how multiplexing is performed at the output links of a node. Multiplexing has been discussed very briefly in Sections 8.1 and 8.2.

The discussion in Section 8.2, in particular, dwelled on the multiplexing of homogeneous fluid-flow on–off sources. The focus was on the overall performance rather than on the performance of individual data streams. That is, not much has been said about the allocation of transmission capacity to a particular session. The performance as perceived by individual sessions was not discussed, although some simple examples were given in Section 8.1 to familiarize the reader with the use of the fluid-flow model as an intuitive tool to aid understanding.

This section is devoted to a more detailed study of the multiplexing issue. In particular, we shall investigate how to schedule the transmission of input traffic on the output channel of a multiplexer in order to guarantee the performance of individual sessions. As shown in Fig. 8.12, the situation at a particular output of an output-buffered switch is the same as that in the multiplexer. Scheduling is a form of open-loop flow control on individual sessions. At the multiplexer, the backlog consists of data from different sources. It is up to the multiplexer to schedule the order in which data will be transmitted.

8.4.1 First-Come-First-Serve Scheduling

The simplest scheduling scheme is perhaps the first-come-first-serve (FCFS) scheme. With FCFS scheduling, the data from different sessions are put into a common buffer and transmitted in a first-come-first-serve manner. This scheduling scheme is "fair" to the extent that the sources are homogeneous and there are no misbehaving sources. However, it cannot guarantee the performance of individual sessions in case there are some "greedy" sessions that pump an excessive amount of data into the multiplexer. A well-behaved session will then suffer a large delay because of the greedy sessions. This is illustrated in Fig. 8.17 using the fluid-flow model.

In Fig. 8.13, there are three sessions. The cumulative curves $A_1(t)$, $A_2(t)$, and $A_3(t)$ are the amount of traffic arrived from sessions 1, 2, and 3, respectively. The capacity of the output channel is C. The arrival rates of both sessions 1 and 2 are C from time

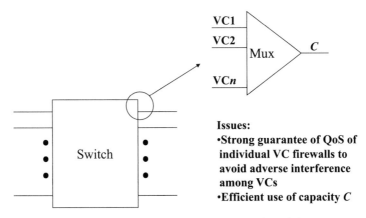

FIGURE 8.12 Multiplexing at the output of a switch.

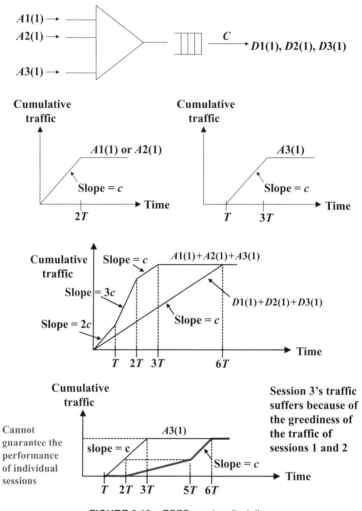

FIGURE 8.13 FCFS service discipline.

0 to $2T$. The arrival rate of session 3 is C from time T to $3T$. Between time 0 and T, the total amount of traffic arriving from sessions 1 and 2 is $2CT$, while the amount of traffic exiting the multiplexer is CT. This leaves a backlog of CT by the time the first bit of session 3 arrives. Therefore, with the FCFS service discipline, session 3's traffic must wait for another T time unit before its first bit will go out of the multiplexer. This service discipline is unfair to session 3 because it is being penalized by the arrivals of other sessions.

An alternative is to maintain a separate queue for each session. Within the queue of a session, data are still served in a FCFS fashion. However, the data from separate queues may not be served in the FCFS manner. An issue then is the order in which data should be taken from the queues for transmission on the output channel in order to avoid a greedy session from hogging the use of the output channel.

8.4.2 Fixed-Capacity Assignment

For illustration, let us first consider the simplistic fixed-capacity scheduling, as illustrated in Fig. 8.14. The output-channel capacity is divided among the sessions. In the

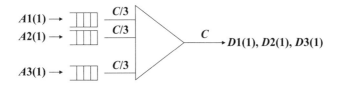

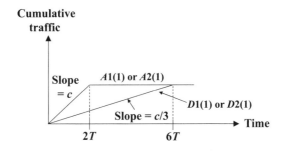

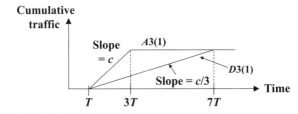

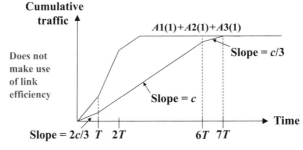

FIGURE 8.14 Fixed-capacity assignment service discipline.

figure, a fixed capacity of $C/3$ is allocated exclusively to each of the three sessions. A session's traffic cannot make use of the capacity already assigned to another session. Conceptually, it is as if the traffic of each session enters a *simple* leaky bucket (i.e., the one in Fig. 8.8(a)) with a drain rate of $C/3$. There is no interference among the sessions. This is the same situation as in that of Fig. 8.11. With the same arrival patterns as in Fig. 8.13, we see from Fig. 8.14 that session 3's traffic is not penalized because of other sessions' traffic.

If the capacity guaranteed to each session is less than or equal to $C/3$, then the above fixed-capacity scheduling can meet the commitment. More generally, it is not necessary that the sessions be allocated equal amount of capacities. The only constraint is that the sum total of the allocated capacities should not exceed C.

Unfortunately, the output channel cannot be used in an efficient manner with exclusive allocation of capacities. For example, in Fig. 8.14 the sum of the arrival rates of sessions 1 and 2 is $2C$ from time 0 to T. The output rate in the same time window is only $2C/3$, which is below the transmission capacity C of the output channel. Meanwhile, the backlogs of sessions 1 and 2 build up. A better scheduling scheme is as follows: when there is no arrival from a particular session and there is no backlog for that session, the capacity allocated to that session should be reallocated and used by other sessions with backlogs. With our example, the capacity of $C/3$ originally assigned to session 3 could have been redistributed to sessions 1 and 2 from time 0 to T. This will reduce the backlogs and delays of both these sessions. At time T, when the traffic of session 3 begins to arrive, this capacity can be returned to session 3. In this way, the capacity already guaranteed to session 4 can still be maintained.

There are many ways in which unused capacities of idle sessions can be redistributed to the active sessions. A multiplexer is said to be *work conserving* if the output-channel capacity is fully used (i.e., data are being transmitted at rate C) when there is backlog (of any session) in the multiplexer. Loosely speaking, a work-conserving multiplexer is "efficient" in the sense that it tries to use the capacity of the output channel whenever it is possible; the only time the output capacity is not fully used is when there is no backlog, and data arrive at a rate smaller than C.

FCFS scheduling is work conserving, while fixed-capacity assignment is not. The FCFS scheduling scheme suffers from the lack of guarantees to individual sessions, while the fixed-capacity assignment scheme is not efficient because it is not work conserving. There are many scheduling schemes that can maintain the capacity guaranteed to each session while being efficient. We shall discuss two general schemes here.

8.4.3 Round-Robin Scheduling

We have used the fluid-flow model to simplify the discussion in the above. In practice, traffic must be transmitted at the output channel in a packet-by-packet manner. Thus, the above discussion, particularly the discussion of the fixed-capacity assignment scheme, is only an approximation to what happens in reality. In the example of the fixed-assignment scheme, what happens in reality is that the queues of the sessions

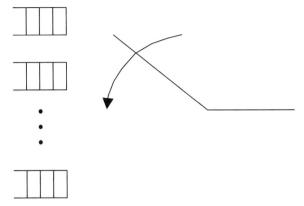

FIGURE 8.15 Round-robin scheduler.

are scanned in a one-by-one fashion. One packet is taken out of each nonempty queue whenever it is scanned. If the queue is empty, no packet is transmitted on the output and the capacity is wasted. A more efficient scheme is to skip directly to a nonempty queue whenever an empty queue is encountered in order to save the time slot. This is the idea underpinning the round-robin scheduling scheme.

The round-robin service discipline is illustrated in Fig. 8.15. The scheduler scans the queues in a round-robin fashion: whenever it encounters a nonempty queue, it outputs the head-of-line packet from the queue and moves on to the next queue; if a queue is empty, it simply moves on to the next queue without wasting a time slot doing nothing. Therefore, it is a work-conserving service discipline.

With N sessions, the worst-case *service* time for a head-of-line packet, T_S, defined as the maximum amount of time it can spend at the head of its queue in the worst case, is N packet times, and this occurs when the queues of all N sessions are nonempty. To see this, consider a packet that moves into the head of its queue immediately after the transmission of the previous head-of-line packet. If all the other sessions have packets, it will take $N - 1$ packet times before the scheduler visits its queue again. Another unit of time is needed to transmit the packet itself. In the round-robin scheme, T_S is also the worst-case cycle time T_C, which is the maximum amount of time before the scanning pattern repeats itself.

There is a minimum performance guarantee to each session in that each session gets a capacity of at least C/N, and the negative effect of a greedy session on other sessions is restrained. It is also possible to construct a more sophisticated round-robin scheduler that takes into account the varying service requirements and priorities of different sessions. For instance, a scheduler may serve more than one packet from a high-priority queue in each *cycle* of service when it scans through all queues. Specifically, for each queue i, we may assign a number n_i ($n_i \geq 1$) that represents the maximum number of packets that will be served each time the scheduler serves the queue. One shortcoming is that T_S would then be larger because T_C is. Specifically,

360 NETWORK TRAFFIC CONTROL AND BANDWIDTH ALLOCATION

T_C is given by

$$T_C = \sum_{j=1}^{N} n_j \qquad (8.70)$$

and the T_S for a particular session i is given by

$$T_S = \sum_{j \neq i} n_j + 1. \qquad (8.71)$$

For a session with $n_i = 1$, its T_S is the same as T_C. For a session with $n_i > 1$, T_S is smaller. It can be easily seen that the worst-case service time would be larger when many queues have $n_i > 1$ and/or when some n_i's are large.

To solve this problem, one may introduce mini-cycles within the original cycle. In each mini-cycle, at most one packet will be served from each queue. In the first mini-cycle, all queues will be scanned; in the second mini-cycle, all queues with $n_i \geq 2$ will be scanned; and in general, in the jth mini-cycle, all queues with $n_i \geq j$ will be scanned. The last mini-cycle within a cycle is the n_{max}th mini-cycle, where $n_{max} = \max_i n_i$, and after that a new scanning cycle begins. For those sessions with $n_i = 1$, the worst-case service time is still $\sum_{i=1}^{N} n_i$, but these are presumably the low-priority sessions. In general, for a session l with $n_l = k$, the worst-case service time can be calculated as follows.

With the aid of Fig. 8.16, we examine how long in the worst case a packet has to wait before the scheduler scans its queue again. The kth mini-cycle is the last mini-cycle within the current cycle that the queue will be served. In the kth mini-cycle, suppose that immediately after the scheduler finishes with serving queue l, a packet from queue l moves to the head of line. The question is what is the worst-case service time for this packet. The scheduler will not scan queue l again until the next overall cycle begins. We assume that within each mini-cycle, the queues with lower indices

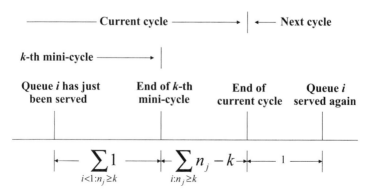

FIGURE 8.16 Computation of the worst-case service time of queue L in a round-robin scheduler.

are scanned before the higher one. The time till the end of the kth mini-cycle after a packet from queue l is served is $\sum_{i>l:n_i\geq k} 1$, or the number of sessions with $n_i \geq k$ and $i > l$. The time from the end of the kth mini-cycle till the end of the current cycle is $\sum_{i:n_i>k}(n_i - k)$. Thus, the time from the end of the service of the packet from queue l till the end of the current cycle is

$$\sum_{i>l:n_i\geq k} 1 + \sum_{i:n_i>k} (n_i - k). \tag{8.72}$$

From the beginning of the next cycle until the completion of the service of the next packet from queue l is l (i.e., there are $(l - 1)$ queues that will be scanned before queue l and it takes one packet time to transmit the packet of queue l). Thus, for session l, the worst-case service time is

$$T_S = \sum_{i>l:n_i\geq k} 1 + \sum_{i:n_i>k} (n_i - k) + l. \tag{8.73}$$

It can be seen that the introduction of the mini-cycles reduces the worst-case service time (see Problem 8.10). However, the mini-cycle scheme can be further improved. For illustration, consider a session l with $n_l = 2$, and suppose that there are many sessions with $n_i > 2$. The worst-case service time is incurred by a packet that moves to the front of the queue l immediately after the scheduler serves the queue in the second mini-cycle. The packet must wait for a long time before the next cycle begins. Certainly, the worst-case service time can be reduced if instead of the second mini-cycle, we serve queue l at a later time within the overall cycle. In particular, if the scheduler can make sure that the intervals between scanning times of queue l are equally separated, then the worst-case service time can be reduced.

For simplicity, let us first examine the problem of scheduling the service of a queue with approximately equal scanning intervals when all the sessions always have packets to send. We will retain the concept of a repeating scanning pattern in every cycle but do away with the mini-cycles. When all sessions always have packets to send, the cycle time is $T_C = \sum_{i=1}^{N} n_i$, and a queue l with $n_l = k$ must be served approximately every T_C/k time slots to minimize the T_S.

There are two ways in which it is not possible to schedule the service interval to be exactly T_C/k. The first is due to the discrete nature of time slots: the scanning interval cannot be exactly T_C/k if T_C/k is not an integer. To approximate, we may let the interval be $\lceil T_C/k \rceil$ sometimes and $\lfloor T_C/k \rfloor$ at other times. The second scenario is more subtle and it occurs even when T_C/k is an integer for all sessions. Consider an example in which the number of sessions $N = 3$, and $n_1 = 1$, $n_2 = 2$, and $n_3 = 3$. The cycle time is 6, and session 3 is to be served once every other time slot. It can be easily seen that if we serve session 3 every other slots and session 1 once every six slots, it is then not possible to serve session 2 once every three slots. In general, the scheduler must perform some computation to approximate the periodic service of once every T_C/k slots for each session, and with a large N, the task would not be simple.

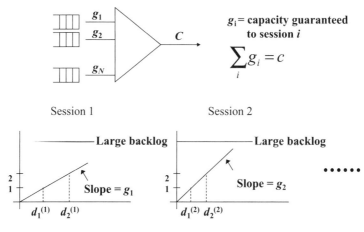

FIGURE 8.17 Fluid-flow model.

Let us consider a systematic way to compute the scanning pattern. We shall first examine the scheduling problem using the fluid-flow model again. With the fluid-flow service model, the scheduling problem is basically a way to determine how the channel capacity C can be partitioned and distributed to the sessions. Let g_i be the capacity assigned to session i. A high-priority session or a session with high input traffic can be assigned a higher g_i. The sum total of the capacities assigned to all sessions should not exceed C. That is, $\sum_{i=1}^{N} g_i \leq C$.

Let us again examine the simple scenario in which all sessions always have packets waiting to be transmitted. Session i should be able to transmit g_i bits per second on the output channel. Consider the fluid-flow model as depicted in Fig. 8.17. We picture data as being infinitely divisible fluid and that data from different sessions may "flow" out of the multiplexer *simultaneously*. The data from session i will flow out at the rate of g_i bits per second. This will be the "ideal" way to distribute the output-channel capacity. Of course, in practice, the multiplexer must transmit packets one at a time and two sessions cannot be served simultaneously. As can be seen, the fluid-flow model that distributes different amounts of capacities to different sessions is really simple. The problem is how to extend that to the packet-by-packet model. A strategy is to use "ideal" fluid-flow service model to help us compute the packet-by-packet service schedule. In other words, we can use the departure times of data in the fluid-flow model as a guide of how packets in the packet-by-packet model are to be served, as explained below.

Each packet has a certain number of bits. We define the time at which the last bit leaves the multiplexing system as the departure time. Each packet in the fluid-flow model has a departure time. We can schedule the transmission of packets in the packet-by-packet model according to these departure times. That is, the packets with earlier departure times in the fluid-flow system will be served earlier. In other words, the departure times under the fluid-flow model will be simulated and used to schedule

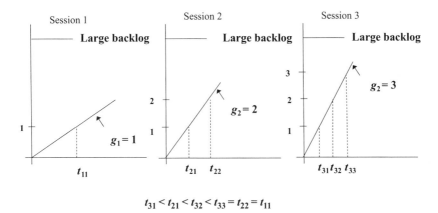

FIGURE 8.18 Departure time: time at which the last bit leaves.

the transmission of packets in the actual system. This is illustrated in Fig. 8.18. There are three sessions in the examples, with $g_1 = 1$, $g_2 = 2$, and $g_3 = 3$. The jth packet of session i is denoted by (i, j), and the departure times of the packet under the fluid-flow model are denoted by t_{ij}. As shown, the transmission schedule should accordingly be

$$(3, 1), (2, 1), (3, 2), (3, 3), (2, 2), (1, 1), \ldots \tag{8.74}$$

Note that $t_{11} = t_{22} = t_{33}$ and we have a tie for three packets, in which case the actual ordering can be determined in an arbitrary fashion. Also, the departure time of a packet in the "real" schedule is different from its departure time in the "simulated" schedule in the fluid-flow model. Let us assume g_i and C are expressed in terms of packets per second. Then $t_{31} = 1/3$ s. However, its actual departure time in the "real" multiplexer is $t = 1/6$ s, ahead of t_{31}. The departure time of a packet in the actual packet-by-packet system is therefore different from that in the simulated fluid-flow model.

The above schedule determines the order of transmission in a cycle of the round-robin scheduling scheme. Corresponding to (8.74) is the order of transmission: session 3, session 2, session 3, session 3, session 2, and session 1. The scanning pattern keeps repeating itself. Note that the transmission times of session 3's packets are sometimes separated by zero time slots, sometimes by one time slot, and sometimes by two time slots. This is somewhat different from the ideal of one time slot, and it is due to the arbitrary tie breaking of simultaneous departures.

Now, in practice some of the queues may not have packets to send when they are scanned. In this case, as mentioned before, the multiplexer simply looks to the next queue in the scanning pattern for a packet to transmit. In this way, the busy sessions are being served more often when some of the other sessions are idling. In fact, a busy session i receives more than its allocated share of capacity g_i when there are some idle sessions.

8.4.4 Weighted Fair Queueing

The fluid-flow model is used to construct the scanning pattern in a cycle in the above. Weighted fair queueing is another work-conserving service discipline. It also uses the departure times of packets under the fluid-flow service model to schedule the actual transmission of the packets. The concept of a repetitive scanning cycle is dropped altogether. The system chooses the packet with the smallest departure time for transmission rather than relying on the scanning-cycle pattern to decide whom to serve next.

For explanation, we first consider the scenario in which all sessions have backlog packets in the multiplexer. With reference to the example in Fig. 8.19, we now explain the calculation of the departure time of a packet. Let $a_n^{(i)}$ be the arrival time of the nth packet from session i. The arrival time is defined to be the time by which the last bit of the packet has arrived. Let $d_n^{(i)}$ be the departure time of this packet under the fluid-flow model. Suppose that the capacity g_i is expressed in terms of packets per second. Then

$$d_n^{(i)} = \max[a_n^{(i)}, d_{n-1}^{(i)}] + 1/g_i. \tag{8.75}$$

Note that $d_n^{(i)} = a_n^{(i)} + 1/g_i$ if the $(n-1)$th packet has already departed by the time the nth packet has arrived, and $d_n^{(i)} = d_{n-1}^{(i)} + 1/g_i$ if the $(n-1)$th packet is still in the multiplexer by the time the nth packet has arrived.

This scheme and the round-robin scheme differ when not all sessions are active and some sessions do not have backlogs in the multiplexer. In the round-robin scheme, the scanning cycle is retained and the scheduler will simply skip over the inactive queues to an active queue. The weighted fair queueing does not use a repetitive scanning pattern. Instead, the departure times are adjusted when some queues are inactive. In other words, the departure times are calculated in a different manner from (8.75) when some queues are empty. Let $B(t)$ denote the set of busy sessions (i.e., sessions with backlog in the fluid-flow multiplexer) at time t. In the weighted fair queueing scheme, the service rate of queue i at time t is

$$g_i'(t) = g_i \times \frac{g}{\sum_{i \in B(t)} g_i}, \tag{8.76}$$

where $g = \sum_{i=1}^{N} g_i$. For our purpose, we may assume that $g = C$. If $g < C$, we can introduce for conceptual purposes a dummy session that is assigned a capacity $C - g$ so that all the capacity C will be used.

Consider the nth arriving packet from session i again. Let $T_h = (\max[a_n^{(i)}, d_{n-1}^{(i)}], d_n^{(i)}]$ be the time interval during which this nth packet is at the head of the queue.

Weighted fair queueing

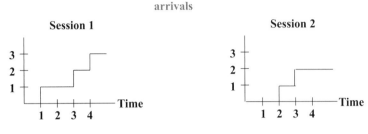

- $d_j^{(1)}$ are sorted. Service performed according to the order
- $d_1^{(1)}, d_1^{(2)}, d_2^{(1)}, d_2^{(2)}, d_3^{(1)}$

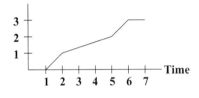 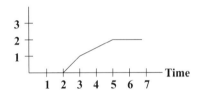

Usually, $\hat{d}_j^{(i)} \le d_j^{(i)}$, $\hat{d}_j^{(i)} - d_j^{(i)} \le \dfrac{1}{c}$

(b)

FIGURE 8.19 (a) Fluid-flow service: packet-by-packet arrivals; (b) packet-by-packet service arrival.

Suppose that $B(t)$ does not change during T_h. Then

$$d_n^{(i)} = \max[a_n^{(i)}, d_{n-1}^{(i)}] + 1/g_i'(t), \qquad (8.77)$$

where $t \in T_h$. Equation (8.77), however, is valid only when $B(t)$ does not change during T_h. If $B(t)$ changes during this time (either due to the arrival of new packets to some empty queues or due to the emptying of some backlogged queues), $g'_i(t)$ will change accordingly. But we know that the packet must accumulate the service equivalent to one packet before it can depart. The time needed to accumulate this service is

$$T_h \text{ such that } \int_{t \in T_h} g'_i(t) dt = 1, \qquad (8.78)$$

where we have assumed that $g'(t)$ has been normalized to have unit in packet per unit time. One way to interpret the above result is the following. At any time, the departure time of a head-of-line packet in a queue will be calculated assuming $B(t)$ does not change during its stay at the head of the queue. However, if $B(t)$ changes before it departs, then the departure time is recalculated, taking into account the amount of service that has already been accumulated so far and the amount of remaining service that is required before the packet can depart. The computation complexity can be quite formidable because it is possible that the departure time of a packet has to be recomputed again and again.

The computation complexity can be solved somewhat by making the observation that the departure times in the fluid-flow system are used to order the transmission of packets in the actual packet-by-packet system. As indicated earlier, they are not the actual departure times of the packets in the real system. In the real system, the scheduler will simply serve the packet with the earliest departure time. It is quite possible that while transmitting this packet (say packet A), another packet (say packet B) has arrived into the empty queue of another session and that this packet should have an earlier departure time under the fluid-flow model (see Problem 8.11). In this case, there is a discrepancy between the departure order in *fluid-flow weighted fair queueing* and that in *packet-by-packet fair queueing*, which corresponds to the actual system. But given that packet B has not arrived when the scheduler serves packet A, in order not to waste the output-channel capacity, the scheduler must go ahead and serve packet A anyway. The actual system does its best to "track" the simulated system. We shall come back to examine the maximum "extra" delay suffered by a packet under the packet-by-packet model relative to the fluid-flow model.

Returning to the issue of computation complexity of the departure times of the fluid-flow model. The observation that the actual departure times of packets are not important and that only the order of departure times is important for scheduling the service of packets in the real system allows us to simplify the computation somewhat. Let $d_A, d_B, \ldots$ be the departure times of packets $A, B, \ldots$ calculated using the method as described previously. Without loss of generality, suppose that

$$d_A \leq d_B \leq \cdots \qquad (8.79)$$

Suppose we have another way of calculating another set of departure times $d'_A, d'_B, \ldots$ for the packets such that the order is preserved:

$$d'_A \leq d'_B \leq \cdots \qquad (8.80)$$

As far as ordering the service of packets in the real packet-by-packet system is concerned, the new way of calculating the departure times is equivalent to the old way because the order of service remains the same. In particular, the departure times do not need to be the time as measured by a real clock. We now describe a method in which the departure time of a packet needs to be calculated only once. Instead of adjusting its departure time whenever $B(t)$ changes, we adjust the rate at which a "virtual" clock ticks.

Consider a real clock and a virtual clock. The real clock advances at a constant rate; the rate at which the virtual clock advances, however, may change with time. For the time being, let us say the advancement rate of the virtual clock changes in an arbitrary manner, except that it must be positive (i.e., the reading of the virtual clock cannot become smaller and smaller). Now, instead of labeling the departure times $d_A, d_B, \ldots$ of packets $A, B, \ldots$ with the readings of the real clock, suppose that we label them with the readings of the virtual clock at real times $d_A, d_B, \ldots$. Let the readings of the virtual clock be $V(d_A), V(d_B), \ldots$. Certainly

$$V(d_A) \leq V(d_B) \leq \cdots \tag{8.81}$$

In other words, the readings of the virtual clock are just as good as far as ordering packets is concerned.

We now specify the advancement rate of the virtual clock. Let $V(t)$ be the reading of the virtual clock at real time t. Then, the rate at which the virtual clock advances is

$$\frac{dV(t)}{dt} = \frac{g}{\sum_{i \in B(t)} g_i} \tag{8.82}$$

Note that the virtual time advances faster (slower) when there are fewer (more) busy sessions. This corresponds to the fact that the service rate received by a busy session is higher (lower) when there are fewer (more) busy sessions, and therefore, the virtual departure time of a packet will be reached earlier (later).

Recall that each packet needs to receive a unit of service before it can depart in the original method. Let us see how much virtual time is needed before a unit of service is accumulated. From (8.76) and (8.78),

$$\int_{\max[a_n^{(i)}, d_{n-1}^{(i)}]}^{d_n^{(i)}} g_i \times \frac{g}{\sum_{i \in B(t)} g_i} dt = 1. \tag{8.83}$$

The above and (8.82) imply that

$$\int_{V(\max[a_n^{(i)}, d_{n-1}^{(i)}])}^{V(d_n^{(i)})} g_i \, dV = 1. \tag{8.84}$$

This gives us

$$V(d_n^{(i)}) = \max[V(a_n^{(i)}), V(d_{n-1}^{(i)})] + 1/g_i, \qquad (8.85)$$

which is the formula for the virtual departure time calculation. Note that this formula is similar to (8.75), with the difference that both the arrival times and the departure times of packets are now virtual times with reference to a virtual clock. In particular, the service rate g_i in the new method is constant.

In the original method, the service rate varies with time according to $g_i'(t)$ and time advances at a constant rate. In the virtual-time method, the service rate is fixed but the time advancement rate varies. The advantage of the virtual-time method is that the virtual departure time of a packet can be immediately computed according to (8.85); this is not possible in the real-time method, because the real departure time needs to be adjusted according to the number of busy sessions in the future. Once the virtual departure time of the packet is computed, it can then be sorted together with other existing packets and their transmission ordered; there is no need to recompute the departure time of a packet again and again.

We have thus reduced the problem to that of constructing or simulating a virtual clock with the advancement rate given by (8.82). Since the advancement rate depends on $B(t)$, the set of busy sessions must be tracked. A subtlety is that $B(t)$ is still a function of the real time, and therefore, we still need to know when B changes in real time, (as opposed to in virtual time). The overall updating of the virtual time and the computation are given as follows.

Define the *busy period of the system* as the time during which there is backlog in the fluid-flow multiplexer. Whenever the system is idle with no packets in any queue, both the real time and virtual time are set to be zero. Therefore, the beginning of any busy period is always time 0. We will see how virtual time is to be updated during the busy period. Since the set B may change only when there is an arrival or when there is a departure, we only need to update the virtual times at this instance. An arrival instant is always known since it is marked by an input packet to the system. (*Note*: Arrival instants in the simulated fluid-flow model are the same as those in the actual packet-by-packet system and therefore the arrival instants in the actual system can be used for updating purposes in the simulated model.)

When the nth packet of session i arrives at real time $a_n^{(i)}$, the following computation is performed.

Computation of Virtual Departure Time

1. The virtual time is updated:

$$V(a_n^{(i)}) = \frac{dV}{dt} \times (a_n^{(i)} - t_{\text{last}}) + V(t_{\text{last}}), \qquad (8.86)$$

where t_{last} is the last time (real time) when the virtual time was updated and dV/dt was the effective virtual clock advancement rate between t_{last} and $a_n^{(i)}$.

2. The parameter t_{last} is updated:

$$t_{\text{last}} = a_n^{(i)}. \tag{8.87}$$

3. Equation (8.85) is used to compute the virtual departure time of the packet. The virtual departure time is used to schedule the transmission of the packet and will be stored.
4. If the queue of session i is empty when the packet arrives, then B will change after the arrival. In this case, update B and the virtual clock advancement rate according to (8.82).

The update of the virtual clock advancement rate due to a departure is more problematic compared to that due to an arrival. Unlike the arrival instant, the departure instant of a packet in the simulated fluid-flow model may not be the same as the actual departure instant of the packet in the actual packet-by-packet system, and therefore, we cannot use the actual departure time as the time to perform the update. In other words, the departure time under the fluid-flow model must be used for the purpose of updating. However, the stored virtual departure time of the packet does not tell us when to update the virtual clock advancement rate; it is only used to order packets for transmission purposes; the *real* time must be used. This means that, at any time, we must know the real time at which the next departure will occur in the fluid-flow model in order that we can adjust B should it change at this instant. Let t_{next} be the real time at which the next packet will depart if there are no more packet arrivals from now to t_{next}. The value of t_{next} can be computed from the virtual time $V(t_{\text{next}})$ of the packet scheduled to depart next, by the following relationship:

$$V(t_{\text{next}}) = V(t_{\text{last}}) + (t_{\text{next}} - t_{\text{last}})\frac{dV}{dt}. \tag{8.88}$$

Or equivalently,

$$t_{\text{next}} = t_{\text{last}} + \frac{[V(t_{\text{next}}) - V(t_{\text{last}})]}{dV/dt}. \tag{8.89}$$

If indeed that there is no arrival between t_{last} and t_{next}, then at t_{next} the following will be performed.

Update of Virtual Clock Rate Due to Departures

1. Check and see if the departure of the packet at t_{next} will leave its associated queue empty. If so, B will change after the departure, and so update B and the virtual clock advancement rate according to (8.82).
2. Update the virtual time: $V(t_{\text{last}}) = V(t_{\text{next}})$.
3. Update the last update time for the virtual time: $t_{\text{last}} = t_{\text{next}}$.

370   NETWORK TRAFFIC CONTROL AND BANDWIDTH ALLOCATION

4. If the system still has backlog packets after the departure of the above packet, compute the real departure time t_{next} of the next packet according to (8.89).

If there is an arrival between t_{last} and t_{next}, there are two possibilities that must be taken into account: (1) this new packet may have a smaller virtual departure time than the packet originally scheduled to depart next, especially if the service rate assigned to the session of the new packet is high; (2) the value of B may change (hence dV/dt) due to this arrival. In either case, after computing the virtual departure time of this new packet according to the procedure "Computation of Virtual Departure Time," recompute t_{next} with virtual departure time of the packet scheduled to depart next (whether it is the original or the new packet) according to (8.89).

To summarize, we have described in the above a way of computing the virtual departure times of packets for the purpose of ordering the packets for transmission. The method obviates the need to recompute the departure times of a packet again and again. Implementation-wise, with the virtual departure times, arriving packets from all sessions can be inserted into a common queue according to the values of their departure times. Of course, simultaneous arrivals and simultaneous departures are still possible in the fluid-flow model. For simultaneous departures, arbitrary tie breaking can be performed to schedule the packet to transmit next in the actual packet-by-packet system.

There is another subtlety. As described above, in the fluid-flow model, a new arrival may have a departure time earlier than all those packets already in the system. In this case, the new packet will still depart earlier than the previous head-of-line packet in the fluid-flow model. In the packet-by-packet model, however, the previous head-of-line packet is already under service when the new packet arrives and the new packet must wait until the previous packet has been totally transmitted before it can depart. This means that it is possible for the departure time of a packet in the packet-by-packet system to be later than its departure time in the fluid-flow system. Nevertheless, we shall see in the following that the worst-case increase in departure time is still quite minimal when packet lengths are small.

Consider an arbitrary packet k. Let its departure time (real) in the fluid-flow system be d_k and its departure time in the packet-by-packet system be $\hat{d}_k$. It can be shown that

$$\hat{d}_k - d_k \leq \frac{1}{g}. \tag{8.90}$$

In other words, the departure time "penalty" for the packet-by-packet system is at most $1/g$. Before explaining the result, let us consider an example shown in Fig. 8.20. There are two sessions, and $g_1 = g_2 = 0.5$. The system is not busy until the arrival of a packet from session 1 at time $a_1^{(1)} = 0$. Another packet arrives from session 2 almost immediately after that at time $a_1^{(2)} = \epsilon$, where ϵ is an arbitrarily small positive number. In the fluid-flow system, the departure times of both the first and second packets are approximately 2. In the packet-by-packet model, however, the departure time of the first packet is 1 and the departure time of the second packet is 2. It can be

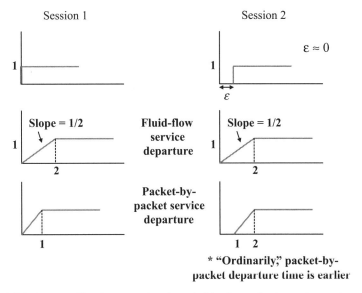

FIGURE 8.20 The departure time "penalty" for the packet-by-packet systems.

seen from this example that the departure time of a packet is usually smaller in the packet-by-packet system than in the fluid-flow system.

An exception is when a packet, say packet A, arrives just immediately after the service of another packet, say packet B, and the virtual departure time of A is smaller than that of B. In this case the virtual departure time of A would be delayed by one slot in the packet-by-packet system, and this is one situation in which (8.90) can be satisfied equally.

We now show (8.90) in general. Since both the fluid-flow and packet-by-packet systems are work conserving, their busy periods coincide. Consider a particular busy period. Let the starting time of the busy period be time zero. Suppose that we order the packets according to their departure times in the packet-by-packet system. Let p_k be the kth packet leaving the packet-by-packet multiplexer and let $\hat{d}_k$ be its departure time and a_k be its arrival time. The arrival time of p_k in the fluid-flow system is also a_k, and let d_k be its departure time.

Consider those packets that leave the packet-by-packet system before p_k does. Let m, $m < k$, be the largest integer such that

$$d_m > d_k,$$
$$d_k \geq d_{k-1} \geq \cdots \geq d_{m+1}. \tag{8.91}$$

This is illustrated in Fig. 8.21. That is, $p_{m+1}, p_{m+2}, \ldots, p_k$ leave the packet-by-packet system after p_m, but they leave the fluid-flow system before p_m. Such m may or may not exist. For the time being, we will assume such m exists and consider the case in which such m does not exist later.

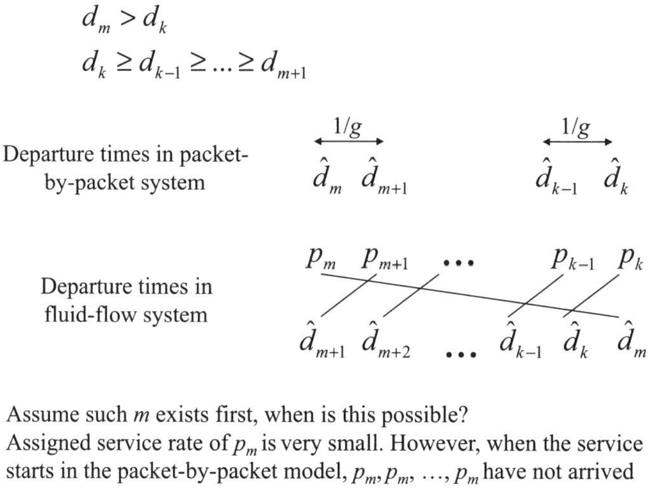

$$d_m > d_k$$
$$d_k \geq d_{k-1} \geq \ldots \geq d_{m+1}$$

Departure times in packet-by-packet system: $\hat{d}_m \; \hat{d}_{m+1}$ (spacing $1/g$) ... $\hat{d}_{k-1} \; \hat{d}_k$ (spacing $1/g$)

Departure times in fluid-flow system: $p_m, p_{m+1}, \ldots, p_{k-1}, p_k$ mapping to $\hat{d}_{m+1}, \hat{d}_{m+2}, \ldots, \hat{d}_{k-1}, \hat{d}_k, \hat{d}_m$

Assume such m exists first, when is this possible?
Assigned service rate of p_m is very small. However, when the service starts in the packet-by-packet model, $p_m, p_m, \ldots, p_m$ have not arrived

FIGURE 8.21 Departure times in (a) fluid-flow systems, (b) packet-by-packet systems.

In the packet-by-packet system, p_m starts getting served at time $\hat{d}_m - 1/g$. Now, the arrival times of $p_{m+1}, p_{m+2}, \ldots, p_k$ must be after $\hat{d}_m - 1/g$; otherwise, by (8.91), they would have been served in the packet-by-packet system before p_m, which contradicts our assumption.

Let us now look at the fluid-flow system. Packets $p_{m+1}, p_{m+2}, \ldots, p_k$ also arrive at the same time. In particular, these $(k - m)$ packets must also arrive during the time interval $(\hat{d}_m - 1/g, d_k)$. Since they also depart during this time interval, we must have

$$\left\{ d_k - \left(\hat{d}_m - \frac{1}{g}\right) \right\} \times g \geq k - m. \tag{8.92}$$

The left-hand side is the amount of service performed during this interval, and the right-hand side is the minimum amount of service needed so that these $(k - m)$ packets can arrive and leave the system within this time interval.

Let us now examine the packet-by-packet system. With reference to Fig. 8.22, we must have

$$(\hat{d}_k - \hat{d}_m) = \frac{(k - m)}{g}, \tag{8.93}$$

since the interdeparture times of successive leaving packets are always $1/g$. From (8.92) and (8.93), we get the result we want: (8.90).

Now, suppose that p_m does not exist such that (8.91) is satisfied. Then, none of $p_1, \ldots, p_{k-1}$ leaves the fluid-flow system after p_k does. In the fluid-flow system, the amount of work performed in the interval $(0, d_k]$ is

$$d_k g \geq k, \tag{8.94}$$

since by time d_k (including time d_k), k packets would have left the fluid-flow system. In the packet-by-packet system, the number of packets that have left the system by time $\hat{d}_k$ is also k, and we have

$$\hat{d}_k = \frac{k}{g}. \tag{8.95}$$

From (8.94) and (8.95), we therefore have

$$\hat{d}_k \leq d_k. \tag{8.96}$$

8.4.5 Delay Bound in Weighted Fair Queueing with Leaky-Bucket Access Control

In the subsequent discussion, we shall see how the packet delay in the weighted fair queueing system can be bounded if the arrivals from sessions are regulated by leaky buckets. These delay bounds are necessary for us to provide deterministic delay guarantees for each data session in the system.

A leaky-bucket scheme is shown in Fig. 8.9(a). Tokens here are generated at a fixed rate, ρ, and packets can be released into the network only after removing the required number of tokens from the token bucket that contains at most σ bits worth of tokens. There is no bound on the number of packets that can be buffered. The traffic is constrained to leave the bucket at a maximum rate of $C > \rho$.

Let $A_i(\tau, t)$ denote the amount of traffic from session i that leaves the leaky bucket and enters the network in time interval $(\tau, t]$. Then,

$$A_i(\tau, t) \leq \min\{(t - \tau)C_i, \sigma_i + \rho_i(t - \tau)\}, \quad \text{for all } t \geq \tau \geq 0, \tag{8.97}$$

for every session i.

We say that the arrival function A_i conforms to (σ_i, ρ_i, C_i), or $A \sim (\sigma_i, \rho_i, C_i)$. This arrival constraint is attractive because it restricts the traffic in terms of three simple parameters: average sustainable rate (ρ_i), burstiness (σ_i), and peak rate (C_i).

$A_i(0, t)$ is plotted in Fig. 8.22. Let $l_i(t)$ be the amount of tokens in the session i token bucket at time t. Assuming that the session starts out with a full bucket of tokens, if $L_i(t)$ is the total number of tokens accepted at the session i bucket in the interval $(0, t]$ (it does not include the full bucket of tokens that session i starts out with, and does not include arriving tokens that find the bucket full), then

$$L_i(t) = \min_{0 \leq \tau \leq t} \{A_i(0, \tau) + \rho_i(t - \tau)\}. \tag{8.98}$$

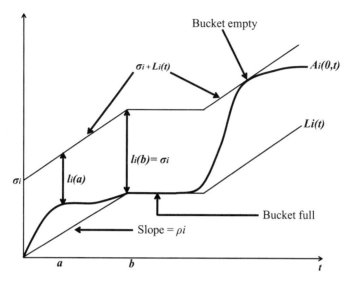

FIGURE 8.22 Arrival curve.

Since the token generation rate is ρ_i, we have

$$L_i(t) - L_i(\tau) \leq \rho_i(t - \tau), \tag{8.99}$$

for all $\tau \leq t$.

Thus, $l_i(t)$ is expressed as follows:

$$l_i(t) = \sigma_i + L_i(t) - A_i(0, t). \tag{8.100}$$

From (8.100) and (8.99), we have

$$A_i(\tau, t) \leq l_i(\tau) + \rho_i(t - \tau) - l_i(t). \tag{8.101}$$

In the following, we analyze the worst-case performance of weighted fair queueing systems for sessions that operate under leaky-bucket constraints. That is, the session traffic is constrained as in (8.97).

Let there be N sessions in the system. The incoming traffic from session i is $A_i \sim (\sigma_i, \rho_i, C_i)$ for $i = 1, 2, \ldots, N$. We assume that the system is empty before time zero. The server is work conserving (i.e., it is never idle if there is work in the system) and operates at the fixed rate of 1.

Let $S_i(\tau, t)$ be the amount of session i traffic served in the interval $(\tau, t]$. Note that $S_i(0, t)$ is continuous and nondecreasing for all t (see Fig. 8.23). It depends on the amount of traffic from all the sessions, that is, the arrival function $A_1, \ldots, A_N$. The session i backlog at time τ is defined to be

$$Q_i(\tau) = A_i(0, \tau) - S_i(0, \tau). \tag{8.102}$$

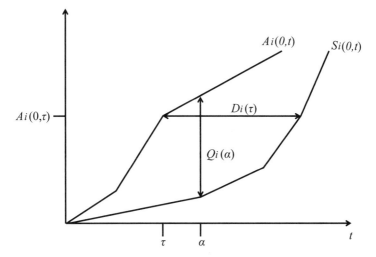

FIGURE 8.23 The session service curve $S_i(0, t)$.

The session i delay at time τ is denoted by $D_i(\tau)$, which is the amount of time that it would take for the session i backlog to clear if no session i bits were to arrive after time τ. Thus,

$$D_i(\tau) = \inf\{t \leq \tau : S_i(0, t) = A_i(0, \tau)\} - \tau. \tag{8.103}$$

From Fig. 8.23, we see that $D_i(\tau)$ is the horizontal distance between curves $A_i(0, t)$ and $S_i(0, t)$ at the ordinate value of $A_i(0, \tau)$.

Clearly, $D_i(\tau)$ depends on $S_i(0, t)$, that is, the arrival functions $A_1, \ldots, A_N$. It is interesting to compute the maximum delay over all time, and over all arrival functions that are consistent with (8.97). Let D_i^* be the maximum delay for session i. Then,

$$D_i^* = \max_{(A_1,\ldots,A_N)} \max_{\tau \geq 0} D_i(\tau). \tag{8.104}$$

Similarly, we define the maximum backlog for session i, Q_i^*:

$$Q_i^* = \max_{(A_1,\cdots,A_N)} \max_{\tau \geq 0} Q_i(\tau). \tag{8.105}$$

Given $g_1, \ldots, g_N$ for a weighted fair queueing server of rate $g = 1$ and given (σ_j, ρ_j), $j = 1, \ldots, N$, we are going to determine D_i^* and Q_i^* for every session i. Let σ_i^τ be the sum of the number of tokens left in the bucket and the session backlog at the server for session i at time $\tau \geq 0$, then

$$\sigma_i^\tau = Q_i(\tau) + l_i(\tau), \tag{8.106}$$

where $l_i(\tau)$ is defined in (8.100). If $C_i = \infty$, we can consider σ_i^τ as the maximum amount of session i backlog at time τ^+ over all arrival functions $A_i, \ldots, A_N$ up to time τ.

Recall (8.101)

$$A_i(\tau, t) \leq l_i(\tau) + \rho_i(t - \tau) - l_i(t). \tag{8.107}$$

Substituting for l_i^τ and l_i^t from (8.106),

$$Q_i(\tau) + A_i(\tau, t) - Q_i(t) \leq \sigma_i^\tau - \sigma_i^t + \rho_i(t - \tau). \tag{8.108}$$

Now note that

$$S_i(\tau, t) = Q_i(\tau) + A_i(\tau, t) - Q_i(t). \tag{8.109}$$

Combining (8.108) and (8.109), we have the following useful result:

$$S_i(\tau, t) \leq \sigma_i^\tau - \sigma_i^t + \rho_i(t - \tau), \tag{8.110}$$

for every session i, $\tau \leq t$.

Let us define a system busy period to be a maximal interval B such that for any $\tau, t \in B$, $\tau \leq t$:

$$\sum_{i=1}^{N} S_i(\tau, t) = t - \tau. \tag{8.111}$$

Suppose $[t_1, t_2]$ is a system busy period, that is, $B = [t_1, t_2]$. Since the system is work conserving, we have

$$\sum_{i=1}^{N} Q_i(t_1) = \sum_{i=1}^{N} Q_i(t_2) = 0. \tag{8.112}$$

Thus,

$$\sum_{i=1}^{N} A_i(t_1, t_2) = \sum_{i=1}^{N} S_i(t_1, t_2) = t_2 - t_1. \tag{8.113}$$

Substituting from (8.97) and rearranging terms, we have

$$t_2 - t_1 \leq \frac{\sum_{i=1}^{N} \sigma_i}{1 - \sum_{i=1}^{N} \rho_i}. \tag{8.114}$$

It is the upper bound on the length of a system busy period.

This result shows that all system busy periods are bounded whenever $\sum_{i=1}^{N} \rho_i < 1$. Since session delay is bounded by the length of the largest possible system busy period, the session delays are bounded as well.

Let a session i busy period be a maximal interval B_i contained in a single system busy period, such that for all $\tau, t \in B_i$:

$$\frac{S_i(\tau, t)}{S_j(\tau, t)} \geq \frac{g_i}{g_j}, \ j = 1, 2, \ldots, N. \tag{8.115}$$

Note that it is possible for a session to have zero backlog during its busy period. However, if $Q_i(\tau) > 0$, then τ must be in a session i busy period at time τ.

For every interval $[\tau, t]$ that is in a session i busy period,

$$S_i(\tau, t) \geq (t - \tau)g_i. \tag{8.116}$$

Session i is defined to be *greedy* starting at time τ, if

$$A_i(\tau, t) = \min\{C_i(t - \tau), l_i(\tau) + (t - \tau)\rho_i\}, \text{ for all } t \geq \tau. \tag{8.117}$$

This means that the session uses as many tokens as possible (i.e., sends at maximum possible rate) for all times $\geq \rho$. At time ρ, session i has $l_i(\tau)$ tokens left in the bucket, but it is constrained to send traffic at a maximum rate of C_i. Thus, it takes $l_i^\tau/(C_i - \rho_i)$ time units to deplete the tokens in the bucket. After this, the rate will be limited by the token arrival rate ρ_i.

Define A_i^τ as an arrival function that is greedy starting at time τ (see Fig. 8.24). From the figure (and from (8.97)), we see that if a system busy period starts at time

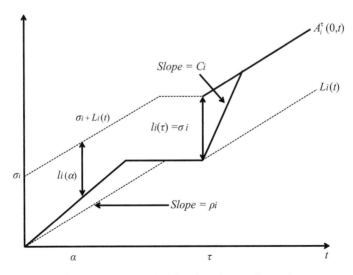

FIGURE 8.24 An arrival function of a greedy session.

zero, then

$$A_i^0(0, t) \geq A(0, t), \text{ for all } A \sim (\sigma_i, \rho_i, C_i), t \geq 0. \tag{8.118}$$

Suppose that $C_j \geq g$ for every session j, where g is the rate of a server. Then, for every session i, D_i^* and Q_i^* are achieved (not necessarily at the same time) when every session is greedy starting at time zero, the beginning of a system busy period.

This means that the all-greedy regime can maximize the delay as well as the backlog for every session in the weighted fair queueing system. In actuality, it seems reasonable that if a session sends as much traffic as possible at all times, it is going to impede the progress of packets arriving from the other sessions. This gives a simple scenario for investigating the worst-case behavior. To prove this result, the easiest way is to show it is true for the case when $C_i = \infty$ for all i first and then based on this analysis to yield the results for the cases with finite link capacities. Since the proof is very complicated and cannot provide any insight to the readers, we do not include it here.

Define e_1 as the first time at which one of the sessions, say $L(1)$, ends its busy period in an all-greedy system. That is, in the interval $[0, e_1]$, each session i is in a busy period and is served at rate $g_i / \sum_{k=1}^{N} g_k$, assuming that $\sigma - i > 0$ for all i. Since session $L(1)$ is greedy after 0, it follows that

$$\rho_{L(1)} < \frac{g_i}{\sum_{k=1}^{N} g_k}, \tag{8.119}$$

where $i = L(1)$. Now each session j still in a busy period will be served at rate

$$\frac{(1 - \rho_{L(1)})g_j}{\sum_{k=1}^{N} g_k - g_{L(1)}} \tag{8.120}$$

until a time e_2 when another session, say $L(2)$, ends its busy period.

Similarly, for each k, we have

$$\rho_{L(k)} < \frac{(1 - \rho_{L(j)})g_i}{\sum_{j=1}^{N} g_j - \sum_{j=1}^{k-1} g_{L(j)}}, \quad k = 1, 2, \ldots, N, \quad i = L(k). \tag{8.121}$$

As shown in Fig. 8.25, the slopes of the various segments that comprise $S_i(0, t)$ are $s_1^i, s_2^i, \ldots$. From (8.121), we get

$$s_k^i = \frac{\left(1 - \sum_{j=1}^{k-1} \rho_{L(j)}\right) g_i}{\sum_{j=1}^{N} g_j - \sum_{j=1}^{k-1} g_{L(j)}}, \quad k = 1, 2, \ldots, L(i). \tag{8.122}$$

It can be seen that $s_k^i, k = 1, 2, \ldots, L(i)$, forms an increasing sequence. Any ordering of sessions that meets (8.121) is known as a *feasible ordering*.

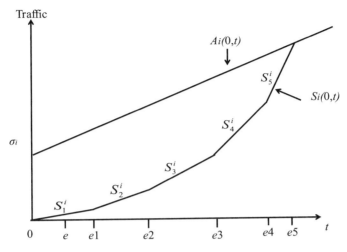

FIGURE 8.25 The departure curve of session i.

Note that we only require that $0 \leq e_1 \leq e_2 \leq \cdots \leq e_N$, allowing for several e_i to be equal. Session $L(i)$ in this case has exactly one busy period, the interval $[0, e_i]$. Also, we are only interested in $t \leq e_{L(i)}$, since the session i buffer is always empty after this time.

In the following, we shall determine bounds for the maximum delay D_i^* and the maximum backlog Q_i^*. Suppose Q_i^* is achieved at time t, and let τ be the first time before t when there are no session i bits backlogged in the server. Then by (8.116), we have

$$Q_i^* \leq (\sigma_i + \rho_i(t-\tau)) - g_i(t-\tau)$$
$$\leq (\sigma_i + \rho_i(t-\tau)) - g_i(t-\tau)$$
$$= \sigma_i. \qquad (8.123)$$

An arriving session i bit will be after at most Q_i^* session i bits have been served. Since these backlogged bits are served at a rate of at least g_i,

$$D_i^* \leq \frac{Q_i^*}{g_i} \leq \frac{\sigma_i}{g_i}. \qquad (8.124)$$

However, the bounds obtained above are quite loose since the assumption is that the session i is served with a constant rate g_i till its backlog is cleared. In reality, the service rates of backlogged sessions increase as more and more sessions complete their backlogged periods, where a backlogged period for session i is any period of time during which packets belonging to that session are continuously queued in the system. This is because as each session completes its backlog, the bandwidth released is distributed among the still backlogged sessions in proportion to their weights.

8.5 CLOSED-LOOP FLOW CONTROL

One problem with open-loop flow control is that network bandwidth may not be used efficiently. Access control is needed to prevent a session from injecting excessive amount of data into the network. The data rate allowed by each session is prenegotiated upfront and it does not take into account that the available network bandwidth as well as the desirable bandwidth by the user sessions may vary dynamically with time.

If the "natural" data rates of all sessions are more or less constant (e.g., such as that for digitized voice), open loop will work well because there is no need to give the sessions more bandwidth, even when there is unused bandwidth in the network. By contrast, if each session would like to pump in as much data as possible by making use of unused bandwidth in the network, open-loop control does not work well. Two example scenarios are as follows: (1) file transfer in which it is desirable to transfer a large file to the destination as fast as possible; and (2) a VBR (variable bit-rate) coded video in which it is desirable to exploit any unused bandwidth in the network to increase image quality.

A reasonable way to provision the network bandwidth is to divide sessions into different service types. For services that do not have highly varying data rates that require strong guarantee for bandwidth availability during transmission, the open-loop control can be used. In this situation, the services would not attempt to use more than their allocated share of bandwidth and the network would not assign more bandwidth than preallocated to the services.

For services that do not require strong guarantee on bandwidth availability but that may desire extra bandwidth in a time-varying way (e.g., file transfers that are initiated in a sporadic fashion), we may use closed-loop control to provision bandwidth. This kind of service is sometimes referred to as the available bit-rate service (ABR).

For the ABR service, there must be a way to inform the sessions if extra bandwidth is available in the network and if so how much of it is available. In other words, the network traffic conditions must be fed back to the sessions so that they know how much traffic to inject into the network. And this is the reason why closed-loop control is needed so that excess unused bandwidth can be utilized.

For simplicity, let us begin by looking at the interactions between two successive links of an ABR VC. Consider Fig. 6.23 in which we show the source node connecting to a switch in the network. When the output link of the VC on the switch does not suffer congestion, its buffer is relatively empty. The source should not be prevented from pumping in data to make off the unused bandwidth on the output link. This output link, however, may be shared by several ABR VCs, and when many of them start to pump in data, it is possible that the input rate to the output link is higher than its output rate. The buffer occupancy starts to build up, and unless something is done to limit the data from the sources, the buffer may eventually overflow, resulting in data loss. If the higher level protocol at the sources attempts to retransmit data because of this data loss, the congestion will tend to become even worse. Therefore, it is better to prevent the data loss to begin with and take actions before the buffer begins to overflow.

We can set a threshold on the buffer occupancy. When the threshold is exceeded, the sources of the ABR VCs that pass through the link will be instructed to cut down their rates. Two issues that must be resolved are

- What should be the value of the threshold in order to prevent data loss.
- How should the source data rates be reduced when congestion occurs.

PROBLEMS

8.1 Use the fluid-flow model in this question. Suppose you want to multiplex two on–off sources using a statistical multiplexer with output capacity $C = 3$. Suppose that the peak rate of one of the sources is 1 and that of the other source is 3.

(a) For the arrival rates shown in Fig. 8.1 for the two sources, plot the amount of total traffic that has arrived at and departed from the multiplexer.

(b) When there is backlog for one source, the whole capacity c will be dedicated to the transmission of its traffic. When both sources have backlogs, the multiplexer will serve source A with rate 1 and source B with rate 2. Plot the arrived and departed traffic for each source.

(c) Suppose the average "on" time for both sources is 1 and the average "off" time for both sources is λ. Both the on and off times are exponentially distributed. Draw a four-state Markov chain to represent the state transition of the system. State 00 means both sources are off, state 01 means source A is off and source B is on, and so on. What is the expected number of on sources at any time?

(d) With the service discipline in part (b) and the arrival processes in part (c), what is the expected amount of time needed to completely clear the traffic of source B that arrives during a single on period?

8.2 Explain why $G(\bar{x})$ derived in (8.59) in the text is an upper bound on the overflow probability of a finite buffer of size $\bar{x}$. Use a simple example and draw a fluid-flow diagram in your explanation. (*Hint*: Consider two systems, one with infinite buffer and one with finite buffer, and both systems are subjected to the same arrival traffic pattern. Argue that at any time, the system with infinite buffer has a higher buffer occupancy than the one with finite buffer. Argue that if the buffer occupancy in the infinite-buffer system is more than $\bar{x}$, the finite-buffer system does not necessarily overflow; when the finite-buffer system overflows, however, the buffer occupancy in the infinite-buffer system is greater than $\bar{x}$.)

8.3 Section 8.2 analyzes an infinite-buffer multiplexing system with on–off fluid flow sources. This problem explores the solution when the buffer size B is finite. The eigenvalues and eigenvectors are the same as before, but the coefficients a_i's must be matched to different boundary conditions.

(a) Explain why the terms with positive eigenvalues cannot be ignored anymore. Argue that

$$F_j(0) = 0 = \sum_{i=0}^{N} a_i(\Phi_i)_j, \quad \lfloor C \rfloor + 1 \le j \le N.$$

(b) Explain why $F_j(x)$ at boundary $x = B$ may have a discontinuity (i.e., $F_j(B_-) \ne F_j(B)$). For what j's are there discontinuities?

(c) Argue that

$$F_j(B_-) = \binom{N}{j} \lambda^j \left(\frac{1}{1+\lambda}\right)^N = \sum_{i=0}^{N} a_i(\Phi_i)_j, \quad 0 \le j \le \lceil C \rceil - 1.$$

(d) Parts (a) and (b) give $N + 1$ equations for $N + 1$ unknowns a_i, $0 \le i \le N$, when C is not an integer. For C an integer, there are only N equations. For an additional equation, argue that

$$\{(N - C)\lambda + C\} F_C(0) = \{N - (C - 1)\} \lambda F_{C-1}(0),$$

which gives

$$\sum_{i=0}^{N} a_i [\{(N - C)\lambda + C\}(\Phi_i)_C - \{N - (C - 1)\}\lambda(\Phi_i)_{C-1}] = 0.$$

(e) Express the overflow probability in terms of $F_j(B_-)$, $0 \le j \le N$, when C is not an integer as well as when C is an integer.

8.4 Continuing from the previous problem, this problem examines the loss probability in a finite-buffer system. Consider an infinitely long stretch of time T. The total amount of traffic arriving at the system is approximately $\bar{r}T$, where $\bar{r} = N\lambda/(1+\lambda)$ is the average arrival rate.

(a) Argue that amount of loss traffic is approximately

$$\sum_{j>C}^{N}(F_j(B) - F_j(B_-))(j - C)T.$$

(b) Express $F_j(B)$ in terms of N and λ.
(c) Express the loss probability in terms of $F_j(B_-)$, λ, and N.

8.5 This problem derives the standard probability result in the second equality of (8.65). Express $\int_0^\infty xf(x)dx$ as $\lim_{b\to\infty} \int_0^b xf(x)dx$.
   (a) By integration by parts, show that
$$\int_0^b xf(x)dx = \int_0^b F(b) - F(x)dx.$$
   (b) Argue that
$$\lim_{b\to\infty} \int_0^b F(b) - F(x)dx = \int_0^\infty 1 - F(x)dx.$$

8.6 Using the fluid-flow traffic model, this problem considers the stochastic analysis of a dual-leaky-bucket system with input traffic from an on–off source. We shall use the same traffic model as described in Section 8.2 and assume that the data buffer in the dual-leaky-bucket size is infinite. A similar analysis as in Section 8.2 can be used to study the buffer occupancy in the data buffer of the leaky bucket. Let $Q_D(t)$ be the buffer occupancy of the data bucket and $Q_T(t)$ be the buffer occupancy of the token bucket at time t. The state of the system is defined by three parameters: whether source is on or off, $Q_D(t)$, and $Q_T(t)$.
   (a) Argue that both $Q_D(t)$ and $Q_T(t)$ cannot be nonzero at the same time.
   (b) Argue that because of the above, we can define a parameter $Q(t)$ such that $Q(t) = \sigma + Q_D(t)$ when $Q_D(t) \neq 0$ and $Q(t) = \sigma - Q_T(t)$ when $Q_T(t) \neq 0$, and describe the state with just two parameters: whether the source is on or off, and $Q(t)$. Argue that $Q_D(t)$ and $Q_T(t)$ can be derived from $Q(t)$ if it is known.
   (c) Interpret $Q(t)$ as the buffer occupancy of a fictitious queue with a server serving at the rate of ρ. Argue that the same analysis as in Section 8.2 can then be used to study $Q(t)$. (*Note*: It is actually simpler here since we have only one source.)

8.7 In the three-leaky-bucket system, argue that if the arrival rate is never more than C, then the third leaky bucket is redundant. That is, the departure curve would have been the same in the dual-bucket system.

8.8 In the three-leaky-bucket system, argue that if $\rho = C$, we can replace the system with a simple leaky-bucket system with bucket size $\beta + \sigma$ as far as the departed traffic is concerned. What is the service rate of the simple leaky bucket? (*Hint*: Argue that it is not possible for the third leaky bucket to be empty while the first leaky bucket is not empty.)

8.9 Consider the situation in Fig. 8.11 in the text. Instead of a dual-leaky-bucket traffic shaper, we have a three-leaky-bucket traffic shaper. Assuming that $C > \rho' > \rho$, draw a diagram similar to that in Fig. 8.11(c). What is the worst-case backlog and delay suffered by the traffic at the output of the switch?

8.10 For the round-robin scheduling scheme that serves n_i packets from session i in a continuous fashion before moving on to the next queue (i.e., the one without mini-cycles), derive the worst-case service time for a session l with $n_l = k$. How does this compare with the worst-case service time in the scheduler with mini-cycles?

8.11 Give an example in which the order of service in the simulated fluid-flow weighted fair queueing is different from that in the real packet-by-packet weighted fair queueing. (*Hint*: Consider the arrival of a packet into an empty queue of session i with a very high ρ_i.)

9

PACKET SWITCHING AND INFORMATION TRANSMISSION

All communication networks comprise transmission systems and switching systems, even though their designs are usually treated as two separate issues. Communication channels are generally disturbed by noise from various sources. In circuit-switched networks, reliable communication requires error-tolerant transmission of bits over noisy channels. In packet-switched networks, however, not only can bits be corrupted with noise, but resources along connection paths are also subject to contention. Thus, quality of service (QoS) is determined by buffer delays and packet losses. The theme of this chapter is to show that transmission noise and packet contention actually have similar characteristics and can be tamed by comparable means to achieve reliable communication. The following analogies between switching and transmission are identified:

1. Buffering against contention is a process that is similar to the error correction of noise corrupted signals. A signal-to-noise ratio that represents the carried load of packet switches can be deduced from the Boltzmann model of packet distribution.
2. When deflection routing is applied to Clos networks, the loss probability decreases exponentially, which is similar to the exponential behavior of the error probability of binary symmetric channels with random channel coding. In information theory, this result is stated as the noisy channel coding theorem.
3. The similarity between Hall's condition of bipartite matching and expander graph manifests the resemblance between nonblocking route assignments and error-correcting codes. An extension of the Sipser–Speilman decoding algorithm of expander codes to route assignments of Benes networks is given to illustrate their correspondence.

Principles of Broadband Switching and Networking, by Tony T. Lee and Soung C. Liew
Copyright © 2010 John Wiley & Sons, Inc.

4. Scheduling in packet switching serves the same function as noiseless channel coding in digital transmission. The smoothness of scheduling, like source coding, is bounded by entropy inequalities.

5. The sampling theorem of bandlimited signals provides the cornerstone of digital communication and signal processing. Recently, the Birkhoff–von Neumann decomposition of traffic matrices has been widely applied to packet switches. With respect to the complexity reduction of packet switching, we show that the decomposition of a doubly stochastic traffic matrix plays a similar role to that of the sampling theorem in digital transmission.

We conclude that packet switching systems are governed by mathematical laws that are similar to those of digital transmission systems as envisioned by Shannon in his seminal 1948 paper, *A Mathematical Theory of Communication*.

9.1 DUALITY OF SWITCHING AND TRANSMISSION

All communication networks comprise transmission systems and switching systems, even though their designs are usually treated as two separate issues. Communication channels are generally disturbed by noise from various sources. In circuit-switched networks, resources are dedicated to connections and reliable communication only requires error-tolerant transmission of bits over noisy channels. In packet-switched networks, however, not only can bits be corrupted with noise, but resources along connection paths are also subject to contention. Despite the great achievements of information theory in dealing with transmission noise [Sha48,Mac03,Mas84], there are still many networking problems in modern communication systems that information theory is unable to provide solutions for. A particular problem that has come to the fore is the delay that is induced by contention in packet-switched networks, the solution for which lies in extending the theory to go beyond the bit rate of transmission [EpHa98].

The theme of this chapter is to show that transmission noise and packet contention actually have similar mathematical characteristics and can be tamed by comparable means to achieve reliable communication. The source information of the transmission channel is a function of time, and errors are corrected by coding, which expands the signal space. For switching systems, source information is a space function $f(i) = j$, for $i = 1, 2, \ldots, N$, from inputs V_I to outputs V_O. The function $f(i)$ represents a set of connection requests. Packets lost in contention are usually buffered or deflected to stretch out time and defer their requests. As shown in Fig. 9.1, the processes of transmission and switching are antisymmetric with respect to time and space and they are both governed by the law of probability.

In transmission systems, the fundamental QoS parameter is bit rate, or channel capacity, which can be pinpointed by signal-to-noise ratio. In packet-switched networks, QoS parameters, buffer delay, and packet loss are all determined by loading. In order to provide a common ground for comparison, the carried load of a packet switch is

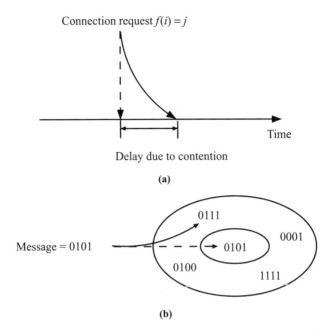

FIGURE 9.1 The parallel characteristics between packet switching and digital transmission.

converted into a pseudo signal-to-noise ratio (PSNR) induced by Boltzmann's model of packet distribution. Based on this correspondence between carried load and PSNR, we show that a packet-switched Clos network with random routing can be modeled as an abstract channel with additive Gaussian noise.

The schematic diagrams of a transmission channel and a three-stage Clos network [Clo53] are shown in Fig. 9.2. The input modules in the first stage and output modules in the third stage of the Clos network correspond to transmitters and receivers, respectively, of the transmission system. The number of central modules is the bandwidth of the Clos network. The disturbances due to internal contentions in the middle stage of a packet-switched Clos network mimic the noise of a channel. In this chapter, we demonstrate that various routing schemes applied to a packet-switched Clos network to cope with contentions are comparable to coding schemes of a transmission channel. The routing schemes of packet switching and their counterparts in digital transmission are listed in Fig. 9.3. Depending on the implementation of routing schemes, and similar to the separation of channel coding and source coding in information theory, the Clos network can be modeled as either a noisy channel with nonblocking routing, analogous to channel coding, or a noiseless channel with route scheduling, analogous to source coding, as explained below.

The noisy channel model of the Clos network. We first compare the deflection routing of packet switching with the random coding over noisy channel. Errors introduced by noises in transmission are subdued by redundant bits, while packet contentions can be relieved by redundant links in switching. When deflection routing is applied

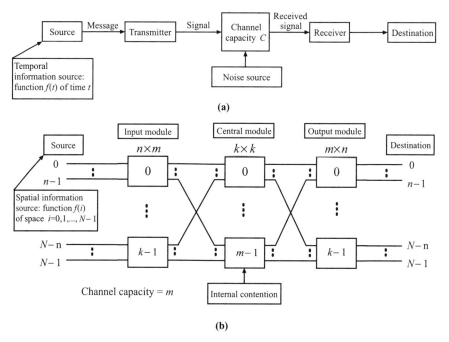

FIGURE 9.2 Comparison of transmission systems and switching systems.

to Clos networks, the loss probability decreases exponentially with respect to the network length, which is similar to the exponential behavior of the error probability of binary symmetric channels with random channel coding. In information theory, this result is stated as the noisy channel coding theorem [Sha48,Mac03,Ham86,Abr63].

In contrast to deflection routing, nonblocking routing in the Clos network will avoid internal contentions completely by conflict-free route assignments according to

Packet-switched Clos network	Transmission channel
o Random routing	o Noisy channel capacity theorem
o Deflection routing	o Noisy channel coding theorem
o Route assignment	o Error-correcting code
o BvN decomposition	o Sampling theorem
o Path switching	o Noiseless channel
o Scheduling	o Noiseless coding theorem

FIGURE 9.3 Analogies between packet switching and information transmission.

input requests [Ben65,ChJi03,Jaj03]. The route assignment algorithms are based on the matching of bipartite graphs. In the error-correcting code, the low-density parity check (LDPC) code developed by Gallager can also be represented by a bipartite graph [Gal62,Tan81], and a subclass called the expander code was constructed by Sipser and Speilman [SiSp96]. The condition on the expander graph is a generalization of Hall's condition on complete matching, which suggests the resemblance between nonblocking route assignments and error-correcting codes. An extension of the Sipser and Speilman decoding algorithm of expander codes to route assignments of Benes networks is given to illustrate their correspondence.

The noiseless channel model of the Clos network. A drastically different approach, called *path switching* (see Chapter 7), to deal with contention in a Clos network is proposed in Ref. [LeLa97]. A set of connection patterns of central modules is determined by the traffic matrix decomposition for this purpose. Path switching periodically uses this finite set of predetermined connections in the central stage of the Clos network to avoid online computation of route assignments. Once the connection patterns of central modules are fixed in each time slot, incoming packets can be scheduled accordingly in the input buffer. Regarding the predetermined connections as a code book, scheduling is a process similar to source coding in digital transmission [Sha48,Mac03,Ham86,Abr63]. The smoothness of scheduling, like noiseless coding, is bounded by entropy inequalities.

The decomposition of traffic matrices, sometimes called the Birkhoff–von Neumann decomposition, has been widely used in SS/TDMA satellite communications [Inu79,BBB87]. Path switching adopts this scheduling scheme in packet switching systems to guarantee the capacity of virtual paths in Clos networks. The same approach can be applied to the crossbar switch, also called the Birkhoff–von Neumann switch, in Ref. [CCH00]. Mathematically, the series expansion and reconstruction of doubly stochastic capacity matrices are similar to the Fourier series expansion and interpolation of sampling theorem of bandlimited signals [Nyq28,Sha49,Whi15,Kot33]. They also serve the same function in terms of complexity reduction of communication systems. The capacity matrix decomposition employed in path switching will reduce the dimension of permutation space of a Clos network from $N!$ to $O(N^2)$ or even lower, while the sampling theorem in transmission reduces the infinite-dimensional signal space of any duration T to a finite number of samples.

The remainder of the chapter is organized as follows. In Section 9.2, we propose the definition of pseudo signal-to-noise ratio of packet switch and prove that the trade-off between the bandwidth and PSNR of a Clos network is the same as that given by noisy channel capacity theorem in transmission. In Section 9.3, we show that the loss probability of the Clos network with deflection routing is similar to the exponential error probability of binary symmetric channels with random coding. In Section 9.4, we demonstrate the correspondence between nonblocking route assignments and error-correcting codes. In Section 9.5, we address the capacity allocation and capacity matrix decomposition issues related to the path switching implemented on a Clos network. In Section 9.6, the entropy inequalities of smoothness of scheduling are derived, and comparisons of scheduling algorithms are discussed. Finally, the conclusion and future research are summarized in Section 9.7.

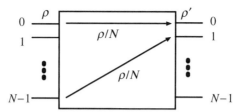

FIGURE 9.4 The carried load of an $N \times N$ crossbar switch.

9.2 PARALLEL CHARACTERISTICS OF CONTENTION AND NOISE

The apparent causes of contention and noise are quite different, but they both limit the performance of communication systems. In transmission, the trade-off between bandwidth and signal-to-noise ratio is given by Shannon–Hartley's noisy channel capacity theorem [Sha49]. In packet switching, the difference between offered load and carried load reflects the degree of contention. In order to provide a common ground to compare contention and noise, a pseudo signal-to-noise ratio is defined to represent the carried load of a packet switch. We show that the packet-switched Clos network with random routing can be mathematically modeled as a noisy channel.

9.2.1 Pseudo Signal-to-Noise Ratio of Packet Switch

Output port contention occurs in a crossbar switch when several packets are destined for the same output in a time slot. Packets lost in contention will be dropped as shown in Fig. 9.4. Consider an $N \times N$ crossbar switch, without any prior knowledge of input traffic. We assume that input loading ρ is homogeneous and output address is uniformly distributed, then the carried load ρ' is the probability that an output is busy in any given time slot:

$$\rho' = 1 - \left(1 - \frac{\rho}{N}\right)^N \xrightarrow{N \to +\infty} 1 - e^{-\rho}. \tag{9.1}$$

The difference between offered load and carried load is caused by packet dropping due to contention at the outputs. Benes considered the number of calls in progress as the energy of the connecting network in the thermodynamics theory of traffic in a telephone system [Ben63]. A similar traffic model was also explored in the broadband network [HuKa95]. If we regard packets as energy quantum, this proposition can be adopted in packet switching and stated as follows.

Proposition 9.1 (Benes). The *signal power* S_p of an $N \times N$ crossbar switch is the number of packets carried by the system, and the *noise power* is equal to $N_p = N - S_p$. Furthermore, both the signal power and noise power are Gaussian random variables when N is large.

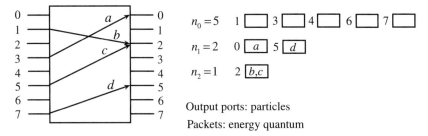

FIGURE 9.5 Boltzmann model of packet distribution.

This proposition can be verified by the Boltzmann model of packet distribution. Consider an $N \times N$ switch as a thermal system. The outputs are particles and the energy level of an output, denoted by ε_i, is equal to the number of packets destined for that output. The distribution of packets over outputs can be determined by maximizing the Boltzmann entropy [KiK80].

Suppose there are M packets from N input ports to N output ports. Let n_i be the number of output ports with energy level ε_i. An example with $N = 8$ and $M = 4$ is shown in Fig. 9.5, where $n_0 = 5, n_1 = 2$, and $n_3 = 1$. The number of possible divisions of N outputs into $r + 1$ distinct energy levels of respective sizes $n_0, \ldots, n_r$ is

$$\frac{N!}{n_0! n_1! \cdots n_r!},$$

and the number of possible divisions of M packets into N distinct outputs of respective sizes $\underbrace{0, \ldots, 0}_{n_0}, \ldots, \underbrace{r, \ldots, r}_{n_r}$ is

$$\frac{M!}{(0!)^{n_0} (1!)^{n_1} \cdots (r!)^{n_r}}.$$

Suppose each input can send one packet at most in any time slot, then the total number of states of the entire switch is given by

$$W = \frac{N!}{(N-M)! M!} \cdot \frac{N!}{n_0! n_1! \cdots n_r!} \cdot \frac{M!}{(0!)^{n_0} (1!)^{n_1} \cdots (r!)^{n_r}} \qquad (9.2)$$

subject to the following constraints:

$$N = n_0 + n_1 + n_2 + \cdots + n_r \qquad (9.3)$$

and

$$M = 0 n_0 + 1 n_1 + 2 n_2 + \cdots + r n_r. \qquad (9.4)$$

The Boltzmann entropy S of the system is given by

$$S = C \ln W, \tag{9.5}$$

where C is the Boltzmann constant. The maximal entropy can be obtained from the following function formed by Lagrange multipliers:

$$f(n_i) = \ln W + \alpha \left(\sum_i n_i - N \right) + \beta \left(\sum_i i n_i - M \right). \tag{9.6}$$

Using Stirling's approximation $\ln x! \approx x \ln x - x$ for the factorials, we have

$$f(n_i) \doteq N \ln N - N - (N - M) \ln(N - M) + (N - M)$$
$$+ N \ln N - N - \sum_i (n_i \ln n_i - n_i) - \sum_i n_i \ln(i!) \tag{9.7}$$
$$+ \alpha \left(\sum_i n_i - N \right) + \beta \left(\sum_i i n_i - M \right).$$

Taking the derivatives with respect to n_i, setting the result to zero, and solving for n_i yields the population number

$$n_i = \frac{e^{(\alpha+\beta i)}}{i!}. \tag{9.8}$$

If the offered load ρ is uniform on each input, then we have

$$\rho = \frac{M}{N} = \frac{\sum_i i n_i}{\sum_i n_i} = e^\beta. \tag{9.9}$$

The probability that there are i packets destined for a particular output has the Poisson distribution:

$$p_i = \frac{n_i}{N} = \frac{\frac{e^{(\alpha+\beta i)}}{i!}}{\sum_i n_i} = e^{-\rho} \frac{\rho^i}{i!}, \quad i = 1, 2, 3, \ldots. \tag{9.10}$$

The carried load is equal to the probability that an output port is busy:

$$\rho' = 1 - p_0 = 1 - e^{-\rho}, \tag{9.11}$$

which is consistent with (9.1). Hence, our proposition on the signal power of switch is verified.

We next show that both the signal S_p and noise N_p of an $N \times N$ crossbar switch are normally distributed under the assumptions of homogeneous input loading ρ and uniform output address.

The signal power S_p is the sum of the following i.i.d. random variables:

$$X_i = \begin{cases} 1, & \text{if output } i \text{ is busy,} \\ 0, & \text{otherwise,} \end{cases}$$

with mean $E[X_i] = \rho'$ and variance $\text{Var}[X_i] = \rho'(1 - \rho')$, where ρ' is the carried load given in (9.1). It follows from the central limit theorem that the signal power $S_p = \sum_{i=1}^{n} X_i$ is normally distributed with mean $E[S_p] = N\rho'$ and variance $\text{Var}[S_p] = N\rho'(1 - \rho')$. The noise power $N_p = N - S_p$ becomes independent of the signal power S_p when N is large. Using a similar argument, the noise N_p is also normally distributed with mean $E[N_p] = N(1 - \rho')$ and variance $\text{Var}[N_p] = N\rho'(1 - \rho')$. It follows that the carried load of a crossbar switch can be converted to *pseudo signal-to-noise ratio*, a concept that is comparable to the signal-to-noise ratio in transmission.

Definition 9.23. The pseudo signal-to-noise ratio of a crossbar switch is the ratio of mean signal power to mean noise power:

$$\text{PSNR} = \frac{E[S_p]}{E[N_p]} = \frac{N\rho'}{N - N\rho'} = \frac{\rho'}{1 - \rho'}. \tag{9.12}$$

It should be noted that the ratio of signal power to noise power SNR in transmission is defined by the ratio of the second moment of signal to that of noise, but PSNR is the ratio of their mean magnitudes here. Next, we will investigate the trade-off between bandwidth and PSNR of the Clos network.

9.2.2 Clos Network with Random Routing as a Noisy Channel

In a three-stage Clos network, each pair of adjacent switch modules is interconnected by a unique link, and each module is a crossbar switch. As shown in Fig. 9.6, the $N \times N$ Clos network with k input/output modules and m central modules is denoted by $C(m, n, k)$ [Clo53,Ben65], where n is the number of input ports or output ports of each input module or output module, where $N = kn$. The dimensions of the modules in the input, middle, and output stages are $n \times m$, $k \times k$, and $m \times n$, respectively. The Clos network $C(m, n, k)$ has the following characteristics:

1. Any central module can only assign to *one* input of each input module and *one* output of each output module.
2. Source address S and destination address D can be connected through any central module.
3. The number of alternative paths between input S and output D is equal to the number of central modules m.

In circuit switching theory, it is known that the Clos network $C(m, n, k)$ is rearrangeably nonblocking if $m \geq n$ and strictly nonblocking if $m \geq 2n - 1$

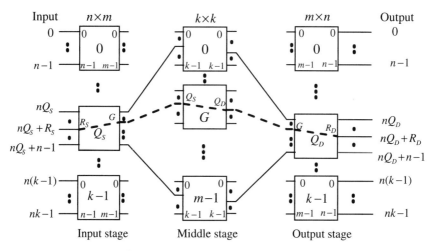

FIGURE 9.6 The Clos network $C(m, n, k)$.

[Ben65,Hui90]. Consider the number of central modules m as the bandwidth, then the trade-off between bandwidth m and PSNR of $C(m, n, k)$ with nonblocking routing is given by

$$\frac{E[S_p]}{E[N_p]} = \frac{n}{m-n}. \qquad (9.13)$$

In a packet-switched Clos network, the maximal data rate of each input module is n packets per time slot. Thus, we have the following.

Definition 9.24. The ratio $\sigma = n/m$ is the *maximal utilization* of a Clos network $C(m, n, k)$.

The route of each packet can be expressed by the numbering scheme of the Clos network [LeTo98]. The switch modules in each stage of the network, as well as the links associated with each module, are independently labeled from top to bottom. According to this numbering scheme, the source address S is represented by the 2-tuple $S(Q_S, R_S)$, indicating that the source S is the link R_S of the input module Q_S, where $R_S = [S]_n$ and $Q_S = \lfloor S/n \rfloor$ are the remainder and quotient of S divided by n. Similarly, the destination address D can be represented by the 2-tuple $D(Q_D, R_D)$. The path of an input packet from source S to destination D is determined by the choice of central module. Suppose the path goes through the central module G, then the routing tag is (G, Q_D, R_D) and the path is expressed by

$$S(Q_S, R_S) \to Q_S \xrightarrow{G} G \xrightarrow{Q_D} Q_D \xrightarrow{R_D} D(Q_D, R_D).$$

In a packet-switched Clos network $C(m, n, k)$, contentions may occur in the middle stage even if destination addresses of input packets are all different. With random routing scheme, the trade-off between bandwidth m and PSNR is given as follows.

Theorem 9.1. The maximum data rate of each input module of the packet-switched Clos network $C(m, n, k)$ with random routing is given by

$$n = m \ln \left(1 + \frac{E[S_p]}{E[N_p]}\right). \tag{9.14}$$

Proof. The maximum data rate is achieved when the input loading $\rho = 1$ and destination addresses of input packets are all different in every time slot, in which case input modules and output modules are contention-free. If the central module is selected randomly for each input packet, then the loading on each input link of a central module is given by

$$\sigma = n/m.$$

Since each switch module of the Clos network is a crossbar switch, the carried load on each output link of a central module is

$$\sigma' = 1 - \left(1 - \frac{n/m}{k}\right)^k \xrightarrow{k \to +\infty} = 1 - e^{-n/m}. \tag{9.15}$$

Substituting the following PSNR for the carried load σ' in (9.15),

$$\frac{E[S_p]}{E[N_p]} = \frac{km\sigma'}{km - km\sigma'} = \frac{\sigma'}{1 - \sigma'},$$

and taking logarithm, we obtain (9.14). □

The trade-offs given in (9.13) and (9.14) for constant data rate n packets per time slot of each input/output module are plotted in Fig. 9.7, which shows the improvement of PSNR that can be achieved by nonblocking routing. The same kind of trade-off between bandwidth and SNR in the transmission channel is stated in the Shannon–Hartley theorem [Sha49] on noisy cannel capacity as follows.

Noisy Channel Capacity Theorem. The channel capacity of a bandlimited Gaussian channel in the presence of additive Gaussian noise is given by

$$C = W \log \left(1 + \frac{S}{N}\right), \tag{9.16}$$

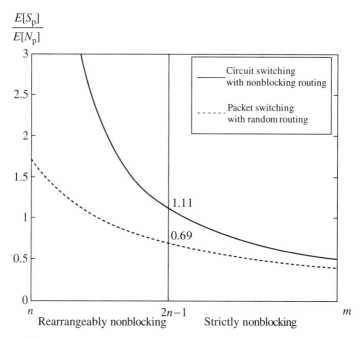

FIGURE 9.7 Trade-off between bandwidth m and PSNR of $C(m, n, k)$.

where C is the capacity in bits per second, W is the bandwidth of the channel in Hertz, and S/N is the signal-to-noise ratio.

With the understanding that the contention is mathematically analogous to noise, the rest of the chapter seeks to investigate the connections between the various routing schemes of the Clos network and the corresponding coding schemes of the transmission channel. We first demonstrate the similarity between deflection routing and random coding in the next section.

9.3 CLOS NETWORK WITH DEFLECTION ROUTING

Packet contention is inevitable in packet switching. One way of solving this problem without having to buffer the losing packet is to use deflection routing. In contrast to "store-and-forward" routing where the network allows buffering, deflection routing, also known as hot potato routing, is a routing scheme without buffering and can only be implemented for packet-switched networks [Tan81]. In case the individual communication links cannot support more than one packet at a time, excessive packets will be transferred to other available links. Conceptually, deflection routing utilizes idle or redundant links of the network as temporary storages. Redundancy is built into the switch design so that deflected packets can use extra stages to correct the deviated routes.

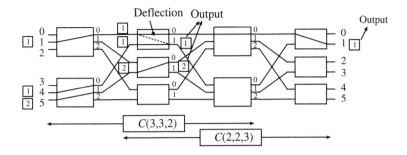

FIGURE 9.8 Cascaded Clos network with deflection routing.

9.3.1 Cascaded Clos Network

A cascaded Clos network is constructed by a sequence of alternate three-stage $C(n, n, k)$ and $C(k, k, n)$ networks, as illustrated in Fig. 9.8. Each output link of a switch module is connected to a module in the next stage and an output concentrator (not shown in Fig. 9.8). A packet will be sent to the concentrator if its destination address matches the output numbering, otherwise it will continue with the remaining journey. The loss probability can be made arbitrarily small by providing a large enough number of stages.

In the cascaded Clos network, the destination address D of a packet can be expressed by either $D = nQ_1 + R_1$ in the $C(n, n, k)$ network or $D = kQ_2 + R_2$ in the $C(k, k, n)$ network, as shown in the example given in Fig. 9.8, and therefore, the routing tag in the packet header comprises a 4-tuple (Q_1, R_1, Q_2, R_2). A packet can reach its destination D in any two consecutively successful steps, either (Q_1, R_1) or (Q_2, R_2). The Markov chain shown in Fig. 9.9 describes the complete journey of a packet, where p_i and $q_i = 1 - p_i$ are respective probabilities of success and deflection in $C(n, n, k)$ and $C(k, k, n)$.

9.3.2 Analysis of Deflection Clos Network

For the sake of simplicity, we assume that the cascaded Clos network is formed by $C(n, n, n)$ network uniformly. Since we will consider the worst-case deflection

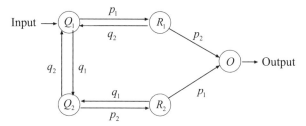

FIGURE 9.9 The Markov chain of deflection routing.

probability when both n and k are sufficiently large, the assumption that $n = k$ does not lose the generality of analysis, and the packet loss probability will be a conservative estimate. Under this presumption, a simplified Markov chain is depicted in Fig. 9.10.

The probability of success p in an $n \times n$ switch module is a function of input loading ρ. Under the homogeneous traffic assumption, we have

$$1 - p\rho = \left(1 - \frac{\rho}{n}\right)^n \stackrel{n \to \infty}{\longrightarrow} e^{-\rho}.$$

Thus, the worst probability of success is given by

$$p = \frac{1 - e^{-\rho}}{\rho}.$$

The probabilities of success p and deflection $q = 1 - p$ versus the loading ρ are plotted in Fig. 9.11(a). Let $G_i(k)$ be the probability that the packet in state i will reach the output state O in exactly k steps. The following equations can be derived from the Markov chain shown in Fig. 9.10:

$$G_O(k) = \begin{cases} 1, & \text{if } k = 0, \\ 0, & \text{otherwise,} \end{cases}$$

$$G_R(k) = pG_O(k-1) + qG_Q(k-1), \quad k = 1, 2, \ldots,$$
$$G_Q(k) = pG_R(k-1) + qG_Q(k-1), \quad k = 1, 2, \ldots. \quad (9.17)$$

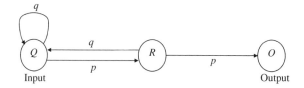

FIGURE 9.10 The simplified Markov chain of deflection routing.

The generating function $G_Q(k)$ is given by

$$G_Q(z) = \sum_{k=0}^{\infty} G_Q(k) z^k = \frac{p^2 z^2}{1 - qz - pqz^2}, \qquad (9.18)$$

from which we obtain the following steady-state probabilities:

$$G_Q(k) = \begin{cases} 0, & k = 0, 1, \\ \frac{p}{vq}(pq)^{(k+1)/2} \cosh(k-1)\theta, & k = 2, 4, 6, \ldots, \\ \frac{p}{vq}(pq)^{(k+1)/2} \sinh(k-1)\theta, & k = 3, 5, 7, \ldots, \end{cases} \qquad (9.19)$$

where $v = \frac{\sqrt{q^2+4pq}}{2}$ and $\theta = \ln \frac{q+\sqrt{q^2+4qp}}{2\sqrt{pq}}$. For a given network length L, a conservative estimate of packet loss probability is given as follows:

$$P_{\text{loss}} \leq \sum_{k=L+1}^{\infty} G_Q(k), \qquad (9.20)$$

which can be explicitly expressed as

$$P_{\text{loss}} \leq \begin{cases} \frac{1}{vq}(pq)^{(L+2)/2} \sinh(L+2)\theta, & \text{for even length } L, \\ \frac{1}{vq}(pq)^{(L+2)/2} \cosh(L+2)\theta, & \text{for odd length } L. \end{cases}$$

When L is large, the logarithm of P_{loss} is linear in L as shown by the curves in Fig. 9.11(c):

$$\ln P_{\text{loss}} \leq m(L+2) + b, \qquad (9.21)$$

where $m = \ln\left(\frac{q+\sqrt{q^2+4pq}}{2}\right)$ and $b = \ln\left(\frac{1}{q\sqrt{q^2+4pq}}\right)$. In the worst case when $\rho = 1$, $p = 0.6321$, $q = 0.3679$, $m = -0.3566$, and $b = 0.9683$.

Theorem 9.2. If the offered load $\rho \leq 1$ on each input of the Clos network with deflection routing, then the loss probability can be arbitrarily small and the carried load ρ' on each output can be arbitrarily close to the carried load ρ.

Proof. It follows from (9.21) that the loss probability in terms of offered load ρ and carried load ρ' is bounded by

$$P_{\text{loss}} = \frac{\rho - \rho'}{\rho} \leq ca^{-L}. \qquad (9.22)$$

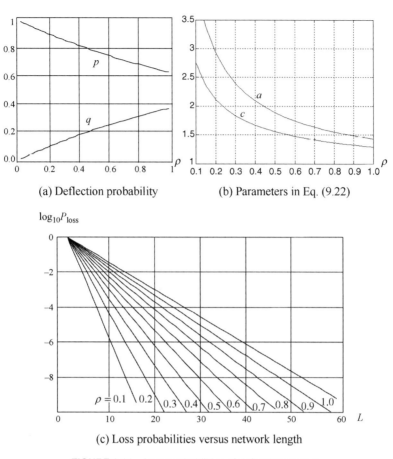

FIGURE 9.11 Loss probabilities of deflection routing.

The two parameters a and c are given by the following functions of deflection probability q:

$$a = \frac{2}{q + \sqrt{q^2 + 4pq}} > 1$$

and

$$c = \frac{\left(q + \sqrt{q^2 + 4pq}\right)^2}{4q\sqrt{q^2 + 4pq}} > 1.$$

The parameters a and c versus the offered load ρ are plotted in Fig. 9.11(b), in which $a = 1.4285$ and $c = 1.2906$ when $\rho = 1$. □

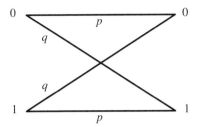

FIGURE 9.12 Binary symmetric channel.

Conversely, if the offered load $\rho > 1$, then the loss probability is unbounded, because the number of packets input to a switch module could be more than the total number of outputs and there is not enough space to deflect packets, which can be lost halfway through.

Compare the deflection Clos network with the binary symmetric channel (BSC) with random coding, as illustrated in Fig. 9.12. Table 9.1 provides the parallels between the above theorem and Shannon's main theorem [Sha48,Mac03,Ham86,Abr63] stated as follows.

Noisy Channel Coding Theorem. Given a noisy channel with capacity C and information transmitted at a rate R, then if $R \leq C$, there exists a coding technique that allows the probability of error at the receiver to be made arbitrarily small. This means it is possible to transmit information without error up to a limit C. However, if $R > C$, the probability of error at the receiver increases without bound.

The capacity of BSC with cross probability $q = 1 - p$ is given by

$$C = 1 + p \log p + (1 - p) \log(1 - p). \tag{9.23}$$

The noisy channel coding theorem states that there exist an encoding function $E : \{0, 1\}^k \to \{0, 1\}^n$ and a decoding function $D : \{0, 1\}^n \to \{0, 1\}^k$, such that the

TABLE 9.1 Comparison of Deflection Clos Network and Binary Symmetric Channel

Deflection Clos Network	Binary Symmetric Channel
Deflection probability $q < 1/2$	Cross probability $q < 1/2$
Deflection routing	Random coding
$\rho \leq 1$	$R \leq C$
Exponential loss probability	Exponential error probability
$\quad P_{\text{loss}} \leq ca^{-L}$	$\quad P_e \leq a^{-n} + c^{-n}$
Complexity increases	Complexity increases
$\quad$ with network length L	$\quad$ with code length n
Equivalent set of outputs	Typical set decoding

error probability that the receiver gets a wrong message is bounded by

$$P_e \leq a^{-n} + c^{-n}, \qquad a > 1, c > 1, \tag{9.24}$$

if the code rate $R = k/n = C - \delta$ for some $\delta > 0$ [Gal65]. Thus, the error probability P_e can be arbitrarily small for a sufficiently large n. Conversely, if the rate $R = k/n = C + \delta$ for some $\delta > 0$, then the error probability is unbounded.

The similar behavior of the deflection routing of the Clos network and the random coding of noisy channel suggests the connection between nonblocking route assignments and error-correcting codes. This analogy can be extended to multistage interconnection networks as well. The deflection routing of the tandem-banyan network proposed in [TKC91] is equivalent to the concept of error detection and retransmission, while the deflection routing of the dual shuffle-exchange network is a distributed error-correction algorithm [LiL94]. These points will be further elaborated on in the next section.

9.4 ROUTE ASSIGNMENTS AND ERROR-CORRECTING CODES

The Clos network $C(m, n, k)$ is rearrangeable if $m \geq n$, meaning that it can realize connections of any permutations between inputs and outputs with the possibility of rearranging the calls in progress. Both Slepian–Duguid's nonblocking theorem [Ben65,Hui90] and Paull's rearrangeable theorem [Pau62] on route assignments of the Clos network are rooted in matching theory, which starts with Hall's marriage theorem that gives the existence condition of a complete matching in a given bipartite graph [Hal35]. In the error-correcting code, the LDPC code can also be represented by a bipartite graph [Gal62,Tan81]. The connection between route assignment and LDPC code in the context of bipartite graph is the main point addressed in this section.

9.4.1 Complete Matching in Bipartite Graphs

A *bipartite graph* $G = (V_L, V_R, E)$ consists of two finite sets of vertices V_L and V_R and a collection of edges E connecting vertices in V_L to vertices in V_R.

Definition 9.3. A *complete matching* in a bipartite graph G is an injective function $f : V_L \to V_R$ so that for every $x \in V_L$, there is an edge in E whose end points are x and $f(x)$.

Definition 9.4. The *neighborhood* of any subset $A \subset V_L$ is defined by

$$N_A = \{b | (a, b) \in E, a \in A\} \subseteq V_R,$$

which are end points of an edge in E whose other end points lie in A.

The necessary and sufficient condition for a bipartite graph to have a complete matching is given by Hall's marriage theorem [Hal35,Wil72]. Because it is the most fundamental theorem in switching theory, we state the theorem and give Rado's elegant proof [Rad67] below.

Theorem 9.3 (Hall). Let $G = (V_L, V_R, E)$ be a bipartite graph, then there exists a complete matching for G if and only if

$$|N_A| \geq |A| \qquad (9.25)$$

for any subset $A \subseteq V_L$.

Proof. It is clear that the condition is essential for the existence of a complete matching, but it remains to show that it is sufficient. Let $V_L = \{1, 2, \ldots, n\}$ and $F = \{N_1, N_2, \ldots, N_n\}$; by contradiction, suppose N_1 contains x, y, then the removal of either x or y violates Hall's condition. There exist $A, B \subseteq \{2, \ldots, n\}$ with the property

$$R_A = N_A \cup (N_1 - \{x\}), \qquad |R_A| \leq |A|,$$
$$R_B = N_B \cup (N_1 - \{y\}), \qquad |R_B| \leq |B|.$$

Then, $|R_A \cup R_B| = |N_{A \cup B} \cup N_1|$ and $|R_A \cap R_B| \geq |N_{A \cap B}|$. It follows that

$$|A| + |B| \geq |R_A| + |R_B| = |R_A \cup R_B| + |R_A \cap R_B|$$
$$\geq |N_{A \cup B} \cup N_1| + |N_{A \cap B}|. \qquad (9.26)$$

On the other hand, from Hall's condition, we have

$$|N_{A \cup B} \cup N_1| + |N_{A \cap B}| \geq |A \cup B| + 1 + |A \cap B| = |A| + |B| + 1. \quad (9.27)$$

Combining (9.26) and (9.27) will lead to

$$|A| + |B| \geq |A| + |B| + 1,$$

a contradiction. Hence, the removal of either x or y does not violate Hall's condition, and a complete matching can be obtained by repeating this procedure until each N_i contains only one element. □

The nonblocking route assignment of the Clos network and the edge coloring of the bipartite graph stated in the following theorem are equivalent consequences of Hall's marriage theorem.

Theorem 9.4. The following statements are equivalent:

1. A regular bipartite graph G with degree n can be edge colored by m colors if $m \geq n$.

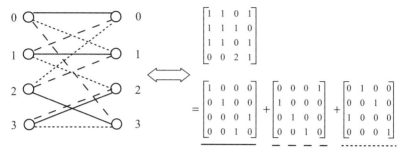

FIGURE 9.13 An edge-colored regular bipartite graph and decomposition of integer doubly stochastic matrix.

2. *Slepian–Duguid*: The Clos network $C(m, n, k)$ is rearrangeably nonblocking if $m \geq n$.

Proof. 1. For any subset $A \subseteq V_L$, the edges terminating on vertices in A must be terminated on vertices in its neighborhood set N_A at the other end; Hall's condition holds because

$$n|N_A| \geq n|A| \Rightarrow |N_A| \geq |A|.$$

The bipartite graph G can be reduced to n complete matching by repeatedly applying Hall's theorem. It is clear that edge-coloring can be satisfied if $m \geq n$. The edge-colored bipartite graph is equivalent to the decomposition of an integer doubly stochastic matrix into permutation matrices. An example to illustrate this point is shown in Fig. 9.13.

2. Consider a set of call requests $\{(S_0, D_0), \ldots, (S_{N-1}, D_{N-1})\}$, in which the destination addresses are all different, and suppose the central module G_i is assigned to the request (S_i, D_i) for $i = 0, \ldots, N - 1$. This set of assignments is nonblocking, or contention-free, if and only if

$$\lfloor S_i/n \rfloor = \lfloor S_j/n \rfloor \Rightarrow G_i \neq G_j$$

and

$$\lfloor D_i/n \rfloor = \lfloor D_j/n \rfloor \Rightarrow G_i \neq G_j$$

for all $i \neq j$. It simply means that if either of the two sources S_i and S_j are on the same input modules, or the two destinations D_i and D_j are on the same output module, then the central modules assigned to (S_i, D_i) and (S_j, D_j) must be different.

The route assignments can be formulated as the edge coloring of a bipartite graph $G(V_L, V_R, E)$, in which vertices of V_L represent input modules and vertices of V_R represent output modules, and each connection request (S_i, D_i) is represented by an edge $(\lfloor S_i/n \rfloor, \lfloor D_i/n \rfloor)$ in E. This bipartite graph is regular with degree n and can be

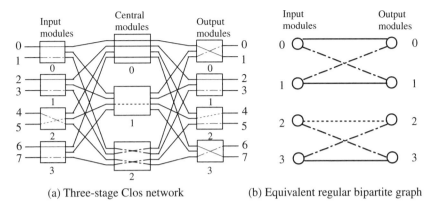

FIGURE 9.14 Correspondence between Clos network and bipartite graph.

colored by m colors if $m \geq n$, such that each color represents a central module and the edge coloring is the set of route assignments. □

An example to show the equivalence between route assignment of the Clos network and edge coloring of the bipartite graph is given in Fig. 9.14. The resulting route assignments of the call request are listed in Table 9.2. The connection between route assignment and error-correcting code will be addressed in the sequel.

9.4.2 Graphical Codes

The LDPC code is a class of linear block code that can be represented by bipartite graphs $G(V_L, V_R, E)$, called Tanner graph [Tan81], in which V_L is the set of variable vertices and V_R the set of constraint vertices. The parity matrix specifies the edge set E. Each variable can assume the value of 0 or 1, and a constraint is satisfied if the sum of all the variables adjacent to it is 0 mod 2. Figure 9.15 illustrates an example of such a bipartite graph that represents the following matrix:

$$\begin{bmatrix} 0 & 1 & 0 & 1 & 1 & 0 & 0 & 1 \\ 1 & 1 & 1 & 0 & 0 & 1 & 0 & 0 \\ 0 & 0 & 1 & 0 & 0 & 1 & 1 & 1 \\ 1 & 0 & 0 & 1 & 1 & 0 & 1 & 0 \end{bmatrix}. \qquad (9.28)$$

TABLE 9.2 The Set of Routing Tags (G, Q, R)

S	0	1	2	3	4	5	6	7
D	1	3	2	0	6	4	7	5
G	0	2	0	2	2	1	0	2
Q	0	1	1	0	3	2	3	2
R	1	1	0	0	0	0	1	1

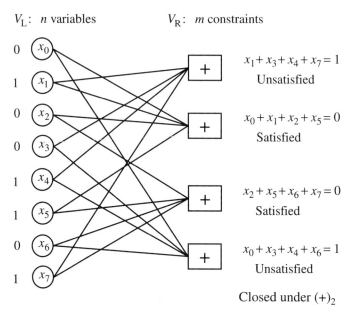

FIGURE 9.15 Low-density parity check code.

A vector $x \in \{0, 1\}^n$ is a code word if and only if x satisfies all constraints. The set of all code words is closed with respect to mod 2 sum and forms a linear subspace of $\{0, 1\}^n$.

A subclass of graphical codes based on expander graphs was constructed by Sipser and Speilman [SiSp96]. The expander code has the following linear time sequential decoding algorithm.

Sipser–Speilman Decoding Algorithm

If there is a variable vertex v such that most of its neighboring constraints are unsatisfied, flip the value of v. Repeat.

The decoding algorithm of expander codes guarantees that the number of unsatisfied vertices will be monotonically decreasing until all constraints are satisfied. The main result is stated in the following theorem.

Theorem 9.5 (Sipser–Speilman). Let $G(V_L, V_R, E)$ be the bipartite graph of size $|V_L| = n$, $|V_R| = m$ that is k-regular on the left. Assume that for any subset $A \subseteq V_L$ of size $|A| \leq \alpha n$, $|N_A| > 3k/4|A|$, then the decoding algorithm will correct up to $\alpha n/2$ errors.

The similarity between Hall's condition (9.25) on complete matching and the condition on expander graphs in the above theorem is quite obvious. The resemblance between these two structural conditions on bipartite graphs suggests the connection between error-correcting code and route assignment of Clos network. The Sipser–

Speilman decoding algorithm is extended to the nonblocking route assignment of Benes network to illustrate their correspondence.

9.4.3 Route Assignments of Benes Network

A Benes network is a multistage connecting network that can be recursively constructed from Clos network $C(2, 2, k)$ [Ben65,Hui90]. A set of call requests is a mapping from inputs to outputs. An example is given in (9.29), which displays a set of connection requests in an 8×8 Benes network. The upper row of the matrix is the ordered inputs from 0 to 7, and the lower row consists of the respective outputs.

$$\pi = \begin{pmatrix} 0 & 1 & 2 & 3 & 4 & 5 & 6 & 7 \\ 1 & 6 & 0 & 5 & 7 & 2 & 4 & 3 \end{pmatrix}. \tag{9.29}$$

The two central modules in $C(2, 2, N/2)$, as shown in Fig. 9.19(a), are labeled, respectively, by 0 and 1. Let x_i be the binary variable associated with the request $(i, \pi(i))$, for $i = 0, \ldots, N-1$. The value of $x_i \in \{0, 1\}$ defines the following route assignment:

$$x_i = \begin{cases} 0, & \text{if module 0 is assigned to } (i, \pi(i)), \\ 1, & \text{otherwise.} \end{cases} \tag{9.30}$$

Since the same central module cannot be assigned to two inputs or two outputs on the same module, the necessary and sufficient condition on nonblocking route assignment can be simply formulated as a set of linear constraints

$$x_i + x_j = 1 \tag{9.31}$$

for all $i \neq j$ such that either two inputs i and j are on the same input module or two outputs $\pi(i)$ and $\pi(j)$ are on the same output module.

As shown in Fig. 9.16, let $V_L = \{x_0, \ldots, x_{N-1}\}$ be the set of variables and V_R be the set of constraints (9.31); a bipartite graph $G(V_L, V_R, E)$ similar to that of LDPC can be constructed to solve for the values of x_i''s, which determine the nonblocking routes of the set of call requests in π. It should be noted that the set of solutions to (9.31) is not closed with respect to mod 2 sum. In fact, two solutions in each connected component of the bipartite graph $G(V_L, V_R, E)$ are complementary to each other, and the total number of solutions of (9.31) should be 2^g, where g is the number of connected components of G. The following route assignment algorithm is a modification of Sipser–Speilman decoding algorithm.

The flip algorithm for nonblocking route assignments:

Step 1 Initially assign $x_0 = 0, x_1 = 1, x_2 = 0, x_3 = 1, \ldots$ to satisfy all input module constraints.

Step 2 In each cycle of the bipartite graph, unsatisfied vertices divide the cycle into line segments. Label them α and β alternately and flip the values of all variables located in α segments.

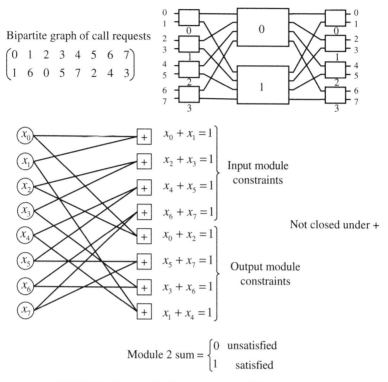

FIGURE 9.16 Nonblocking constraints of Benes network.

The initial values of all variables can be arbitrarily assigned. Compared to the Sipser–Speilman decoding algorithm, the initial assignment can be considered as the received signal, and errors are systematically removed by the flip algorithm to satisfy all constraints.

Theorem 9.6. All constraints are satisfied when the flip algorithm terminates.

Proof. The proof of this theorem is illustrated by the example displayed in Fig. 9.17. First, we want to show that every cycle has an even number of unsatisfied vertices. It is clear that the total number of constraints is even because there is an equal number of input modules and output modules in each cycle. Assume by contradiction that the number of unsatisfied vertices is odd, then the number of satisfied vertices must also be odd. Let u_0 and u_1 be the number of unsatisfied vertices whose neighboring variables both equal 0 and 1, respectively, and let s be the number of satisfied vertices. Then, the number of variables equal to 0 and 1, respectively, should be $u_0 + s/2$ and $u_1 + s/2$, which is impossible if s is odd.

Next, we want to show that all constraints are satisfied when the flip algorithm terminates as shown in Fig. 9.17. For an arbitrary constraint vertex $v \in V_R$, let $x_i, x_j \in V_L$

ROUTE ASSIGNMENTS AND ERROR-CORRECTING CODES

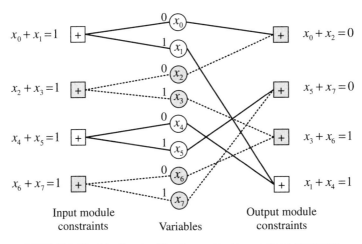

FIGURE 9.17 The flip algorithm for nonblocking route assignments.

be the two neighboring variables of v. If v is unsatisfied, the two variables x_i and x_j should have the same value but are located in segments bearing different labels. Only one located in the α segment will flip the value, and the vertex v should be satisfied when the algorithm terminates.

On the other hand, if the constraint v is initially satisfied, then x_i and x_j flip simultaneously if they are both in an α segment, or else they keep the same value if they are located in a β segment. Hence, the constraint v remains satisfied in either case when the algorithm terminates. □

The route assignments of the set of call requests (9.29) resulting from the flip algorithm are displayed in Fig. 9.18, and the complete assignments shown in Fig. 9.19 can be determined iteratively. If both N and n are of the power of 2, the set of equations (9.31) can also be solved by the parallel algorithm proposed in Ref. [LeLi02] with time complexity on the order of $O(\log^2(N))$.

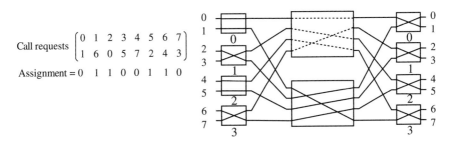

FIGURE 9.18 The connection requests and route assignments.

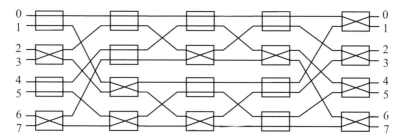

FIGURE 9.19 The complete route assignments of Benes network.

9.5 CLOS NETWORK AS NOISELESS CHANNEL-PATH SWITCHING

Path switching proposed in Ref. [LeLa97] is a compromise of static routing and dynamic routing schemes for a Clos network. A set of connection patterns of the central modules is predetermined according to the traffic between pairs of input and output modules. This set of connection patterns is repeatedly used in a cyclic manner such that online path hunting can be avoided and bandwidth requirements can be satisfied in the long run.

The scheduling of path switching is based on the following relationships between the edge-colored bipartite graph and the connection pattern in the middle stage of the Clos network as depicted in Fig. 9.14 earlier:

1. The number of edges e_{ij} represents the number of packets that can be sent from input module I_i to output module O_j, the *virtual path* V_{ij}, in one time slot.
2. Each color of the bipartite graph corresponds to a central module of the Clos network.

This correspondence suggests the scheduling of switch according to predetermined connection patterns if traffic matrix is expressed as weighted sum of a finite number of permutation matrices as shown in Fig. 9.20.

For any given traffic matrix $T = [\lambda_{ij}]$, where λ_{ij} is the number of packets per time slot from input module I_i to output module O_j, such that $\sum_i \lambda_{ij} < n \leq m$ and $\sum_j \lambda_{ij} < n \leq m$, there exist a finite number F of regular bipartite graphs such that the capacity c_{ij} of the virtual path V_{ij} between I_i and O_j satisfies

$$c_{ij} = \frac{\sum_{t=0}^{F-1} e_{ij}(t)}{F} > \lambda_{ij}, \tag{9.32}$$

where $e_{ij}(t)$ is the number of edges from node i to node j in the tth bipartite graph. The bandwidth requirement $T = [\lambda_{ij}]$ can be satisfied if the system periodically provides connections according to the edge coloring of these F bipartite graphs. Adopting convention in the TDMA system, each cycle is called a *frame* and the period F *frame*

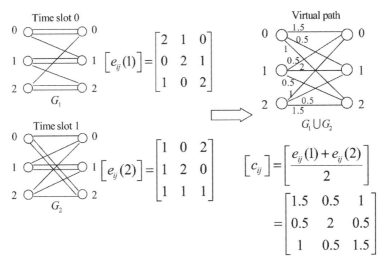

FIGURE 9.20 Capacity of virtual path as average number of edges.

size. The routing information can be stored in the local memory of each input module to avoid the slot-by-slot computation of route assignments [EHY,MeT89].

Since the connection patterns of switch modules in the middle stage are fixed, the virtual path between any pair of input module and output module is contention-free. As shown in Fig. 9.21, input/output modules are scaled to a much smaller size and routing decisions in these modules become independent. Each input module is an independent input-queued switch, and arrival packets can be scheduled in the input buffer according to predetermined routes in every time slot by some matching algorithm similar to the scheduling algorithm for input-queued crossbar switch [McK99], while each output module could be an output-queued switch similar to a knockout switch [YHA87]. Consider the scheduled Clos network as a noiseless channel, then the scheduling of path switching maps central modules and time slots into incoming packets, a process similar to the source coding of transmission [Sha48,Mac03,Ham86,Abr63] if the set of predetermined connection patterns is regarded as a code book. In this section, we address the capacity allocation and traffic matrix decomposition issues, and the smoothness of scheduling will be discussed in the next section.

9.5.1 Capacity Allocation

The capacity allocation problem seeks to find the capacity $c_{ij} > \lambda_{ij}$ for each virtual path V_{ij} between I_i and O_j such that non-overbooking condition $\sum_i c_{ij} = \sum_j c_{ij} = m$ for each input/output module is observed. The problem can be formulated as constrained optimization of some objective function [Str86]. The choice of the objective function depends on the stochastic characteristic of the traffic on virtual paths and the quality-of-service requirements of connections. Following Kleinrock's independency assumption [Klei75,BeG92], each virtual path can be modeled as an $M/M/1$ queue

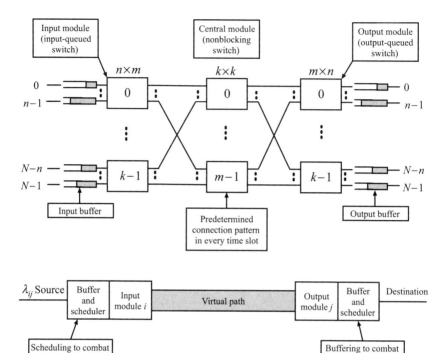

FIGURE 9.21 Noiseless virtual path of Clos network.

with arrival rate λ_{ij} and service rate c_{ij} for all i, j, then the average delay for the packets from input module I_i to output module O_j is given by

$$D_{ij} = \frac{1}{c_{ij} - \lambda_{ij}} \quad (9.33)$$

and the objective is to minimize the total weighted delay [Klei75]

$$\sum_{i,j} \frac{\lambda_{ij}}{c_{ij} - \lambda_{ij}} \quad (9.34)$$

subject to $c_{ij} > \lambda_{ij}$ and $\sum_i c_{ij} = \sum_j c_{ij} = m$. The computation of the optimal allocation is quite involved in general. Nevertheless, the following heuristic algorithm based on Kleinrock's square root rule [Klei75] can always yield suboptimal capacity allocation that is close to the optimal solution.

Theorem 9.7. For any given traffic matrix $T = [\lambda_{ij}]$ with nonzero entries, $\lambda_{ij} > 0$ for all i and j, then

$$\sum_{j=1}^{n} \lambda_{ij} \leq \sum_{j=1}^{n} c_{ij} = m \quad (9.35)$$

and

$$\sum_{i=1}^{n} \lambda_{ij} \leq \sum_{i=1}^{n} c_{ij} = m \qquad (9.36)$$

when the capacity allocation algorithm terminates.

Capacity Allocation Algorithm

Step 1 Initialization: Set $k = 1$ and $c_{ij}^{(0)} = \lambda_{ij}$.

Step 2 Calculate slack capacity of each input module (row) and output module (column).

$$\begin{cases} A_i^{(k-1)} = m - \sum_{j=1}^{n} c_{ij}^{(k-1)} \\ B_j^{(k-1)} = m - \sum_{i=1}^{n} c_{ij}^{(k-1)} \end{cases}$$

Step 3 Remove any saturated row i if $A_i^{(k-1)} = 0$ and column j if $B_j^{(k-1)} = 0$. The algorithm terminates if there is no slack capacity left, otherwise do

$$c_{ij}^{(k)} = c_{ij}^{(k-1)} + \min\left(\frac{A_i^{(k-1)} \sqrt{\lambda_{ij}}}{\sum_{j=1}^{n} \sqrt{\lambda_{ij}}}, \frac{B_j^{(k-1)} \sqrt{\lambda_{ij}}}{\sum_{i=1}^{n} \sqrt{\lambda_{ij}}} \right),$$

and set $k = k + 1$, go to step 2.

Proof. Since $\lambda_{ij} > 0$, it is obvious that

$$c_{ij}^{(k)} > c_{ij}^{(k-1)} \qquad (9.37)$$

in kth iteration of the algorithm for all k. Also, we have

$$\sum_{j=1}^{n} c_{ij}^{(k)} \leq \sum_{j=1}^{n} c_{ij}^{(k-1)} + \sum_{j=1}^{n} \frac{A_i^{(k-1)} \sqrt{\lambda_{ij}}}{\sum_{j=1}^{n} \sqrt{\lambda_{ij}}} = m. \qquad (9.38)$$

The combination of (9.37) and (9.38) assures that $\sum_{j=1}^{n} c_{ij} = m$, and similarly, that $\sum_{i=1}^{n} c_{ij} = m$, when the algorithm terminates. □

Notice that if some entries λ_{ij} are zero, then the strict inequality (9.37) does not hold in general, and there is no guarantee that the algorithm will halt. It is also true for any capacity allocation algorithm that if $\lambda_{ij} = 0$ implies $c_{ij} = 0$, then the doubly stochastic matrix $C = [c_{ij}]$ that satisfies (9.35) and (9.36) may not exist. Nevertheless, this problem can be eliminated in practice if a small amount of capacity $c_{ij} = \epsilon$ allocated to a virtual path with $\lambda_{ij} = 0$ is allowed.

9.5.2 Capacity Matrix Decomposition

The time axis is divided into frames of time slots, and the frame size is denoted by F. Within each frame, the rate requirement will be satisfied by a set of connection patterns, which are predetermined by the decomposition of the capacity matrix $[c_{ij}]$. In a time-slotted system, it is reasonable to assume that all entries c_{ij} are rational numbers with a least common denominator F. Then $[F \cdot c_{ij}]$ is an integer doubly stochastic matrix with constant mF row sum and column sum. According to Theorem 2, the bipartite graph corresponding to the matrix $[F \cdot c_{ij}]$ can be colored by mF colors. The colored bipartite graph can also be expressed by the generalized Birkhoff–von Neumann decomposition [Bir46,LLL83,LLL86] stated in the following theorem.

Theorem 9.8 (Generalized Birkhoff–von Neumann Decomposition). The capacity matrix C has the following expansions:

$$C = \frac{1}{F}\sum_{i=1}^{mF} M_i = \frac{1}{F}\sum_{i=1}^{F} G_i = \sum_{i=1}^{K} \phi_i P_i, \qquad (9.39)$$

such that $G_i = \sum_{j=m(i-1)+1}^{mi} M_j, i = 1, \ldots, F$, and $\sum_{i=1}^{K} \phi_i = 1$.

With respect to the scheduling of path switching, the correspondence between connection patterns in the Clos network $C(m, n, k)$ and these matrices in the series expansions (9.39) is given as follows:

1. Each M_i is a permutation matrix, or complete matching.
2. Each matrix G_i is a sum of m permutation matrices, $M_{m(i-1)+1}, \ldots, M_{mi}$, which represent connection patterns of those m central modules of $C(m, n, k)$ in time slot i of every frame. That is, the matrix G_i is an edge-colored regular bipartite graph with degree m. The combination of permutation matrices in G_i can be arbitrary.
3. Each matrix P_i is a *state* of the switch such that $P_i \neq P_j$ for all $i \neq j$, and $\{P_1, \ldots, P_K\} = \{G_1, \ldots, G_F\}$. The coefficient ϕ_i is the frequency of the state P_i within each time frame. Since the total number of constraints in the doubly stochastic matrix C is equal to $(N-1)^2 + 1$, the number of states is bounded by $K \leq \min\{F, N^2 - 2N + 2\}$.

The scheduling information stored in the memory is linearly proportional to F, and the frame size is limited by the access speed and the memory space of input modules. In the capacity matrix decomposition (9.39), both K and F could be too large in practice, because the number of states K is on the order of $O(N^2)$ and the frame size F is determined by the least common denominator of c_{ij}. The following limitations compromise between complexity and QoS:

1. If the capacity matrix C is bandlimited such that $c_{ij} \leq B/F$, then it is easy to show that $K \leq F \leq BN/m$.

$$[c_{ij}] = \begin{bmatrix} 1.4 & 0.5 & 1.1 \\ 0.6 & 1.8 & 0.6 \\ 1.0 & 0.7 & 1.3 \end{bmatrix}$$

Capacity matrix
(per slot)

$$F = 4 \ [e_{ij}] = \begin{bmatrix} 5 & 2 & 4 \\ 2 & 7 & 2 \\ 4 & 2 & 5 \end{bmatrix}$$

RMS error = 0.104

$$F = 8 \ [e_{ij}] = \begin{bmatrix} 11 & 4 & 8 \\ 4 & 14 & 4 \\ 8 & 5 & 10 \end{bmatrix}$$

RMS error = 0.0608

FIGURE 9.22 Round-off errors with different frame size F.

2. The frame size F can be a constant independent of switch size N if $[F \cdot c_{ij}]$ is rounded off into an integer matrix, and the round-off error on the order of $O(1/F)$ is acceptable. The error can be arbitrarily small if the frame size F is sufficiently large, as shown in the examples given in Fig. 9.22.

Consider the traffic matrix as the aggregate of signals input to the packet switch, then the series expansion of doubly stochastic capacity matrix into permutation matrices (9.39) and the reconstruction of capacity by weighted running sum of scheduled connections are mathematically similar to those of the Nyquist–Shannon sampling theorem of bandlimited signals [Nyq28,Sha49,Whi15,Kot33] input to the transmission channel.

Sampling Theorem of bandlimited signal. If a function $f(t)$ contains no frequencies higher than W cps, it is completely determined by its ordinates at a series of equally spaced sampling intervals of $1/2W$ s, called Nyquist intervals. If $|F(\omega)| = 0$, for $|\omega| \geq 2\pi W$, then

$$f(t) = \sum_{-\infty}^{+\infty} f_n \frac{\sin \pi(2Wt - n)}{\pi(2Wt - n)}$$

where $f_n = f\left(\frac{n}{2W}\right)$ is the nth sample of $f(t)$.

An important common characteristic of both series expansions is the reduction of complexity. Without scheduling, the total number of possible permutations of a packet switch for a given capacity matrix is $N!$. The complexity reduction is achieved by limiting the space to those permutations only involved in the expansion (9.39). Hence, the dimension of permutation space is reduced to $O(N^2)$ by the traffic matrix decomposition. It can further be reduced to $O(N)$ if each connection is bandwidth limited. The complexity of packet-level problems can be significantly reduced if the

416   PACKET SWITCHING AND INFORMATION TRANSMISSION

TABLE 9.3 Comparison of Traffic Matrix Decomposition and Sampling Theorem

	Packet Switching	Digital Transmission
Network environment	Time-slotted switching system	Time-slotted transmission system
Bandwidth limitation	Capacity-limited traffic matrix	Bandwidth-limited signal function
	$\sum_{i=1}^{N} \lambda_{ij} \leq \sum_{i=1}^{N} c_{ij} = m$	$f(t) = \dfrac{1}{2\pi} \int_{-2\pi W}^{2\pi W} F(\omega) d\omega$
	$\sum_{j=1}^{N} \lambda_{ij} \leq \sum_{i=1}^{N} c_{ij} = m$	$F(\omega) = 0, \quad |\omega| \geq 2\pi W$
	$\lambda_{ij} \leq c_{ij}$	
Samples	(0, 1) Permutation matrices	(0, 1) Binary sequences
	BvN decomposition (Hall's marriage theorem)	Fourier series
Expansion	$C = \sum_{i=1}^{Fm} \dfrac{M_i}{F} = \sum_{i=1}^{K} \phi_i P_i$	$f(t) = \sum_{-\infty}^{+\infty} f_n \dfrac{\sin\pi(2Wt - n)}{\pi(2Wt - n)}$
	Frame size $= F = $ lcm of denominators of c_{ij}	Nyquist interval $= T = 1/2W$
Inversion	Reconstruct the capacity by running sum	Reconstruct the signal by interpolation
Complexity reduction	• Reduce number of permutations from $N!$ to $O(N^2)$ • Reduce to $O(N)$ if bandwidth is limited • Reduce to constant F if truncation error of order $O(1/F)$ is acceptable	• Reduce infinite-dimensional signal space to finite number $2tW$ in any duration t
QoS	Capacity guarantee, scheduling, delay bound	Error-correcting code, data compression, DSP

switch fabric is limited to this set of predetermined permutations only, operating in a much smaller subspace, yet the capacity of each connection is still guaranteed. In general, the minimal number of permutations required is determined by the edge coloring of the corresponding bipartite graphs as stated in Theorem 9.4.

In digital transmission, the sampling theorem reduces the infinite-dimensional signal space in any time interval T to a finite number $2TW$ of samples, and the minimal number of samples is determined by the Nyquist sampling rate for limited bandwidth W of the input signal function. The comparisons of the these two expansions and their respective roles in packet switching and digital transmission are listed in Table 9.3.

9.6 SCHEDULING AND SOURCE CODING

The process and function of a scheduling algorithm are similar to those of source coding in transmission. If we regard the set of predetermined connection patterns as

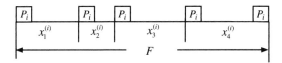

FIGURE 9.23 Interstate time for ith state within a frame of size F.

a code book, scheduling incoming packets in the input buffer according to predetermined connection patterns is the same as encoding the source signal. This point will be elaborated by the construction of a scheduling algorithm based on the Huffman tree. We first show that the smoothness of scheduling, like source coding, is bounded by the source entropy, or the entropy of capacity decomposition. The results can then be generalized to the smoothness of two-dimensional scheduling of tokens assigned to each virtual path of the Clos network.

9.6.1 Smoothness of Scheduling

In light of the fairness of services, the scheduling is expected to be as smooth as possible. In a Clos network with path switching, in addition to guaranteeing capacity for each virtual path, smooth scheduling also reduces delay jitters and alleviates *head-of-line* (HOL) blocking at input buffer [Hui90].

Consider a scheduling of a given capacity decomposition $C = \sum_{i=1}^{K} \phi_i P_i$ with frame size F. Let $X_1^{(i)}, X_2^{(i)}, \ldots, X_{n_i}^{(i)}$ be a sequence of interstate times of state P_i within a frame. It is easy to see from Fig. 9.23 that

$$n_i = \phi_i F, \tag{9.40}$$

$$X_1^{(i)} + \cdots + X_{n_i}^{(i)} = F, \quad \text{for all } i = 1, \ldots, k. \tag{9.41}$$

Since delay jitter is one of the major concerns of scheduling, it is natural to define the smoothness of scheduling by the second moment of interstate time.

Definition 9.5. The smoothness of state P_i is defined by

$$L_i = \log \sqrt{\frac{\sum_{k=1}^{n_i} (X_k^{(i)})^2}{n_i}}, \tag{9.42}$$

and the smoothness of a scheduling is the weighted average given by

$$L = \sum_{i=1}^{K} \phi_i L_i. \tag{9.43}$$

418 PACKET SWITCHING AND INFORMATION TRANSMISSION

The following theorem reveals the fact that the smoothness of scheduling is parallel to the code length of source coding in information theory.

Theorem 9.9. For any scheduling of a given capacity decomposition $C = \sum_{i=1}^{K} \phi_i P_i$ with frame size F, we have

$$\sum_{i=1}^{K} 2^{-L_i} \leq 1 \quad \text{(Kraft's inequality)}, \tag{9.44}$$

and the average smoothness is bounded by

$$L = \sum_{i=1}^{K} \phi_i L_i \geq \sum_{i=1}^{K} \phi_i \log \frac{1}{\phi_i} = H. \tag{9.45}$$

Both equalities hold when $X_k^{(i)} = 1/\phi_i$, for all $i = 1, \ldots, K$, and $k = 1, \ldots, n_i$.

Proof. The inequality (9.44) can be obtained by substituting (9.40) and (9.41) into the Cauchy–Schwartz inequality:

$$\sum_{i=1}^{K} 2^{-L_i} = \sum_{i=1}^{K} \sqrt{\frac{n_i}{\sum_{k=1}^{n_i}(X_k^{(i)})^2}} \leq \sum_{i=1}^{K} \frac{n_i}{F} = \sum_{i=1}^{K} \phi_i = 1.$$

Again, using the Cauchy–Schwartz inequality, we have

$$L = \sum_{i=1}^{K} \phi_i L_i = \sum_{i=1}^{K} \phi_i \log \sqrt{\frac{\sum_{k=1}^{n_i}(X_k^{(i)})^2}{n_i}}$$

$$= \sum_{i=1}^{K} \frac{\phi_i}{2} \log \frac{\sum_{k=1}^{n_i}(X_k^{(i)})^2}{n_i} \geq \sum_{i=1}^{K} \phi_i \log \frac{1}{\phi_i} = H.$$

□

Suppose equalities in the above theorem hold, then the scheduling is obviously optimal. First, we consider a simple example where $K = F$, $\phi_i = 1/F$, and $n_i = 1$ for all i, then $X_1^{(i)} = F$ for all $i = 1, \ldots, F$, in which case the smoothness equals entropy.

$$L = \frac{1}{F} \sum_{i=1}^{F} \log \sqrt{(F^2)} = \log F = H. \tag{9.46}$$

SCHEDULING AND SOURCE CODING

| P_1 | P_2 | P_1 | P_3 | P_1 | P_2 | P_1 | P_4 |

(a) Optimal scheduling

| P_1 | P_1 | P_2 | P_1 | P_1 | P_2 | P_3 | P_4 |

(b) WFQ scheduling

FIGURE 9.24 Examples of Scheduling.

When all weights are reciprocals of the power of 2, for example $\phi_i = 1/2, 1/4, 1/8, 1/8$, an optimal scheduling can be easily found. An example is shown in Fig. 9.24(a), where $F = 8$, $K = 4$, and $n_i = 4, 2, 1, 1$ with respect to state P_i for $i = 1, \ldots, 4$. The scheduling with constant interstate time $X^{(i)} = 2, 4, 8, 8$ satisfies both equalities of Theorem 10.9:

$$H = L = \frac{1}{2} \cdot 1 + \frac{1}{4} \cdot 2 + \frac{1}{8} \cdot 3 + \frac{1}{8} \cdot 3 = 1.75. \tag{9.47}$$

Another scheduling based on weighted fair queueing (WFQ) for the same decomposition is shown in Fig. 9.24(b). The smoothness $L = 1.8758$ is greater than the decomposition entropy $H = 1.75$, and the superiority of the optimal scheduling is quite obvious by comparing Fig. 9.24(a) and (b). An upper bound of smoothness is given in the following theorem.

Theorem 9.10. For any capacity decomposition $C = \sum_{i=1}^{K} \phi_i P_i$, it is always possible to devise a scheduling whose smoothness L is within $1/2$ of decomposition entropy:

$$H \leq L < H + \frac{1}{2}. \tag{9.48}$$

Proof. Consider a random scheduling without frame structure such that the state P_i of each time slot is randomly selected with probability ϕ_i. The interstate time $X^{(i)}$ of state P_i is a geometric random variable with $E[X^{(i)}] = 1/\phi_i$ and $\text{Var}[X^{(i)}] = (1 - \phi_i)/\phi_i^2$. The second moment of $X^{(i)}$ and the smoothness of state P_i are given, respectively, as follows:

$$E[(X^{(i)})^2] = \text{Var}[X^{(i)}] + (E[X^{(i)}])^2 = \frac{2 - \phi_i}{\phi_i^2} \tag{9.49}$$

and

$$L_i = \log \sqrt{E[(X^{(i)})^2]} = \frac{1}{2} \log(2 - \phi_i) + \log \phi_i. \tag{9.50}$$

It is counterintuitive that the smoothness L of random scheduling is not equal to the decomposition entropy H; in fact, we have

$$L = \sum_{i=1}^{K} \phi_i L_i = \frac{1}{2} \sum_{i=1}^{K} \phi_i \log(2 - \phi_i) + H. \tag{9.51}$$

420 PACKET SWITCHING AND INFORMATION TRANSMISSION

It is easy to show that the Kullback–Leibler distance $L - H$ reaches the maximum value when $\phi_1 = \phi_2 = \cdots = \phi_K = \frac{1}{K}$, in which case we have

$$L - H = \frac{1}{2}\sum_{i=1}^{K} \phi_i \log(2 - \phi_i) \leq \frac{1}{2}\log\left(2 - \frac{1}{K}\right) < \frac{1}{2}. \qquad (9.52)$$

The random scheduling is actually no scheduling at all. Hence, a scheduling algorithm, with or without frame, should exist that satisfies the bound given in (9.48). □

It is obvious that Theorems 9.9 and 9.10 are counterparts of the source coding theorem stated as follows.

Noiseless Coding Theorem. Let random variable X take the possible values $x_1, \ldots, x_K$ with respective probabilities $p(x_1), \ldots, p(x_K)$. Then, the necessary and sufficient condition to encode the values of X in binary prefix code (none of which is an extension of another) of respective lengths $L_1, \ldots, L_K$ is

$$\sum_{i=1}^{K} 2^{-L_i} \leq 1 \qquad \text{(Kraft's inequality)}. \qquad (9.53)$$

The average code length is bounded by

$$L = \sum_{i=1}^{K} p(x_i) L_i \geq H(X) = \sum_{i=1}^{K} p(x_i) \log \frac{1}{p(x_i)} \qquad (9.54)$$

and it is always possible to devise an optimal prefix code for X whose average code length L is within 1 of entropy:

$$H(X) \leq L < H(X) + 1. \qquad (9.55)$$

9.6.2 Comparison of Scheduling Algorithms

The scheduling algorithm for an arbitrary set of decomposition weights ϕ_i achieves optimal smoothness if $L = H$, which requires the constant interstate time that is equal to the inverse of weight $X_k^{(i)} = F/n_i = 1/\phi_i$ for all state P_i. Most proposed scheduling algorithms are indeed constructed from the reciprocal of the normalized weights. A comparison of several stereotype algorithms is given below.

9.6.2.1 Weighted Fair Queueing (WFQ) Scheduling Algorithm
The WFQ scheduling algorithm [DKS89] is constructed from the sequence of virtual finish times $1/\phi_i, 2/\phi_i, \ldots$ of each state P_i with the initial finish time $1/\phi_i$. The algorithm is given as follows.

The WFQ Scheduling Algorithm

Select the state with the smallest finish time and increase its finish time by the inverse of its weight. Repeat this process until the frame size is reached.

We will use the running example of a set of five states P_1, P_2, P_3, P_4, and P_5 with respective weights $\phi_1 = 0.5$ and $\phi_2 = \phi_3 = \phi_4 = \phi_5 = 0.125$ to illustrate the performance of the scheduling algorithms presented in this section. The sequence generated by WFQ in each time slot τ within a frame is given in Table 9.4, and the final sequence is $P_1 P_1 P_1 P_1 P_2 P_3 P_4 P_5$.

Although the WFQ algorithm properly utilizes the inverse weights in constructing the scheduling, the above example shows that the sequence generated by WFQ is still far from the achievable optimal smoothness. Possible amendments to WFQ are discussed below.

9.6.2.2 WF²Q Scheduling Algorithm WF²Q is a scheduling algorithm [Zha95] that incorporates the stringent rate requirement of *generalized processor sharing* (GPS)[PaGa93] in WFQ. Let $T_i(\tau)$ be the number of time slots assigned to state P_i up to time τ, then the WF²Q will select the state P_i that satisfies

$$T_i(\tau - 1) < \tau \cdot \phi_i \qquad (9.56)$$

in every time slot $\tau = 1, 2, \ldots$. That is, only the states that have started its service in the corresponding GPS system will be selected. The set of states qualified for selection in time slot τ is defined by

$$Q_\tau = \{P_i \mid T_i(\tau - 1) < \tau \cdot \phi_i\}. \qquad (9.57)$$

Hence, all states are qualified in the first time slot (see Table 9.4), because $T_i(0) = 0$ for all i. The set of qualified states $Q_1 = \{P_1 P_2 P_3 P_4 P_5\}$ is then ordered by their finish time and the state P_1 is selected. In the second time slot, however, $T_1(1) = 1$ and $2 \cdot \phi_1 = 1$, so the state P_1 cannot be selected and $Q_2 = \{P_2 P_3 P_4 P_5\}$. In each time slot, whenever a state is selected, like in WFQ, its finish time will be increased by the inverse of its weight.

TABLE 9.4 An Example of WFQ Algorithm

τ	P_1	P_2	P_3	P_4	P_5	Selection
1	2	8	8	8	8	P_1
2	4	8	8	8	8	P_1
3	6	8	8	8	8	P_1
4	8	8	8	8	8	P_1
5	10	8	8	8	8	P_2
6	10	16	8	8	8	P_3
7	10	16	16	8	8	P_4
8	10	16	16	16	8	P_5

TABLE 9.5 An Example of WF²Q Algorithm

τ	P_1	P_2	P_3	P_4	P_5	Q_τ	Selection
1	2	8	8	8	8	$P_1 P_2 P_3 P_4 P_5$	P_1
2	4	8	8	8	8	$P_2 P_3 P_4 P_5$	P_2
3	4	8	8	8	8	$P_1 P_3 P_4 P_5$	P_1
4	4	16	8	8	8	$P_3 P_4 P_5$	P_3
5	6	16	8	8	8	$P_1 P_4 P_5$	P_1
6	6	16	16	8	8	$P_4 P_5$	P_4
7	8	16	16	8	8	$P_1 P_5$	P_1
8	10	16	16	16	8	P_5	P_5

Table 9.5 provides the whole procedure of WF²Q. The final sequence is $P_1 P_2 P_1 P_3 P_1 P_4 P_1 P_5$, which is better than the sequence produced by WFQ. Another approach to amend WFQ based on the Huffman tree is given next.

9.6.2.3 Huffman Round Robin (HuRR) Algorithm

The Huffman code [Gal78] is the optimal source code constructed from the binary probability tree, called the *Huffman tree*, in a hierarchical manner. When the WFQ is applied to only two states, the average interstate time of each state is very close to the optimum because the two-state Huffman binary tree has only one level.

For example, consider the two states P_1 and P_2 with respective weights $\phi_1 = 0.25$ and $\phi_2 = 0.75$, the WFQ will produce the sequence $P_2 P_2 P_1 P_2$. The interstate time of state P_1 is 4, which is exactly the reciprocal of 0.25, and state P_2 has interstate times of 1 and 2, which are the two integers nearest to $1/0.75$. However, if there are more than two states, the WFQ fails to achieve the optimality because the Huffman tree has multiple levels.

The optimality of Huffman code is closely related to Kraft's inequality, which is the necessary and sufficient condition of prefix source coding. In scheduling, the optimal smoothness $L = H$ also implies $L_i = \log 1/\phi_i$ for all states P_i. Hence, the equality of Kraft's inequality holds:

$$\sum_{i=1}^{K} 2^{-L_i} = 1.$$

Similar to the Huffman code, the HuRR scheduling algorithm proposed in Ref. [ChLe06] is a hierarchical scheduling algorithm synthesized from the Huffman tree.

Consider each state of the decomposition as a symbol and the weight of the state as its probability, then a Huffman tree can be constructed from ordered sequence of probabilities as usual. The probability of the root node is 1, the probability of each leaf node is the probability of the symbol represented by that leafe, and the probability of each intermediate node is the sum of the probabilities of two successors. The HuRR algorithm comprises the following steps.

The HuRR Scheduling Algorithm

Step 1 Initially set the root to be temporary node P_X and $S = P_X \cdots P_X$ to be a temporary sequence.

Step 2 Apply the WFQ to the two successors of P_X to produce a sequence T and substitute T for the subsequence $P_X \cdots P_X$ of S.

Step 3 If there is no intermediate node in the sequence S, then terminate the algorithm. Otherwise, select an intermediate node P_X appearing in S and goto step 2.

The Huffman tree of the previous example is shown in Fig. 9.25; the HuRR scheduling algorithm will generate the following sequences:

$$P_1 P_Z P_1 P_Z P_1 P_Z P_1 P_Z \to P_1 P_X P_1 P_Y P_1 P_X P_1 P_Y$$
$$\to P_1 P_2 P_1 P_4 P_1 P_3 P_1 P_5.$$

In this particular example, the following Huffman code can be generated from the tree depicted in Fig. 9.25:

$$P_1 \leftarrow 0, \ P_2 \leftarrow 100, \ P_3 \leftarrow 101, \ P_4 \leftarrow 110, \ P_5 \leftarrow 111$$

and the logarithm of interstate time of each state is equal to the length of its Huffman code. The smoothness of the sequence of this example is the same as that generated by WF^2Q earlier. However, the HuRR outperforms WF^2Q in general as demonstrated by the comparison given next.

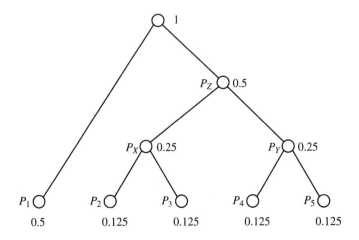

FIGURE 9.25 A Huffman tree of HuRR Algorithm.

TABLE 9.6 The Smoothness Comparison of Scheduling Algorithms

P_1	P_2	P_3	P_4	Random	WFQ	WF^{2}Q	HuRR	Entropy
0.1	0.1	0.1	0.7	1.628	1.575	1.414	1.414	1.357
0.1	0.1	0.2	0.6	1.894	1.734	1.626	1.604	1.571
0.1	0.1	0.3	0.5	2.040	1.784	1.724	1.702	1.686
0.1	0.2	0.2	0.5	2.123	1.882	1.801	1.772	1.761
0.1	0.1	0.4	0.4	2.086	1.787	1.745	1.745	1.722
0.1	0.2	0.3	0.4	2.229	1.903	1.903	1.884	1.847
0.2	0.2	0.2	0.4	2.312	2.011	1.980	1.933	1.922
0.1	0.3	0.3	0.3	2.286	1.908	1.908	1.908	1.896
0.2	0.2	0.3	0.3	2.370	2.016	2.016	1.980	1.971

9.6.2.4 Comparison of Smoothness of Scheduling Algorithms The smoothness of preceding scheduling algorithms is compared with random scheduling in Table 9.6 to show its progressive improvement. It is clear that the sequences generated by HuRR are, in general, closer to the entropy than the other two. The performance of WF^{2}Q is comparable to HuRR, while WFQ is not satisfactory in some cases.

The significance of the Huffman coding scheme lies in the structure of the Huffman tree. By the same token, HuRR, which is also implemented with the Huffman tree, can provide the best performance in terms of smoothness among the scheduling algorithms discussed in this section. However, HuRR has not yet been proven to be the optimal scheduling algorithm. It is clear that the key to construct optimal scheduling is to explore the structure of the smoothness L_i of each state P_i in a frame of size F and Kraft's inequality that is satisfied by L_i, which is mathematically equivalent to the codeword length in the source coding theorem.

9.6.3 Two-Dimensional Scheduling

The smoothness of scheduling described in the preceding section is for a single-server system. However, the capacity matrix decomposition of the packet switch system is a two-dimensional scheduling aimed at multiple inputs and outputs [LiLe00]. The generalization of the smoothness to two-dimensional scheduling is discussed in this section.

Recall that the capacity matrix $C = [c_{ij}]$ of a $C(m, n, k)$ Clos network is a $k \times k$ matrix, where c_{ij} is the average number of tokens assigned to the virtual path V_{ij} between input module i and output module j. If the frame size is F, then the matrix $F \cdot C$ is integer doubly stochastic, such that $\sum_i^k c_{ij} = m, \sum_j^k c_{ij} = m$.

In the following discussion of two-dimensional smoothness, we will consider, without loss of generality, capacity matrix $C = [c_{ij}]$ and $\sum_i^N c_{ij} = 1, \sum_j^N c_{ij} = 1$ of a crossbar switch for the sake of simplicity. The two-dimensional entropies of the capacity matrix C are defined as follows.

Definition 9.6. Let $C = [c_{ij}]$ be a doubly stochastic capacity matrix; the entropy of input module i is defined by

$$\overline{H}_i = -\sum_{j=1}^{N} c_{ij} \log c_{ij}, \tag{9.58}$$

and the set of entropies of input modules is represented by the column vector

$$\overline{H} = \begin{bmatrix} \overline{H}_1 \\ \overline{H}_2 \\ \vdots \\ \overline{H}_N \end{bmatrix} = \begin{bmatrix} \overline{H}_1, \overline{H}_2, \ldots, \overline{H}_N \end{bmatrix}^T. \tag{9.59}$$

Similarly, the entropy of output module j is defined by

$$\underline{H}_j = -\sum_{i=1}^{N} c_{ij} \log c_{ij}, \tag{9.60}$$

and the set of entropies of output modules is represented by the row vector

$$\underline{H} = \begin{bmatrix} \underline{H}_1, \underline{H}_2, \ldots, \underline{H}_N \end{bmatrix}. \tag{9.61}$$

The entropy of the capacity matrix C is defined by

$$H(C) = \sum_{i=1}^{N} \overline{H}_i = \sum_{j=1}^{N} \underline{H}_j = -\sum_{j=1}^{N} \sum_{i=1}^{N} c_{ij} \log c_{ij}. \tag{9.62}$$

The smoothness of a two-dimensional scheduling is defined by the intertoken time in the same manner as the interstate time of single-server scheduling. Let n_{ij} be the number of tokens assigned to the virtual path V_{ij} within a frame F, and $y_1^{(ij)}, y_2^{(ij)}, \ldots, y_{n_{ij}}^{(ij)}$ be the sequence of the intertoken times. We have

$$n_{ij} = c_{ij} F, \tag{9.63}$$

$$y_1^{(ij)} + y_2^{(ij)} + \cdots + y_{n_{ij}}^{(ij)} = F \tag{9.64}$$

for all $i, j = 1, \ldots, N$. It should be noted that the number of tokens assigned to a virtual path V_{ij} in a time slot can be more than one if the number of central modules $m > 1$, in which case the degenerate intertoken time $y_k^{(ij)} = 0$ will be allowed. The smoothness of two-dimensional scheduling is defined as follows.

Definition 9.7. For a given scheduling of the capacity matrix decomposition $C = \sum_{i=1}^{K} \phi_i P_i$, the smoothness of the virtual path V_{ij} is defined by

$$d_{ij} = \log \sqrt{\frac{\sum_{k=1}^{n_{ij}} \left(y_k^{(ij)}\right)^2}{n_{ij}}}. \tag{9.65}$$

The smoothness of input module i is defined by

$$\overline{D}_i = \sum_{j=1}^{N} c_{ij} d_{ij} = \sum_{j=1}^{N} c_{ij} \log \sqrt{\frac{\sum_{k=1}^{n_{ij}} (y_k^{(ij)})^2}{n_{ij}}}, \tag{9.66}$$

and the input smoothness is the column vector

$$\overline{D} = \begin{bmatrix} \overline{D}_1 \\ \overline{D}_2 \\ \vdots \\ \overline{D}_N \end{bmatrix} = \left[\overline{D}_1, \overline{D}_2, \ldots, \overline{D}_N\right]^T. \tag{9.67}$$

Similarly, the smoothness of output module j is defined by

$$\underline{D}_j = \sum_{i=1}^{N} c_{ij} d_{ij} = \sum_{i=1}^{N} c_{ij} \log \sqrt{\frac{\sum_{k=1}^{n_{ij}} (y_k^{(ij)})^2}{n_{ij}}}, \tag{9.68}$$

and the output smoothness is the row vector

$$\underline{D} = \left[\underline{D}_1, \underline{D}_2, \ldots, \underline{D}_N\right] \tag{9.69}$$

The smoothness of the two-dimensional scheduling is defined by

$$D = \sum_{i=1}^{N} \overline{D}_i = \sum_{j=1}^{N} \underline{D}_j = \sum_{j=1}^{N} \sum_{i=1}^{N} c_{ij} d_{ij}. \tag{9.70}$$

The properties of two-dimensional smoothness are similar to that of single-server scheduling. The proof of the following theorem is the same as that of Theorem 9.9.

Theorem 9.11. For capacity matrix decomposition $C = \sum_{i=1}^{K} \phi_i P_i$ with frame size F, any scheduling satisfies the following smoothness inequalities:

1. The matrix $K_r = [2^{-d_{ij}}]$, called *Kraft's matrix*, is doubly substochastic such that
$$\sum_{i=1}^{N} 2^{-d_{ij}} \leq 1, \quad \sum_{j=1}^{N} 2^{-d_{ij}} \leq 1. \tag{9.71}$$

2. For input module $i = 1, \ldots, N$, we have
$$\overline{D}_i \geq \overline{H}_i. \tag{9.72}$$

3. For output module $j = 1, \ldots, N$, we have
$$\underline{D}_j \geq \underline{H}_j. \tag{9.73}$$

4. The overall smoothness is bounded by
$$D \geq H(C). \tag{9.74}$$

The above equalities hold when $y_k^{(ij)} = 1/c_{ij}$, for all $i, j = 1, 2, \ldots, N$ and $k = 1, 2, \ldots, n_{ij}$.

The difference between smoothness and entropy can be considered as the distortion of delay jitter introduced by scheduling. For circuit switch, the capacity matrix C is a permutation matrix, in which case we have $H(C) = D = 0$. Another extreme case is the scheduling of uniform capacity matrix $C = [\frac{1}{N}] = \frac{1}{N}\sum_{i=1}^{N} P_i$ with frame size $F = N$. In this case, the entropy $H(C) = N \log N$ is also equal to the smoothness $D = N \log N$ because intertoken time $y_k^{(ij)} = N$, for all $i, j = 1, 2, \ldots, N$ and $k = 1, 2, \ldots, n_{ij}$ for any scheduling, in which case the entropy $H(C)$ is the maximum because the capacity matrix is completely uniform, and the smoothness of any scheduling cannot make it any worse.

It is clear that if the Kraft matrix
$$K_r = [2^{-d_{ij}}]$$

is doubly stochastic, then the two-dimensional scheduling is optimal. However, the existence of the optimal scheduling that can reach the entropy $H(C)$ of any capacity matrix C is still an open issue. Nevertheless, a rule of thumb is to minimize K, the number of permutation matrices in the decomposition [KKLS05], since the maximal delay bound is on the order of $O(K)$ for most known scheduling algorithms [Sha49,ChLe03].

For example, given a doubly stochastic matrix C
$$C = \begin{bmatrix} 0.75 & 0 & 0.125 & 0.125 \\ 0.125 & 0.5 & 0.375 & 0 \\ 0.125 & 0.125 & 0.5 & 0.25 \\ 0 & 0.375 & 0 & 0.625 \end{bmatrix}.$$

The input entropy $\overline{H}$ and output entropy $\underline{H}$ of matrix C are given, respectively, by

$$\overline{H} = \begin{bmatrix} 1.0613 & 1.4056 & 1.7500 & 0.9544 \end{bmatrix}^T$$

and

$$\underline{H} = \begin{bmatrix} 1.0613 & 1.4056 & 1.4056 & 1.2988 \end{bmatrix}.$$

The entropy of the capacity matrix C is $H(C) = 5.1714$. One possible decomposition is given by

$$8 \cdot C = \begin{bmatrix} 6 & 0 & 1 & 1 \\ 1 & 4 & 3 & 0 \\ 1 & 1 & 4 & 2 \\ 0 & 3 & 0 & 5 \end{bmatrix} = 4 \begin{bmatrix} 1 & 0 & 0 & 0 \\ 0 & 1 & 0 & 0 \\ 0 & 0 & 1 & 0 \\ 0 & 0 & 0 & 1 \end{bmatrix}$$

$$+ \begin{bmatrix} 1 & 0 & 0 & 0 \\ 0 & 0 & 1 & 0 \\ 0 & 1 & 0 & 0 \\ 0 & 0 & 0 & 1 \end{bmatrix} + \begin{bmatrix} 1 & 0 & 0 & 0 \\ 0 & 0 & 1 & 0 \\ 0 & 0 & 0 & 1 \\ 0 & 1 & 0 & 0 \end{bmatrix}$$

$$+ \begin{bmatrix} 0 & 0 & 1 & 0 \\ 1 & 0 & 0 & 0 \\ 0 & 0 & 0 & 1 \\ 0 & 1 & 0 & 0 \end{bmatrix} + \begin{bmatrix} 0 & 0 & 0 & 1 \\ 0 & 0 & 1 & 0 \\ 1 & 0 & 0 & 0 \\ 0 & 1 & 0 & 0 \end{bmatrix}$$

in which $K = 5$ and the permutation matrices are denoted by $\{P_1, P_2, \ldots, P_5\}$, respectively. If the WFQ scheduling algorithm is applied to this set of permutation matrices, the resulting sequence is

$$P_1 P_1 P_1 P_1 P_2 P_3 P_4 P_5,$$

which gives rise to the following two-dimensional scheduled tokens:

a	a	a	a	a	a	b	c
b	b	b	b	c	d	d	d
c	c	c	c	b	b	a	b
d	d	d	d	d	c	c	a

where each row corresponds to an output module and the symbols a, b, c, and d represent the respective tokens assigned to input modules 1–4. The input smoothness of this scheduling is given by

$$\overline{D} = \begin{bmatrix} 1.2084 & 1.6997 & 2.0323 & 1.3118 \end{bmatrix}^T,$$

and the output smoothness is

$$\underline{D} = \begin{bmatrix} 1.2084 & 1.7636 & 1.6997 & 1.5805 \end{bmatrix},$$

which sum up to $D = 6.2522$. When the HuRR scheduling algorithm is applied to the same set of permutation matrices, the resulting sequence is

$$P_1 P_2 P_1 P_4 P_1 P_3 P_1 P_5,$$

which gives rise to the following two-dimensional scheduled tokens:

$$\begin{matrix} a & a & a & b & a & a & a & c \\ b & c & b & d & b & d & b & d \\ c & b & c & a & c & b & c & b \\ d & d & d & c & d & c & d & a \end{matrix}$$

This result is obviously smoother than the previous one as evident by the following smoothness measures:

$$\overline{D} = \begin{bmatrix} 1.1250 & 1.4375 & 1.7902 & 1.0267 \end{bmatrix}^T$$

and

$$\underline{D} = \begin{bmatrix} 1.1250 & 1.4375 & 1.4375 & 1.3794 \end{bmatrix},$$

which sum up to $D = 5.3794$, an improvement consistent with the comparison of single-server scheduling described in the preceding section.

Suppose we consider another capacity matrix decomposition

$$8 \cdot C = \begin{bmatrix} 0 & 0 & 1 & 0 \\ 1 & 0 & 0 & 0 \\ 0 & 1 & 0 & 0 \\ 0 & 0 & 0 & 1 \end{bmatrix} + \begin{bmatrix} 0 & 0 & 0 & 1 \\ 0 & 0 & 1 & 0 \\ 1 & 0 & 0 & 0 \\ 0 & 1 & 0 & 0 \end{bmatrix}$$

$$+2 \begin{bmatrix} 1 & 0 & 0 & 0 \\ 0 & 0 & 1 & 0 \\ 0 & 0 & 0 & 1 \\ 0 & 1 & 0 & 0 \end{bmatrix} +4 \begin{bmatrix} 1 & 0 & 0 & 0 \\ 0 & 1 & 0 & 0 \\ 0 & 0 & 1 & 0 \\ 0 & 0 & 0 & 1 \end{bmatrix}$$

in which the number of permutation matrices $K = 4$ is reduced. Again, applying the HuRR scheduling, the resulting sequence is

$$P_1 P_4 P_3 P_4 P_2 P_4 P_3 P_4,$$

which gives rise to the following two-dimensional scheduled tokens:

$$\begin{matrix} b & a & a & a & c & a & a & a \\ c & b & d & b & d & b & d & b \\ a & c & b & c & b & c & b & c \\ d & d & c & d & a & d & c & d \end{matrix}$$

The smoothness measures of this two-dimensional scheduling are given by

$$\overline{D} = \begin{bmatrix} 1.1250 & 1.4375 & 1.7500 & 1.0267 \end{bmatrix}^T$$

and

$$\underline{D} = \begin{bmatrix} 1.125 & 1.4375 & 1.4375 & 1.3392 \end{bmatrix},$$

which sum up to the overall smoothness $D = 5.3392$. By comparing the output smoothness $\underline{D}$ with that of the previous decomposition, we find some improvement in the fourth output. Similar improvement can be found at input module 3, where the optimal smoothness $\overline{D}_3 = 1.7500 = \overline{H}_3$ is achieved. The overall smoothness $D = 5.3392$ is also better than the previous $D = 5.3794$ but still greater than the entropy of capacity matrix $H(C) = 5.1714$.

Preceding examples indicate that the performance of path switching depends not only on the scheduling algorithm but also on the number of permutation matrices in the capacity matrix decomposition. As we mentioned above that the number of permutation matrices in the case of uniform capacity matrix with maximal entropy is $K = N$, and the worst delay bound is on the order of $O(K)$ for most known scheduling algorithm, it is therefore reasonable to arrive at the conjecture that K is on the order of $O(N)$ for the optimal decomposition of any capacity matrix, even though the known bound of K in the Birkhoff–von Neumann decomposition is $N^2 - 2N + 2$. The Kraft matrix and the smoothness measure introduced in this section will provide the objective of optimal two-dimensional scheduling to be explored in the future, which is expected to be a very hard problem because of the large number of possible combinations.

9.7 CONCLUSION

The parallels between packet switch subject to the contention and transmission channel in the presence of noise are consequences of the law of probability. Input signal to transmission is a function of time, the main theorem on noisy channel coding being based on the law of large number. On the other hand, the input signal to switch is a function of space, and both theorems on deflection routing and smoothness of scheduling are proved on the ground of randomness. In both systems, the random disturbances are tamed by means of similar mathematical tools aimed for reliable communication.

The communication network has gone through two phases of quantization in the past century. The first phase was the quantization of transmission channels, from analog to digital, based on the sampling theorem of bandlimited signal. The second phase is the quantization of the switching system, from circuit switching to packet switching. The comparisons provided in this chapter demonstrate that the scheduling based on capacity matrix decomposition serves the same function as the sampling theorem to reduce the complexity of communication.

The concept of path switching can be further extended to the network level to cope with resource contentions. A connection-oriented subnetwork with predetermined topology and bandwidth can be embedded in the current IP network for supporting QoS of real-time services such as voice or video over IP. The scheduling of path switching in conjunction with the routing scheme of multiprotocol label switching (MPLS) will provide a platform to support different traffic classes of differentiated services (DiffServ) [Law01,TAPF01]. The predetermined paths of label switched subnetwork are similar to the subway system embedded in the public transportation network. It will provide a coherent end-to-end QoS solution to the multilevel resource allocation [LeLa97] issues arising from real-time services. First, the predetermined connection patterns at each router are contention-free at the packet level. Next, the periodic connection patterns are frame based, and the bandwidth assigned to each session is fixed at burst level within each frame. Finally, the stable capacity of each session is guaranteed at call level. The traffic engineering of this Internet subway system could be a challenging networking research area in the future.

Bibliography

[Abr63] N. Abramson, *Information Theory and Coding*, New York: McGraw-Hill, 1963.

[ABF01] M. Allman, H. Balakrishnan, and S. Floyd, "Enhancing TCPs Loss Recovery Using Limited Transmit," *RFC 3042*, 2001.

[ADL95] J. S. Ahn, P. B. Danzig, Z. Liu, and L. Yan, "Evaluation with TCP Vegas: Emulation and Experiment," *ACM SIGCOMM Comput. Commun. Rev.*, Vol. 25, No. 4, 1995, pp. 185–195.

[AhD89] H. Ahmadi and W. Denzel, "A Survey of Modern High-Performance Switch Fabric for Integrated Circuit and Packet Switching," *IEEE J. Select. Areas Commun.*, Vol. 7, No. 7, 1989.

[AKS83] M. Ajtai, J. Komlos, and E. Szemeredi, "An $O(n \log n)$ Sorting Network," *Proceedings of the 15th ACM Symposium on Theory of Computation*, 1983, pp. 1–9.

```
A significant theoretical breakthrough: the first
O(n log n) sorting network. The constant in front of the
n log n, however, is rather large, rendering this net-
work impractical.
```

[AMS81] D. Anick, D. Mitra, and M. M. Sondhi, "Stochastic Theory of a Data-Handling System with Multiple Sources," *Bell Syst. Tech. J.*, Vol. 61, No. 8, 1982, pp. 1871–1894.

```
The original work on the application of the fluid-
flow model that led to much subsequent work by others.
This book attempts to simplify the presentation in the
paper to make the theory more accessible to undergrad-
uate students. Some details have been omitted.
```

[Bar64] P. Baran, "On Distributed Communications Networks," *IEEE Trans. Commun.*, Vol. 12, 1964, pp. 1–9.

```
The first paper on deflection routing (called hot-
potato routing in the paper) in the context of wide-
area communications networks.
```

Principles of Broadband Switching and Networking, by Tony T. Lee and Soung C. Liew
Copyright © 2010 John Wiley & Sons, Inc.

[Bat68] K. E. Batcher, "Sorting Networks and Their Applications," *AFIPS Proceeding of the Spring Joint Computer Conference*, 1968, pp. 307–314.

```
The seminal work by Batcher on the bitonic and odd-
even sorting networks. Although both networks are
O(n log² n), they remain perhaps the most implementable
networks in practice throughout all these years
because of their simple structures.
```

[Bat76] K. E. Batcher, "The Flip Network in STARAN," *Proceedings of the 1976 International Conference on Parallel Processing*, 1976, pp. 65–71.

[BBB87] A. Bertossi, G. Bongiovanni, and M. Bonuccelli, "Time Slot Assignment in SS/TDMA Systems with Intersatellite Links," *IEEE Trans. Commun.*, Vol. 35, No. 6, 1987, pp. 602–608.

[BeG92] D. Bertsekas and R. Gallager, *Data Networks*, 2nd edition, Prentice-Hall, 1992.

```
A widely used graduate-level course book. Highly
analytical compared to other books in the same area.
Mostly about traditional data networks, although
many of the issues remain valid even for broadband
networks.
```

[Bel81] J. Bellamy, *Digital Telephony*, Wiley, 1982.

```
An excellent reference book for modern telephone
systems.
```

[Ben63] V. E. Benes, "A Thermodynamics Theory of Traffic in Telephone Systems," *Bell Syst. Tech. J.*, Vol. 42, 1963, pp. 567–607.

[Ben65] V. E. Benes, *Mathematical Theory of Connecting Networks and Telephone Traffic*, Academic Press, 1965.

```
The first book on switching theory. Highly mathemati-
cal.
```

[BiB88] B. Bingham and H. Bussey, "Reservation-Based Contention Resolution Mechanism for Batcher–Banyan Packet Switches," *Electron. Lett.*, Vol. 24, No. 13, 1988, pp. 772–773.

```
On using a ring contention-resolution scheme for an
input-buffered switch.
```

[Bir46] G. Birkhoff, "Tres observaciones sobre el algebra lineal," *Universidad Nacional de Tucuman Revista (Serie A)*, Vol. 5, 1946, pp. 147–151.

[BOP94] L. S. Brakmo, S. W. O'Malley, and L. L. Peterson, "TCP Vegas: New Techniques for Congestion Detection and Avoidance," *Proceedings of ACM SIGCOMM94*, No. 4, October 1994, pp. 24–35.

[BP95] L. S. Brakmo and L. L. Peterson, "TCP Vegas: End to End Congestion Avoidance on a Global Internet," *IEEE J. Select. Areas Commun.*, Vol. 13, No. 8, 1995, pp. 1465–1480.

[ByL94] J. W. Byun and T. T. Lee, "The Design and Analysis of an ATM Multicast Switch with Adaptive Traffic Controller," *IEEE/ACM Trans. Networking*, Vol. 2, No. 3, 1994, pp. 288–298.

```
Much of the discussion in Section 6.1.2 is based on
this work.
```

BIBLIOGRAPHY

[Can71] D. Cantor, "On Non-Blocking Switching Networks," *Networks*, Vol. 1, 1971.

```
The original paper on Cantor Networks.
```

[CCH00] C. S. Chang, W. J. Chen, and H. Y. Huang, "Birkhoff–von Neumann input buffered crossbar switches," *INFOCOM 2000: Proceedings of the 19th Annual Joint Conference of the IEEE Computer and Communications Societies*, Vol. 3, IEEE, 2000, pp. 1614–1623.

[Cha91] H. J. Chao, "A Recursive Modular Terabit/Second ATM Switch," *IEEE J. Select. Area Commun.*, Vol. 9, No. 8, 1991, p. 1161.

[ChJi03] H. Chao, Z. Jing, and S. Liew, "Matching Algorithms for Three-Stage Bufferless Clos Network Switches," *IEEE Commun. Mag.*, Vol. 41, No. 10, 2003, pp. 46–54.

[ChL91] C. J. Chang and C. J. Ling, "Overflow Controller in Copy Network of Broadband Packet Switch," *Electron. Lett.*, Vol. 27, No. 11, 1991, pp. 937–939.

[ChLe03] M. C. Chan and T. Lee, "Statistical Performance Guarantees in Large-Scale Cross-Path Packet Switch," *IEEE/ACM Trans. Networking*, Vol. 11, 2003, pp. 325–337.

[ChLe06] M. T. Choy and T. T. Lee, "Huffman Fair Queueing: A Scheduling Algorithm Providing Smooth Output Traffic," *ICC 06*, IEEE, June 2006.

[ChLi91] H. J. Chao and S. C. Liew, "Architecture Design for ATM Statistical Multiplexers," *Int. J. Digital Analog Commun. Syst.*, Vol. 4, No. 4, 1991, pp. 239–248.

[ChLi95] T. M. Chen and S S. Liu, *ATM Switching Systems*, Artech House, 1995.

```
A book that attempts to include control and manage-
ment functions in the examination of the ATM switching
system. It contains useful information on the SONET
and ATM standards for interested readers. Its focus
is more about how rather than why things are done in a
certain way.
```

[Cho91] C.-H. Chow, "On Multicast Path Finding Algorithm," *Conference Record, IEEE Infocom'91*, pp. 1274–1283.

```
On finding a multicast tree for a multicast connection
over wide-area networks.
```

[ChS91] J.S.-C. Chen and T. E. Stern, "Throughput Analysis, Optimal Buffer Allocation, and Traffic Imbalance Study of a Generic Nonblocking Packet Switch," *IEEE J. Select. Areas Commun.*, Vol. 9, No. 3, 1991, pp. 439–449.

[CJ89] D. M. Chiu and R. Jain, "Analysis of Increase and Decrease Algorithms for Congestion Avoidance in Computer Network," *Comput. Networks ISDN Syst.*, Vol. 17, 1989.

[Clo53] C. Clos, "A Study of Non-Blocking Switch Network," *Bell Syst. Tech. J.*, Vol. 32, 1953, pp. 406–424.

```
The seminal paper by Clos on Clos networks.
```

[DeD87] M. De Prycker and M. Desomer, "Performance of a Service Independent Switching Network with Distributed Control," *IEEE J. Select. Areas Commun.*, Vol. 5, 1987, pp. 1293–1301.

[DeP93] M. De Prycker, *Asynchronous Transfer Mode: Solution for Broadband ISDN*, 2nd edition, Ellis Horwood, 1993.

```
A good reference book on details of the ATM standard
and reasons why ATM was originally proposed.
```

[Des88]   A. Descloux, "Contention Probabilities in Packet Switching Networks with Strung Input Processes," *Proceedings of the ITC 12*, 1988.

[DGH87]   C. Day, J. Giacopelli, and J. Hickey, "Applications of Self-Routing Switches to LATA Fiber Optic Networks," *Proceedings of ISS '87*, March 1987.

[DiJ81]   D. M. Dias and J. R. Jump, "Analysis and Simulation of Buffered Delta Network," *IEEE Trans. Comput.*, Vol. C-30, 1981, pp. 273–282.

[DKS89]   A. Demers, S. Keshav, and S. Shenker, "Analysis and Simulation of a Fair Queueing Algorithm," *Proceedings of the ACM Sigcomm*, 1989, pp. 3–12.

[DyS95]   D. E. McDysan and D. L. Spohn, *ATM: Theory and Application*, McGraw-Hill, 1995.

```
A reference book that is suitable for network managers
and engineers who want to have an overall but not nec-
essary in-depth (i.e., research-level) understanding
of ATM networking.
```

[EcH88]   A. E. Eckberg and T. C. Hou, "Effects of Output Buffer Sharing on Buffer Requirements in an ATDM Packet Switch," *Proceedings of INFOCOM '88*.

```
An analysis of the advantage of having the outputs of
the switch share a common buffer.
```

[EHY]   K. Eng, M. Hluchyj, and Y. Yeh, "A Knockout Switch for Variable-Length Packets," *IEEE J. Select. Areas Commun.*, Vol. 5, No. 9, 1987.

```
The original paper on the Knockout switch.
```

[EKY]   K. Y. Eng, M. J. Karol, and Y. S. Yeh, "A Growable Packet (ATM) Switch Architecture: Design Principles and Applications," *Conference Record, Globecom '89*, pp. 32.2.1–32.2.7.

[EpHa98]   A. Ephremides and B. Hajek, "Information Theory and Communication Networks: An Unconsummated Union," *IEEE Trans. Inform. Theory*, Vol. 44, 1998, pp. 2416–2434.

[FH99]   S. Floyd, and T. Henderson, "The NewReno Modification to TCP¡¯s Fast Recovery Algorithm," *RFC 2582*, April 1999.

[Fu01]   C. P. Fu, "TCP Veno: End-to-End Congestion Control over Heterogeneous Networks," Ph.D. Dissertation, The Chinese University of Hong Kong, August 2001.

[Gal62]   R. Gallager, "Low-Density Parity-Check Codes," *IEEE Trans. Inform. Theory*, Vol. 8, No. 1, 1962, pp. 21–28.

[Gal65]   R. Gallager, "A Simple Derivation of the Coding Theorem and Some Applications," *IEEE Trans. Inform. Theory*, Vol. 11, No. 1, 1965, pp. 3–18.

[Gal78]   R. Gallager, "Variations on a Theme by Huffman," *IEEE Trans. Inform. Theory*, Vol. 24, No. 6, 1978, pp. 668–674.

[Ger98]   M. Gerla, E. Leonardo, F. Neri, and P. Panalti, "Minimum Distance Routing in Bidirectional Shufflenet," *IEEE INFOCOM '98*.

[GHM91]   J. N. Giacopelli, J. J. Hickley, W. S. Marcus, W. D. Sincoskie, and M. Littlewood, "Sunshine: A High Performance Self-Routing Broadband Packet Switch Architecture," *IEEE J. Select. Areas Commun.*, Vol. 9, No. 8, 1991, p. 1289.

```
On a Batcher-banyan switch architecture.
```

[GAC86] P. Gonet, P. Adams, and J. P. Courdreuse, "Asynchronous Time-Division Switching: The Way to Flexible Broadband Communication Networks," *Proceedings of the IEEE 1986 International Zurich Seminar on Digital Communications*, Zurich, Switzerland, March 1986.

The people who first proposed ATM.

[Haj88] K. Hajikano, K. Murakami, E. Iwabuchi, O. Isono, and T. Kobayashi, "Asynchronous Transfer Mode Switching Architecture for Broadband ISDN," *Conference Record, IEEE ICC'88*, pp. 911–915.

[Haj89] B. Hajek, "Bounds on Evacuation Time for Deflection Routing," *Distributed Computing*, Springer-Verlag, Vol. 5, 1991, pp. 1–6.

[Hal35] P. Hall, "On Representatives of Subsets," *J. London Math. Soc.*, Vol. 10, 1935, pp. 26–30.

[Ham86] R. W. Hamming, *Coding and Information Theory*, Englewood Cliffs, NJ: Prentice-Hall, 1986.

[HBM91] J. F. Hayes, R. Breault, and M. K. Mehmet-Ali, "Performance Analysis of a Multicast Switch," *IEEE Trans. Commun.*, Vol. 39, No. 4, 1991, pp. 581–587.

[HlK88] M. G. Hluchyj and M. J. Karol, "Queueing in High-Performance Packet Switching," *IEEE J. Select. Areas Commun.*, Vol. 6, No. 9, 1988, pp. 1587–1597.

Analysis on output-buffered and input-buffered switches.

[HuA87] Y. N. J. Hui and E. Arthurs, "A Broadband Packet Switch for Integrated Transport," *IEEE J. Select. Areas Commun.*, Vol. 5, No. 8, 1987, pp. 1264–1273.

The original paper on the three-phase algorithm applied on a Batcher-banyan input-buffered switch.

[Hui88] J. Hui, "Resource Allocation for Broadband Networks," *IEEE J. Select. Areas Commun.*, Vol. 6, No. 9, 1988.

[Hui90] J. Hui, *Switching and Traffic Theory for Integrated Broadband Networks*, Kluwer Academic Publishers, 1990.

A good reference book for circuit and fast-packet switching as well as the mathematical tools used to analyze broadband network performance. Requires the readers to have a certain level of mathematical maturity. Suitable textbook for a graduate-level course.

[HuK84] A. Huang and S. Knauer, "Starlite: A Wideband Digital Switch," *Proceeding of Globecom '84*, pp. 121–125.

The Batcher-banyan switch structure was originally proposed here.

[HuKa95] J. Y. Hui and E. Karasan, "A Thermodynamic Theory of Broadband Networks with Application to Dynamic Routing," *IEEE J. Select. Areas Commun.*, Vol. 13, 1995, pp. 991–1003.

[Ino79] H. Hiroshi, *Digital Integrated Communications Systems*, University of Tokyo Press, 1979.

[Inu79] T. Inukai, "An Efficient SS/TDMA Time Slot Assignment Algorithm," *IEEE Trans. Commun.*, Vol. 27, No. 10, 1979, pp. 1449–1455.

[Jac88]   V. Jacobson, "Congestion Avoidance and Control," *ACM SIGCOMM 88*, August 1988.

[Jaj03]   A. Jajszczyk, "Nonblocking, Repackable, and Rearrangeable Clos Networks: Fifty Years of the Theory Evolution," *IEEE Commun. Mag.*, Vol. 41, No. 10, 2003, pp. 28–33.

[JaG93]   A. Jajszczyk and G. Jekel "A New Concept—Repackable Networks," *IEEE Trans. Commun.*, Vol. 41, No. 8, 1993, pp. 1232–1237.

[Jain89] R. Jain, "A Delay-Based Approach for Congestion Avoidance in Interconnected Heterogencous Computer Networks," *ACM Comput. Commun. Rev.*, Vol. 19, No. 5, 1989.

[KaJ91]   M. Kawarasaki and B. Jabbari, "B-ISDN Architecture and Protocol," *IEEE J. Select. Areas Commun.*, Vol. 9, 1991, pp. 1405–1415.

[KeK79]   P. Kermani and L. Kleinrock, "Virtual Cut-Through: A New Computer Communication Switching Technique," *Comput. Networks*, Vol. 3, 1979, pp. 167–286.

[Kes91]   S. Keshav, "A Control-Theoretic Approach to Flow Control," *SIGCOMM91*, September 1991.

[KHM87]   M. J. Karol, M. G. Hluchyj, and S. P. Morgan, "Input vs. Output Queuing on a Space Division Packet Switch," *IEEE Trans. Commun.*, Vol. 35, No. 12, 1987, pp. 1347–1356.

[KiK80]   C. Kittel and H. Kroemer, *Thermal Physics*, New York: W. H. Freeman, 1980.

[KiL88]   H. S. Kim and A. Leon-Garcia, "Performance of Buffered Banyan Networks Under Nonuniform Traffic Patterns," *INFOCOM '88*, New Orleans, March 1988.

[KKLS05] I. Keslassy, M. Kodialam, T. V. Lakshman, and D. Stiliadis, "On Guaranteed Smooth Scheduling for Input-Queued Switches," *IEEE/ACM Trans. Networking*, Vol. 13, No. 6, 2005, pp. 1364–1375.

[Kle75]   L. Kleinrock, *Queueing Systems Theory*, Vol. 1, Wiley, 1975.

[Klei75] L. Kleinrock, *Queueing Systems, Computer Applications*, Vol. 2, New York: Wiley, 1975.

[Knu81]   D. E. Knuth, *The Art of Computer Programming, Sorting and Searching*, Addison-Wesley, Vol. III, 1981.

[Kot33]   V. A. Kotelnikov, "On the Carrying Capacity of the Ether and Wire in Telecommunications," *Material for the First All-Union Conference on Questions of Communication, Izd. Red. Upr. Svyazi RKKA, Moscow (Russian)*, 1933.

[KrH90]   A. Krishna and B. Hajek, "Performance of Shuffle-Like Switching Networks with Deflection," *Proceedings of IEEE INFOCOM'90*, San Francisco, CA, pp. 473–480.

[KrS83]   C. P. Kruskal and M. Snir, "The Performance of Multistage Interconnection Networks for Multiprocessors" *IEEE Trans. Comput.*, Vol. 32, No. 12, 1983.

[KuJ86]   M. Kumar and J. R. Jump, "Performance of Unbuffered Shuffle-Exchange Networks" *IEEE Trans. Comput.*, Vol. 35, No. 6, 1986, pp. 573–577.

[Kuw89]   H. Kuwahara, et al., "A shared buffer memory switch for an ATM exchange," *Conference Record, ICC'89*, Vol. 1, pp. 118–122.

[Lai95]   C. W. Lai and S. C. Liew, "A Framework for Statistical Multiplexing onto a Variable-Bit Rate Output Channel," *IEEE INFOCOM '95*, pp. 394–401.

[LaP80] D. H. Lawrie and D. A. Padua, "Analysis of Message Switching with Shuffle-Exchanges in Multiprocessors," *Proceedings of the Workshop on Interconnection Networks for Parallel and Distributed Processing*, 1980, pp. 116 – 123. Also reprinted in Ref. [WuF84].

[Law01] J. Lawrence, "Designing Multiprotocol Label Switching Networks," *IEEE Commun. Mag.*, Vol. 39, No 7, 2001, pp. 134–142.

[Law75] D. Lawrie, "Access and Alignment of Data in an Array Processor," *IEEE Trans. Comput.*, Vol. C-24, No. 12, 1975, pp. 1145–55.

[Lea86] C. T. Lea, "The Load-Sharing Banyan Network," *IEEE Trans. Comput.*, Vol. 35, 1986, pp. 1025–1034.

[Lee88] T. T. Lee, "Non-Blocking Copy Networks for Multicast Packet Switching," *IEEE J. Select. Areas Commun.*, Vol. 6, No. 9, 1988, pp. 1455–1467.

[Lee90] T. T. Lee, "A Modular Architecture for Very Large Packet Switches," *IEEE Trans. Commun.*, Vol. 6, No. 9, 1990, pp. 1455–1467.

[Lee93] T. T. Lee, "Parallel Communications for ATM Network Control and Management," *IEEE Globecom '93*.

[LeL94] T. T. Lee and S. C. Liew, "Broadband Packet Switches based on Dilated Interconnected Networks," *IEEE Trans. Commun.*, Part I of 3 parts, Vol. 42, No. 2/3/4, 1994, pp. 732–744.

[LeLa97] T. T. Lee and C. H. Lam, "Path Switching: A Quasi-Static Route Scheme for Large-Scale ATM Switches," *IEEE J. Select. Areas Commun.*, Vol. 15, No. 5, 1997, pp. 914–924.

[LeLi02] T. T. Lee and S. Y. Liew, "Parallel Routing Algorithms in Benes–Clos Networks," *IEEE Trans. Commun.*, Vol. 50, No 11, 2002, pp. 1841–1847.

[LeTo98] T. T. Lee and P. P. To, "Non-Blocking Routing Properties of Clos Networks," *DIMACS: Series in Discrete Mathematics and Theoretical Computer Science*, American Mathematical Society, 1998, pp. 181–195.

[LeWi00] A. Leon-Garcia and I. Widjaja, *Communication Networks*, McGraw-Hill, 2000.

[Lie89] S. C. Liew, "Comments on 'Fundamental Conditions Governing TDM Switching Assignments in Terrestrial and Satellite Networks'," *IEEE Trans. Commun.*, Vol. 37, No. 2, 1989, pp. 187–189.

[Lie94a] S. C. Liew, "Performance of Various Input-Buffered and Output-Buffered ATM Switch Design Principles Under Bursty Traffic: Simulation Study," *IEEE Trans. Commun.*, part II of 3 parts, Vol. 42, No. 2/3/4, 1994, pp. 1371–1379.

[Lie94b] S. C. Liew, "Multicast Routing in 3-Stage Clos ATM Switching Networks," *IEEE Trans. Commun.*, part II of 3 parts, Vol. 42, No. 2/3/4, 1994, pp. 1380–1390.

[Lie95] S. C. Liew, "A General Packet Replication Scheme for Multicast in Interconnection Networks," *IEEE INFOCOM '95*, pp. 394–401.

[Liew99] S. Y. Liew, "Real Time Scheduling in Large Scale ATM Cross-Path Switch," Ph.D. Dissertation, The Chinese University of Hong Kong, July 1999.

[LiL89] S. C. Liew and K. W. Lu, "A 3-Stage Interconnection Structure for Very Large Packet Switches," *Int. J. Digital Analog Cabled Syst.*, Vol. 2, 1989, pp. 303–316.

[LiL91] S. C. Liew and K. W. Lu, "Comparison of Buffering Strategies for Asymmetric Packet Switch Modules," *IEEE J. Select. Areas Commun.*, 1991, pp. 428–438.

[LiL94] S. C. Liew and T. T. Lee, "$N \log N$ Dual Shuffle-Exchange Network with Error-Correcting Routing," *IEEE Trans. Commun.*, part I of 3 parts, Vol. 42, No. 2/3/4, 1994, pp. 754–766.

[LiL95] S. C. Liew and T. T. Lee, "SKEL: A Fundamental Property desirable in ATM Switches for Simple Traffic Management: Illustration with Generic Output-Buffered and Input-Buffered Switches," *Perform. Evaluation*, Vol. 25, No. 4, 1996, pp. 247–266.

[LiLe89] S.-Q. Li and M. J. Lee, "A Study of Traffic Imbalances in a Fast Packet Switch," *Conference Record, ICC'89*, pp. 538–545.

[LiLe00] S. Y. Liew and T. T. Lee, "Bandwidth Assignment with QoS Guarantee in a Class of Scalable ATM Switches," *IEEE Trans. Commun.*, Vol. 483, 2000, pp. 377–380.

[LLL83] J. Lewandowski, J. Liu, and C. Liu, "SS/TDMA Time Slot Assignment with Restricted Switching Modes," *IEEE Trans. Commun.*, Vol. 31, No. 1, 1983, pp. 149–154.

[LLL86] J. L. Lewandowski, C. L. Liu, and J. W. S. Liu, "An Algorithmic Proof of a Generalization of the Birkhoff–von Neumann Theorem", *J. Algorithms*, Vol. 7, 1986, pp. 323–330.

[Mac03] D. J. C. MacKay, *Information Theory, Inference, and Learning Algorithms*, Cambridge: Cambridge University Press, 2003.

[Mas84] J. Massey, "Information Theory: The Copernican System of Communications," *IEEE Commun. Mag.*, Vol. 22, No. 12, 1984, pp. 26–28.

[Max87] N. F. Maxemchuk, "Routing in the Manhattan Street Network," *IEEE Trans. Commun.*, Vol. 35, No. 5, 1987, pp. 503–512.

[Max89] N. F. Maxemchuk, "Comparison of Deflection and Store and Forward Techniques in the Manhattan Street Network and Shuffle Exchange Networks." *Proceedings of INFOCOM'89*, April 1989, pp. 800-809.

[McK99] N. McKeown, "The iSLIP Scheduling Algorithm for Input-Queued Switches," *IEEE/ACM Trans. Networking*, Vol. 7, No. 2, 1999, pp. 188–201.

[MeT89] R. Melen and J. S. Turner, "Nonblocking Networks for Fast Packet Switching," *Proceedings of IEEE INFOCOM'89*, April 1989, pp. 548–557.

[MeT93] R. Melen and J. S. Turner, "Nonblocking Multirate Distribution Networks," *IEEE Trans. Commun.*, Vol. 41, No. 2, 1993.

[MM96] M. Mathis and J. Mahdavi, "Forward Acknowledgment: Refining TCP Congestion Control," *Proceedings of SIGCOMM96*, August 1996.

[MMFR96] M. Mathis, J. Mahdavi, S. Floyd, and A. Romanow "TCP Selective Acknowledgement Options," *RFC 2018*, October 1996.

[New88] P. Newman, "A Fast Packet Switch for the Integrated Services Backbone Network," *IEEE J. Select. Areas Commun.*, Vol. 6, No. 9, 1988, pp. 1468–1479.

[Nyq28] H. Nyquist, "Certain Topics in Telegraph Transmission Theory", *Trans. AIEE*, Vol. 47, 1928, pp. 617–644.

[Oie89] Y. Oie, M. Murata, K. Kubota, and H. Miyahara, "Effect of Speedup in Nonblocking Packet Switch," *Conference Record, ICC'89*, Vol. 1, pp. 410–415.

[OpT71] D. C. Opferman and N. T. Tsao-Wu, "On a Class of Rearrangeable Switching Networks," *Bell Syst. Tech. J.*, Vol. 50, 1971, pp. 1579–1618.

[PaGa93]   A. Parekh and R. Gallager, "A Generalized Processor Sharing Approach to Flow Control in Integrated Services Networks: The Single-Node Case," *IEEE/ACM Trans. Networking*, Vol. 1, No. 3, 1993, pp. 344–357.

[Pau62]   M. C. Paull, "Reswitching of Connection Networks," *Bell Syst. Tech. J.*, Vol. 41, 1962, pp. 833–855.

[Par94]   C. Partridge, *Gigabit Networking*, Addison-Wesley, 1994.

[Pat81]   J. H. Patel, "Performance of Processor–Memory Interconnections for Multiprocessors," *IEEE Trans. Comput.*, Vol. 30, 1981, pp. 771–780.

[Pat88]   A. Pattavina, "Multichannel Bandwidth Allocation in a Broadband Packet Switch," *IEEE J. Select. Areas Commun.*, Vol. 6, No. 9, 1988, pp. 1489–1499.

[Pea77]   M. C. Pease, "The Indirect Binary n-Cube Microprocessor Array," *IEEE Trans. Comput.*, Vol. C-26, 1977, pp. 458–473.

[PG99]   C. Parsa and J.J. Garcia-Luna-Aceves, "Improving TCP Congestion Control over Internet with Heterogeneous Transmission Media," *Proceeding of IEEE ICNP99*.

[Rad67]   R. Rado, "On the Number of Systems of Distinct Representatives of Sets," *J. London Math. Soc.*, Vol. 42, 1967, pp. 107–109.

[Rob93]   T. G. Robertazzi (Ed.), *Performance Evaluation of High Speed Switching Fabrics and Networks: ATM, Broadband ISDN, and MAN Technology*, Collection of Papers, IEEE Press, 1993.

[Sak90]   Y. Sakurai, N. Ido, S. Gohara, and N. Endo, "Large-Scale ATM Multi-Stage Switching Network with Shared Buffer Memory Switches," *Proceedings of ISS '90*, Vol. 4, pp. 121–126.

[SaZ95]   S. Sibal and J. Zhang "On a Class of Banyan Networks and Tandem Banyan Switching Fabrics," *IEEE Trans. Commun.*, Vol 43, No. 7, 1995, pp. 2231–2240.

[Sch87]   M. Schwartz, *Telecommunication Networks, Protocols, Modeling, and Analysis*, Addison-Wesley, 1987.

[Sha48]   C. E. Shannon, "A Mathematical Theory of Communication," *Bell Syst. Tech. J.*, Vol. 27, 1948, pp. 379–423 (Part I), pp. 623–656 (Part II).

[Sha49]   C. E. Shannon, "Communication in the Presence of Noise," *Proceedings of Institute of Radio Engineers*, Vol. 37, January 1949, pp. 10–21.

[Sha50]   C. E. Shannon, "Memory Requirements in a Telephone Exchange," *Bell Syst. Tech. J.*, Vol. 29, 1950, pp. 343–349.

[Sch87]   M. Schwartz, *Telecommunication Networks Protocols, Modeling and Analysis*, Addison-Wesley, 1987.

[Sie85]   H. J. Siegel, *Interconnection Networks for Large-Scale Parallel Processing*, Lexington Books, 1985.

[SiSp96]   M. Sipser and D. Spielman, "Expander Codes," *IEEE Trans. Inform. Theory*, Vol. 42, No. 6, 1996, pp. 1710–1722.

[Sta95]   W. Stallings, *ISDN and Broadband ISDN with Frame Relay and ATM*, Prentice-Hall, 1995.

[Sto71]   H. S. Stone, "Parallel Processing with the Perfect Shuffle," *IEEE Trans. Comput.*, Vol. C-20, 1971, pp. 153–161.

[Str86]   G. Strang, *Introduction to Applied Mathematics*, Wellesley, MA: Wellesley-Cambridge Press, 1986.

[Suz89] H. Suzuki, H. Nagano, T. Suzuki, T. Takeuchi, and S. Iwasaki, "Output-Buffer Switch Architecture for Asynchronous Transfer Mode," *Conference Record, IEEE ICC'89*, pp. 99–103.

[Tan81] A. S. Tanenbaum, *Computer Networks*, Englewood Cliffs, NJ: Prentice-Hall, 1981.

[Tan81] R. M. Tanner, "A Recursive Approach to Low Complexity Codes," *IEEE Trans. Inform. Theory*, Vol. 27, No. 5, 1981, pp. 533–547.

[TAPF01] P. Trimintzios, I. Andrikopoulos, G. Pavlou, P. Flegkas, D. Griffin, P. Georgatsos, D. Goderis, Y. T. Jocns, L. Georgiadis, C. Jacquenet, and R. Egan, "A Management and Control Architecture for Providing IP Differentiated Services in MPLS-Based Networks," *IEEE Commun. Mag.*, Vol. 39, No 5, 2001, pp. 80–88.

[TKC91] F. Tobagi, T. Kwok, and F. Chiussi, "Architecture, Performance, and Implementation of the Tandem Banyan Fast Packet Switch," *IEEE J. Select. Areas Commun.*, Vol. 9, No. 8, 1991, pp. 1173–1193.

[ToK91] F. A. Tobagi and T. Kwok, "The Tandem Banyan Switching Fabric: A Simple High-Performance Fast Packet Switch," *Proceedings of IEEE INFOCOM '91*.

[Tur86a] J. S. Turner, "New Directions in Communications (or Which Way to the Information Age)," *IEEE Commun. Mag.*, Vol. 4, No. 10, 1986, pp. 8–15.

[Tur86b] J. S. Turner, "Design of a Broadcast Switching Network," *Proceedings of IEEE INFOCOM '86*, pp. 667–675.

[Tur86c] J. S. Turner, "Design of an Integrated Service Packet Network," *IEEE J. Select. Areas Commun.*, 1986.

[Tur88] J. Turner, "Design of a Broadband Packet Switching Network," *IEEE Trans. Commun.*, Vol. 36, No. 6, 1988.

[Tur93] J. Turner, "A Practical Version of Lee's Multicast Switch Architecture," *IEEE Trans. Commun.*, Vol. 41, No. 8, 1993, pp. 1166–1169.

[WC91] Z. Wang and J. Crowcroft, "A New Congestion Control Scheme: Slow Start and Search (Tri-S)," *ACM Comput. Commun. Rev.*, 1991.

[WFL80] C. L. Wu and T. Y. Feng, "On a Class of Multistage Interconnection Networks," *IEEE Trans. Comput.*, Vol. C-29, 1980, pp. 694–702.

[Whi15] E. T. Whittaker, "On the Functions Which are Represented by the Expansions of the Interpolation Theory," *Proc. R. Soc. Edinburgh*, Vol. 35, 1915, pp. 181–194.

[Wil72] R. J. Wilson, *Introduction to Graph Theory*, New York: Academic Press, 1972.

[WuF84] C. L. Wu and T. Y. Feng, *Tutorial: Interconnection Networks for Parallel and Distributed Processing*, IEEE Computer Society Press, 1984.

[YHA87] Y. S. Yeh, M. G. Hluchyj, and A. S. Acampora, "The Knockout Switch: A Simple, Modular Architecture for High-Performance Packet Switching," *IEEE J. Select. Areas Commun.*, Vol. 5, No. 8, 1987, pp. 1274–1283.

[Zha95] H. Zhang, "Service Disciplines for Guaranteed Performance Service in Packet-Switching Networks," *Proc. IEEE*, Vol. 83, No. 10, 1995, pp. 1374–1399.

[ZOK93] W. De Zhong, Y. Onozata, and J. Kaniyil, "A Copy Network with Shared Buffers for Large-Scale Multicast ATM Switching," *IEEE/ACM Trans. Networking*, Vol. 1, No. 2, 1993, pp. 157–165.

INDEX

1-to-R expander, 123, 124
2-tuple numbering scheme, 83, 85
ABR service, 244, 299, 325, 380
Absorption state, 170
Accept phase, 117, 118
Acknowledgement phase, 73, 74
Active link, 164, 262
Active token, 164, 165
Activity bit(s), 71, 72, 212
Adaptation delay, 281, 283, 285
Adaptation layer, 295, 297–299, 314
Adaptation process, 281
Adaptation processing delay, 281
Adaptation waiting delay, 281
Additive Gaussian noise, 387, 395
Address assignment, 133
Address comparison, 132
Address generator, 133, 134, 135, 144
Address interval, 208, 210, 212–214, 219, 221, 230, 231
Aggregated route-base IP switching (ARIS), 306
Air medium, 5
Air space, 2, 5
Allocation, 7, 244, 245–246, 256, 299, 323–324, 355, 358, 389, 411–413, 431
Alternating path algorithm, 247
Alternative routes, 137, 324
Arrival rate, 37, 39, 163, 245, 246, 270, 329–336, 346, 349, 355–358, 377, 412

Asymptotic complexity order, 154, 158
Asynchronous transfer mode(ATM), 10, 275
  ATM adaptation layers (AALs), 297
  cell header, 325
  network, 280, 286–287, 295, 297–303, 312, 313, 325, 329
Automatic repeat request (ARQ), 285
Available bit rate (ABR), 244, 299, 325
Average data rate, 327

Backlogged packet(s), 97, 98, 113, 163
Back-pressure mechanism, 325
Bandlimited signal(s), 386, 389, 415, 431
Bandpass filter, 5
Bandwidth, 5, 7, 9, 138, 224, 235, 236, 239, 244, 245, 256, 268–270, 276–280, 287, 293–295, 299, 301, 324, 327, 346, 352–354, 380
  allocation, 299, 324
  effective, 244, 256, 268–270
  external, 224
  fragmentation, 278
Banyan network(s), 49–60, 69, 71–75, 88, 89, 122–127, 132, 133, 135, 137, 138–139, 141, 143–144, 151–155, 158, 173, 174, 183, 187, 191, 193–196, 207–208, 210–214, 217–220, 231, 233, 268
  baseline network, 29, 31, 32, 49, 50, 81
  baseline switch, 16, 17, 47

Principles of Broadband Switching and Networking, by Tony T. Lee and Soung C. Liew
Copyright © 2010 John Wiley & Sons, Inc.

443

Banyan network(s) (*continued*)
   dilated banyan network, 138, 139, 141, 143
   dilated reverse banyan network, 219
   feedback bidirectional shuffle-exchange network, 166, 175, 177
   feedback shuffle-exchange network, 156, 158, 162
   feedforward shuffle-exchange network, 159, 160
   parallel banyan network(s), 122, 137, 138, 144, 193, 268
   reverse banyan concentrator, 132–135
   reverse banyan network(s), 59, 211
   reverse baseline network, 29, 31, 32, 81
   reverse omega network(s), 81, 84–86, 231–233
   reverse shuffle-exchange network, 49, 50
   shuffle-exchange network, 49–51, 54, 154–158, 220, 226
   tandem-banyan network, 151–154, 158, 180
Bar state, 16, 17, 71–72, 136, 207, 223
Batcher network
   Batcher–banyan knockout switch, 133, 135, 187
   Batcher–banyan network, 73, 122, 132, 138, 144, 187, 237
   Batcher–banyan switch, 132, 154, 187, 192, 193, 196, 197, 268
   Batcher–R-banyan network, 122
   Batcher sorting network, 69, 71, 191, 193, 196, 268
   Batcher–truncated banyan network, 125
   dual shuffle-exchange network, 156, 158, 175, 176
   modular Batcher–binary–banyan switch, 192
Benes network, 28–34, 81, 266
Bernoulli process, 95
Bernoulli trials, 225
Best-effort, 301, 306
Binary–banyan expansion network, 194
Binary numbers, 54, 156, 194, 208
Binary symmetric channel(s), 385, 388, 389, 401
Binary tree(s), 31, 33, 34, 191–196, 124, 136, 216

Binomial distribution, 335
Bipartite graph, 115, 116, 236
Bipartite matching, 385
Bipartite multigraph(s), 236, 239–240
Birkhoff-von Neumann decomposition, 386, 389, 414, 430
Bitonic sequence, 65–67
Bitonic sorter, 65–69, 70
Blocking probability, 15, 39, 278, 291–292, 324
Boltzmann entropy, 391
Boltzmann model, 385, 391
Boolean interval splitting algorithm, 208, 213
Buffer, 43, 45–48, 54, 73, 75, 96, 101–103, 112–115, 118, 123, 151, 159, 160, 180, 184–190, 195, 221, 224, 250, 260, 282–283, 297, 317
   buffer delay(s), 385, 386
   buffer occupancy, 327, 335–336, 344, 346, 380
   buffer size, 102, 260, 269–270, 338, 344–347, 382
   buffer-sharing approach, 186
   infinite buffer(s), 101, 113, 335, 346
   input buffering, 47, 48, 224
   input queue(s), 75, 96, 97, 101, 104, 106, 108–110, 112–114, 119–121, 128, 131, 160–164, 187, 197, 259, 282, 411, 412
   internal buffer(s), 47, 195
   output-buffered switch, 47, 48, 112, 224, 282, 353
Bursty traffic, 12, 27, 52, 76, 348
Bus-based switch, 206
Busy period, 9, 103, 104, 107–109, 368, 371, 376–379
Busy state, 331
Busy stream, 331–332
Bypass, 156–158, 179–184, 308
Bypass mechanism(s), 158, 179, 183, 184

Call admission control, 324
Call admission process, 324
Call level, 244, 268
Call's duration, 324
Call splitting, 214–216, 219
Cantor Switching Network, 32, 33

INDEX **445**

Capacity(s), 5, 7, 35, 43, 58, 97, 118, 119, 121, 213, 214, 217, 222, 224, 235–242, 244–249, 255–259, 268–270, 294, 362, 378
  capacity allocation, 27, 244–246, 256, 268
  capacity assignment matrix, 236, 247
  capacity graph, 239, 240, 242, 247
  capacity matrix decomposition, 389, 414, 424, 426, 429–431
Carried load, 96, 99, 102, 112, 113, 153, 162, 197, 198
Carrier frequencies, 5
Cascaded Clos network, 397
Cauchy–Schwartz inequality, 418
CBR traffic, 348
Cells, 9, 238, 244, 255, 280–289, 291–293, 297–308
  cell-level control, 323–324
  cell loss priority (CLP), 313
  cell size, 283–285
  cell switching, 43, 235, 238, 239, 283
  cell switch router (CSR), 303
  cell-traffic model, 326–331
Central controller, 48, 49, 51, 60, 133, 236, 237
Centralized route assignment, 235
Central limit theorem, 393
Central switch, 3, 4
Channel coding theorem, 385, 388, 401
Channel-grouping principle, 121, 122
Channel-path switching, 410
Circuit-switched network(s), 9, 15, 35
Circuit switching, 8, 43, 44, 48, 184, 235–236, 239, 275, 283
Clos networks, 20–23, 26–30, 37, 39, 75, 79, 81, 205, 228, 229
  three-stage network, 20–23, 26–28, 37–39
Closed-form, 29, 53, 129, 140, 326, 344
Closed-loop control, 325, 380
Coaxial cable, 2, 5
Combinatoric property(s), 54
Communication channel(s), 10, 276
Communication session, 9, 126
Community-of-interest phenomenon, 5
Comparator, 61–67, 69–72
Comparison network, 61–63, 66, 67

Complexity, 18, 20, 31, 45–47, 120, 125, 135, 136, 138, 141, 144, 366, 154, 155, 158, 180, 183, 187, 188, 196, 206, 220, 236, 246–247, 249, 254–255, 264, 266–269, 278–279, 301, 303–305
Concentrated inputs, 54, 83, 84, 87, 124, 210
Concentration, 5, 6, 7, 54, 55, 59, 61, 72, 83–87, 89, 133, 217, 233
Concentrator, 6–9, 59, 126, 127, 132–136, 143, 144, 151, 152, 158, 187–190, 196, 210, 211, 217, 219, 267–268
Conectivity, 6, 27
Conflict-free route assignment(s), 388
Conflicts, 43, 47–49, 51, 57, 59, 72–74, 83, 115, 137, 197
Connection-level control, 324
Connection matrix, 21, 22, 23, 25, 30, 31
Connection-oriented, 9, 286, 298–301, 311
Congestion points, 325
Consecutive, 75, 86, 230, 233, 234
Constant bit-rate (CBR), 298
Contention, 44–47, 51, 54, 61, 71, 73–75, 89, 96, 97, 101–103, 113–115, 119–120, 122, 131, 132, 135, 141, 151, 154, 157, 159–162, 164–166, 171, 174, 187–188, 196, 197, 207, 219, 224, 226–227, 236, 237, 260, 264, 282
  contention problem, 44, 45, 119, 151, 157, 282
  contention-resolution, 47, 73, 74, 99, 102, 103, 110, 113, 114, 119, 120, 122, 132, 138, 219, 224, 249
  mechanism, 224
  policy, 110, 160, 164, 165
  process, 99, 120, 138
  shortest-distance priority contention resolution, 166
Contiguous outputs, 208
Continuous-time Markov chain, 332
Convergence sublayer (CS), 297
Convergence sublayer indicator (CSI), 314
Copy network, 208, 213, 220, 228
  copy index (CI), 220
  copying function, 207

Copy network (*continued*)
 copy number (CN), 211
 cyclic running-adder network, 215, 217, 219
Cost saving, 6
Cross and bar states, 207
Cross state, 16, 48, 72, 213, 223
Crossbar switch, 16, 17, 20, 194
Cross-path switch(es), 205, 236, 237, 239, 244, 246, 249, 250, 254–259, 266
Cross-plane switching, 171
Crosspoint(s), 16, 18, 19, 20, 28, 29, 31
Cyclically shifted monotone sequence, 219
Cyclic redundancy code (CRC), 313
Cyclic running-adder network, 215, 217, 219

Datagram network, 286
Data link layer, 285–286, 297
Data loss, 295, 381
d-dilated-Banyan network, 139
Deadlock, 222–224, 227, 228
 deadlock-free, 223
 deadlock prevention, 222
Decoding process, 211
Decomposition, 28, 66–69, 191, 196, 231, 240
Deflection
 deflection Clos network, 397, 401
 deflection distance, 154, 155, 158, 166, 171, 172, 174
 deflection penalty(s), 155, 158, 164, 166, 176
 deflection probability, 160, 164, 171, 173–175, 181
 deflection routing, 151–184, 220
Degeneration, 231
Delay, 10–11, 36, 49, 71–72, 95, 102–106, 110, 112, 113, 119–120, 130–131, 152, 160–163, 173, 186–189, 217, 222, 247, 244, 246, 247, 268–270, 277–286, 294–301
 delay at the fictitious queues, 103
 delay jitter, 268, 282
 delay of an input-buffered switch, 103
 delay of an output-buffered switch, 112
 input delay, 282
 queueing delay, 187

Demultiplexer, 6, 39, 195
Departure curve, 327–332, 349–354, 379
Departure rate, 163, 329, 330, 336, 349–351
Destination address, 8, 10, 81, 122, 123, 156–158, 160, 167, 170, 177, 184, 188, 193–195, 236
Deterministic routing policy, 223
Diagonal structure, 341
Dilated banyan network, 138, 139, 141, 143
Dilated reverse banyan network, 219
Dilation degree, 139, 140, 143, 219
Dilation principle, 138
Direct connection, 4
Direction for packet routing, 208
Distance, 154, 155, 158, 160, 162, 166, 169–174, 181, 183, 313
Distinct and monotonic, 54, 210
Distributed control, 236, 291
Distributed queue dual bus (DQDB), 312
Distribution network(s), 188, 189
Divide-and-conquer, 64, 193
Downstream, 324–325
DS-1, 276–278, 287
DS-3, 276–278
Dual shuffle-exchange network, 156, 158, 175, 176
Dual-bucket system, 351
Dual-leaky-bucket system, 352
Dummy address encoder, 211, 212, 218
Dummy packets, 60, 61, 72
Dummy routing bit(s), 184
Duplication, 208, 224, 301
Dynamic routing, 205, 235, 236

Edge coloring, 236, 238–240, 249
Edge-colored bipartite graph, 404, 410
Effective bandwidth, 244, 256, 268–270
Effective fanout, 226
Efficient use of link capacity, 354
Eigenvalue, 340–346
Eigenvector, 340–346
Elastic buffer, 297
Encoding process, 211
End office, 4
End-to-end delay, 10, 282, 297–299
End-to-end flow control, 325
Energy quantum, 390
Entropy, 386, 389–392, 417–420, 424–430
 entropy inequality(s), 386, 389

Equilibrium probability densities, 338
Error correction, 167, 169, 170, 171, 175, 177–179, 313–315
Error-correction routing bit(s), 167, 169, 170, 171, 177–179
Error function, 225
Error mode, 170
Error-tolerant transmission, 385, 386
Even distribution, 331
Even-indexed, 70
Expander graph(s), 385, 389, 406
Expansion
 expansion factor, 249–255, 259–262, 264, 266
 expansion network, 191, 193–196
External bandwidth, 224
External link, 119, 120, 224
 external link rate, 119
Externally nonblocking, 83, 84
External rearrangement(s), 27
External time slot, 224

Failure probability, 225
Feasible ordering, 378
Feedback bidirectional shuffle-exchange network, 166, 175, 177
Feedback flow control, 325
Feedback shuffle-exchange network, 156, 158, 162
Feedback network
 feedback bidirectional shuffle-exchange network, 166, 175, 177
 feedback shuffle-exchange network, 156, 158, 162
Feedforward flow control, 325, 354
Feedforward shuffle-exchange network, 159, 160
Feedforward network
 feedforward shuffle-exchange network, 159, 160
Fictitious queue(s), 96, 97, 101–103, 106, 108–110, 112, 113, 127, 130
Filter(s), 5, 135, 136, 138, 152, 206
Finite-state machine, 154, 155, 169
First-come-first-serve(FCFS), 329, 355
First-in-first-out, 96, 102
Fixed-capacity scheduling, 357, 358
Fixed-length, 43, 95, 275, 280, 283, 286, 308

Fixed-length packet switching, 275, 280
Fixed traffic intensity, 345
Fluid-flow
 fluid-flow model, 326, 327, 331, 335, 339, 348, 351, 352, 355, 358, 362–364, 366, 368–370
 fluid-flow service, 330, 362, 364, 365
 fluid-flow weighted fair queueing, 366
Forwarding equivalence classes (FEC), 310
Forwarding information base (FIB), 308, 310
Fourier series expansion, 389, 416
Frame(s), 5–8, 35–37, 39, 43, 184, 240, 242, 247, 249, 263, 275–280, 300, 306–308, 410, 414–421, 424–431
Frequency band(s), 5
Frequency-division multiplexing(FDM), 5, 6
Fundamental design principle(s), 15, 95

Gaussian channel, 395
Gaussian noise, 387, 395
Generalized connection networks (GCN), 233
Generalized interval splitting algorithm, 222, 227, 230–232, 234
Generalized knockout principle, 187–189, 191
Generalized processor sharing, 421
Generalized reduced gradient method (GRG), 246
Generating function, 95, 98, 100, 101, 104, 105, 109, 127–129, 181, 262–264, 341, 399
Generic flow control (GFC), 312
Grant phase, 117, 118
Group identification, 206
Grouping network, 188–190
Group size, 46–48, 53, 112, 121, 129, 189, 240, 259, 264–266, 269

Half-cleaner, 67, 68
Half-Clos networks, 81, 82, 86
Hall's condition, 385, 389, 403, 404, 406
Header(s), 7, 8, 44, 45, 50, 71–74, 132–134, 159, 160, 166, 170, 184, 189, 194, 195, 206–209, 211–213, 219–222, 230, 231, 280, 281, 283–286, 289, 297, 299, 300, 305, 306, 311–313, 317–319, 325, 397

448 INDEX

Header error control (HEC), 313
Header translation process, 44
Head-of-line (HOL), 75, 95–104, 112–115, 127, 129, 196, 258, 259, 329, 359, 366, 370
   head-of-line (HOL) blocking, 196, 258, 259
Heavy-traffic, 187, 214
Hierarchical structure, 5
Hierarchy, 4, 5, 8, 276, 277, 287, 289, 293, 294, 312
Highly blocking, 207
High-speed network, 325
High-speed packet switch, 205
Hop-by-hop routing, 286, 289
Huffman round rabin algorithm(HURR), 422
Huffman tree, 417, 422–424
Hungarian algorithm, 247

Identity matrix, 340
Idle period, 9, 103, 109
Idle state, 331
Idle stream, 331, 332
Independent, 10, 21, 95, 98, 100–106, 114, 125, 127, 131, 154, 158, 160, 188, 192, 210, 225, 246, 250, 251, 255, 257, 262, 263, 280, 304, 313, 332, 335, 393, 394, 411, 415
   independent random variable(s), 101
   link-independent, 57, 210
Index reference (IR), 212, 219
Individual data streams, 355
Infinite buffer(s), 101, 113, 335, 346
Information stream, 5, 35
Information theory, 385–388, 418
In-plane switching, 171
Input broadcast packet(s), 210
Input-buffered switch, 47, 48, 96, 102, 103, 112–115, 118, 119
Input buffering, 47, 48, 224
Input capacity(s), 7
Input-channel group(s), 139
Input delay, 282
Input faireness, 214
Input offered load, 226
Input-output mapping(s), 19, 38, 39, 50–54
Input-output path, 210
Input-output tree(s), 207, 209, 210

Input queue(s), 75, 96, 97, 101, 104, 106, 108–110, 112–114, 119–121, 128, 131, 160–164, 187, 197, 259, 282, 411, 412
Input-stage module, 77, 229, 237, 244, 254, 255, 258–260, 263, 267
Input traffic intensity, 345
Integrated broadband network, 1
Integrated network, 2, 10–12, 323
Integrated switch routers (ISRs), 306
Interconnection network(s), 45, 48, 49, 81, 86, 195, 196, 207, 231
Internal buffer(s), 47, 195
Internal contention(s), 47, 387, 388
Internal link, 76, 138, 219
Internally nonblocking, 48, 49, 54, 59, 60, 83–85, 91, 95, 96, 115, 119, 137, 180, 195
Internal time slot, 224
Inverse multiplexing, 278, 280
Isomorphic, 49–51, 54, 155

Kleinrock's inependency assumption, 411
Knockout concentrator(s), 187, 188, 190
Knockout principle, 131, 132, 135, 137, 187–189, 191
Knockout switch, 133, 135, 149, 187, 190, 196, 206, 237, 251, 267, 268, 411
Kraft's inequality, 418, 420, 422, 424
Kraft's matrix, 427
Kullback–Leibler distance, 420

Label distribution prptocol (LDP), 310
Label-forwarding information base (LFIB), 310
Label information base (LIB), 310
Label switch router (LSR), 309
Label switched path (LSP), 311
Lagrange multiplier(s), 392
Latin square, 240, 241
Leaky-bucket analogy, 328
Leaky-bucket constraint(s), 374
Least significant bit(s), 51, 58, 59, 88, 156, 177, 179
Linear difference equation, 161, 172, 181
Linear differential equation, 340
Link-independent, 57, 210
Link-to-link flow control, 325
Little's law, 113, 161, 173, 346

INDEX   **449**

Local address, 44, 194, 250
Local-area computer networks, 1
Local-area networks (LANs), 300, 312
Local office, 4, 5
Logical channel(s), 5, 6, 36
Logical IP subnets (LISs), 301
Look-ahead, 113, 114, 196, 249, 251, 258–260
Look-ahead contention resolution, 113, 114, 196, 249
Looping algorithm, 30, 31
Losing packet, 73, 119, 120, 132, 135, 136, 151, 396
Loss probability, 46, 47, 51–53, 131, 132, 135, 138–141, 143, 151–154, 159, 180, 183, 185, 186, 188–190, 213, 222, 225, 226, 236, 249–252, 254, 257, 260–262, 264–267, 269, 323, 324, 385, 388, 389, 397–399, 401
Loss system(s), 45–47, 50, 51, 96, 131, 132, 137–139, 183, 187, 213
Lost contention, 47, 54, 73, 75, 96, 97, 114, 131, 187, 282
Low-density parity check (LDPC) code, 389
Lower branch, 210
Low-priority session(s), 360
Low-priority traffic, 348

Manhattan-street network, 227
Marked packet, 151, 152, 219
Markov chain, 102, 170, 181, 332, 333, 335, 397, 398
Markovian on-off source model, 269
Master packet, 213, 217, 219, 222, 227
Matching problem, 115,116
Maximal utilization, 394
Maximum capacity(s), 97, 118
Maximum data rate, 395
Maximum throughput, 53, 96, 101, 102, 111, 112, 114, 118, 119, 127, 129, 159, 161–163, 173, 198
m-colorable, 239
Memory I/O, 184, 185
Merger, 64, 65, 69–71
Message mode, 316, 318
Middle-stage module, 21–24, 31–35, 76–77, 79–81, 86, 228–231
Mini-cycle, 360, 361

Modest-size switch(es), 151, 187, 188
Modular architecture, 191–193, 196
Modular Batcher–binary–banyan switch, 192
Modulated binomial process, 98
Monotonic, 54, 55, 62–66, 74, 77–89, 124, 132, 133, 141, 210, 217, 228–230, 232, 406
Multicast capability, 135, 206
Multicast group, 206
Multicasting algorithm, 220
Multicast network
  branching state, 213
  broadband integrated network, 2, 10, 323
  broadband network, 1, 2, 205
  broadcast banyan network, 208, 210–214, 217, 220, 231, 233
  broadcast channel number (BCN), 211
  broadcast clos network, 228–235
  broadcast knockout switch, 268
  broadcast omega network, 231–233
  broadcast routing, 138
  broadcast switch, 230
  served copy number (SCN), 216
  truck number (TNs) translators, 211
  truncated banyan network, 123, 124
Multicast switching, 205, 268
  multicast capability, 135, 206
  multicast group, 206
  multicasting algorithm, 220
Multimedia service(s)(QoS), 1, 205
Multiple-access communication network(s), 102
Multiplexed channel, 6
Multiplexer(s), 6–10, 39, 122, 123, 126, 127, 133, 137, 151, 177, 179, 180, 190, 191, 195, 284, 294, 312, 327, 329, 330, 332, 335, 355, 356, 358, 362–364, 368, 371
Multiplexing indentifier (MID), 316
Multiprotocol label switching (MPLS), 308, 431
Multirate, 205, 235, 236, 242, 278, 279
  multirate circuit switching, 235, 236
  multirate switching, 278, 279
  multirate traffic, 242
Multirate circuit switching, 235, 236
Multirate switching, 278, 279
Multirate traffic, 242

Multistage crossbar switch, 194
Multistage interconnection networks
 (MINs), 81, 86, 87, 231, 402
Multistage network
 multistage crossbar switch, 194
 multistage interconnection networks
  (MINs), 81, 86, 87, 231, 402

Network resource(s), 9, 323, 324
Network time protocol (NTP), 305
Next hop resolution protocol (NHRP), 302
Noiseless channel model, 389
Noiseless coding theorem, 388, 420
Noise power, 390, 393
Noisy channel capacity theorem, 388–390, 395
Noisy channel coding theorem, 385, 388, 401
Nonblocking condition, 39, 54, 75, 83, 85, 195, 210, 220, 231, 232
Nonblocking property(s), 16, 21, 34, 195, 228
Nonconflicting, 48, 61, 73, 122
Numbering scheme, 76, 81–85, 194, 195, 394
Nyquist–Shannon sampling theorem, 415

Occurrence of congestion, 325
Odd–even merger, 70, 71
Odd–even sorting network, 70
 odd–even merger, 70, 71
 odd-indexed, 70
Offered load, 95, 96, 102, 112, 119, 131, 140, 153, 162, 163, 189, 226, 242, 245, 246, 390, 392, 399–401
Omega network, 49, 50, 81–89, 231–233
One-packet buffer, 159, 160
On-off source, 9, 244, 269, 326, 327, 332, 333, 349, 355
Open-loop control, 325, 354, 380
Operations, administration, and management (OAM), 312
Optical fiber, 2, 5
Order-preserving property, 61–65, 79, 87
Out-of-sequence, 286
Output address, 44, 45, 50, 51, 54, 56, 59, 71–74, 83, 120–122, 124, 125, 133, 139, 140, 152, 158, 177, 207, 208, 210, 211, 213, 220, 221, 231, 232, 390, 392

Output-buffered switch, 47, 48, 112, 224, 282, 353
Output capacity(s), 7, 119, 121, 259, 330, 335, 345, 346, 354, 358
Output capacity expansion, 119, 121
Output contention, 45, 47, 95, 260
Output interval, 208, 210–212
Output link, 43, 48, 55, 112, 119, 120, 123, 126, 156, 158, 167, 172, 174, 177, 179, 185, 226, 230, 244, 256, 258, 284, 286, 287, 289, 291, 292, 307, 308, 354, 380, 395, 397
Output offered load, 226
Output-stage module, 76, 77, 229
Overflow, 54, 113, 185, 187, 213–217, 219, 222, 224–227, 270, 325, 332, 344, 346, 381
Overhead penalty, 284

Packet collision, 194, 195, 217, 219
Packet conflict, 73, 125, 220
Packet contention, 45, 120, 151, 207, 385–387, 396
Packet level, 244, 256, 268, 269, 415, 431
Packet loss(es), 46–48, 53, 120, 139, 140, 141, 155, 159, 180, 185–190, 213, 222, 236, 250, 251, 257, 260, 385, 386, 398, 399
Packet replication, 207, 208, 211, 222, 227–233, 255, 258, 267
Packet routing, 195, 208
Packet slicing, 180, 120, 121
Packet switching, 7, 30, 43–45, 48, 49, 61, 71–73, 184, 235, 236, 275, 276, 280, 283, 386–392, 396, 416, 431
Packet-based, 280
Packet-by-packet fair queueing, 366
Packet-loss probability, 46, 47, 53, 131–133, 139, 140, 141, 155, 159, 180, 185, 186, 188–190
Packet-switched network(s), 8, 285, 385, 386, 396
Parallel algorithm, 247, 409
Parallel banyan network(s), 122, 137, 138, 144, 193, 268
Parallel iterative matching(PIM), 115, 116
Parallel network(s), 137
 parallel algorithm, 247, 409

parallel banyan network(s), 122, 137, 138, 144, 193, 268
parallel iterative matching(PIM), 115, 116
parallelization approach, 185
Partially busy state, 331
Partial service, 214
Partial splitting, 228
Path switching, 205, 235–237, 239, 254, 388, 389, 410, 411, 414, 417, 430, 431
Paull's rearrangeable theorem, 402
Penalty, 158, 164, 166, 176, 187, 278, 284, 370
Perfect shuffle, 155, 156, 158, 196
Performance, 48, 95, 102, 113, 120, 125, 151, 154, 159, 171, 174, 175, 177, 187, 196, 197, 225
Performance of simple switch design, 95
Point-to-multipoint, 205, 255
Point-to-point, 89, 205–210, 224, 230, 267
Poisson approximation, 197
Poisson process, 98, 100, 111, 249
Policing function, 324
Polling, 133
Pre-agreed rate, 324, 325
Preprocessing, 220, 224
Primary office, 5
Primary route, 178, 179
Probe phase, 73
Projection, 324
Propagation delay, 281, 282
Pseudo signal-to-noise ratio (PSNR), 387

Quality of service(QoS), 53, 323
Quasi-static, 236, 239, 295
Quasi-static routing, 236
Queueing delay, 187
Queueing theory, 98, 105

Random access memory (RAM), 36, 184, 185
Random channel coding, 385, 388
Random number generator, 223
Random routing, 137, 138, 223
    random routing policy, 223
Random-walk model, 183
Rank-based assignment algorithm, 78, 228

Rate-versus-time, 327
Real-time communication services, 112
Real-time service, 11
Rearrange, 16–18, 23, 24, 43, 48, 49
Rearrangeably nonblocking, 17, 18, 21, 23, 26–28, 32, 39, 48
Recursiveness properties, 87
Redundancy, 151
Redundant packets, 138
Regular mode, 170, 171
Reliable communication, 385, 386, 431
Replicating, 210
Replication principle, 137
Replication process, 220–225, 227, 232
Request phase, 117, 118
Reservation mechanism, 224
Residual time, 106, 107
Resource reservation protocol (RSVP), 305
Resource sharing, 7–12, 15
Restriction, 208
Reverse banyan concentrator, 132–135
Reverse banyan network(s), 59, 211
Reverse baseline network, 29, 31, 32, 81
Reverse omega network(s), 81, 84–86, 231–233
Reverse shuffle-exchange network, 49, 50
Rouche's theorem, 128
Round-off penalty, 284
Round-robin, 5, 126
Round-robin fashion, 5, 126, 127, 184
Route assignment, 76–81, 228, 233–239, 249, 266, 389, 402–407
Routing algorithm, 49, 51, 60
Routing bits, 51, 166–174, 222
Routing cells, 324
Routing function, 207
Routing tag, 159, 160, 167–171, 175, 179, 183, 184, 230, 394, 397
Running-adder address generator, 133, 144
Running-adder network, 211–219, 233
Running sum(s), 133–135, 415, 416
Running-sum header, 134

Sampling theorem, 386, 389, 415, 416, 431
Saturated, 97, 111, 118, 161–163, 413
Saturation throughput, 161, 163, 173

Scheduling algorithm, 329, 416–424, 428–430
Segmentation and reassembly, 297
Self-routing, 31, 49, 61, 73, 75, 77, 81, 85, 87, 89, 160, 166, 191, 195, 196, 207, 231
Sequence number (SN), 314, 315
Sequence number protection (SNP), 314
Sequence-preservation, 286
Sequential linear approximation algorithm, 246
Served copy number (SCN), 216
Service guarantee, 354
Service quality, 323
Service times, 103, 104, 106, 108
Shaded adders, 216
Shannon–Hartley theorem, 395
Shannon's main theorem, 401
Shared-buffer memory switching, 185
Sharing of resourses, 5
Shifter, 59
Shifting concentrator, 126, 127, 133
Shifting starting point, 217
Shortest-distance priority contention resolution, 166
Shuffle-exchange network, 49–51, 54, 154–158, 220, 226
Shuffle link, 166, 167, 174–178
Shuffle plane(s), 167, 169, 174, 178
Signal power, 390–393
Signal-to-noise ratio, 385–390, 393, 396
Simple leaky bucket, 348–353
Sipser–Speilman, 385, 406
Sipser–Speilman decoding algorithm, 406–408
Slot-by-slot, 236
Smoothing buffer, 282, 283
Smoothness, 386, 389, 417–431
Sorting network, 60–73, 89
 sort-banyan network, 54, 55, 71–74
 sorted and concentrated, 210
Source data rates, 381
Source routing, 286
Space-division switch, 16, 35–39, 43
Space-domain, 277
Space domain circuit switching, 16
Speedup, 120–125, 224
 speedup principle, 119

speedup scheme, 185
Splitting, 209–232
Sporadic fashion, 380
SS/TDMA satellite communications, 389
Starting point (SP) field, 219
State-transition diagram, 154, 155, 158, 171, 172
Static
 static routing, 205, 236
 statistical multiplexing, 7
 statistical multiplexing gain, 235, 249, 259
 statistically stable, 239
Stirling's approximation, 392
Stirling's formula, 20, 142
Stochastic capacity matrix(s), 415, 425
Stochastic situation, 332
Store-and-forward networks, 10
Store-and-forward routing, 396
Stream mode, 297
Strictly nonblocking, 16, 18, 21, 23, 26, 27, 32–34, 39, 319
Subchannel, 7
Subnetwork, 30, 31, 51, 54–56, 79–81, 124, 431
Subscriber loop, 4
Subsequent switch, 44, 47
Sustainable output rate, 162
Switch
 switch dimensions, 45
 switching, 2
 switching centre, 4
 switching facilities, 3, 15
 switching system(s), 416, 431
 switch level, 244
Synchronous transfer mode (STM), 275

Tag information base (TIB), 308
Tag switching, 306
Tag switch router (TSR), 308
Talkspur, 9
Tandem-banyan network, 151–154, 158, 180
Tanner graph, 405
Taylor series, 53
 approximation, 153
TDM, 6, 35–37, 184
TDMA, 410

Telephone network(s), 1, 3, 4, 6, 8, 15, 35, 276, 285, 287
Telephone session, 9
Thermal system, 391
Thermodynamics theory, 390
Three-leaky-bucket system, 351, 352
Three-phase scheme, 73, 74, 122, 132
Three-phase sort-banyan network, 114
Three-stage expansion network, 194
Three-stage network, 20–23, 26–28, 37–39
Threshold, 381
Throughput, 10, 47, 48, 51–54, 74, 75, 96, 101, 110–120, 127, 159–165, 236–260, 284
   throughput of an input-buffered switch, 96
   throughput of an internally nonblocking loss system, 96
Time assigned speech interpolation(TASI), 9
Time slot(s), 5, 35–39, 43, 47, 59, 103, 105, 106, 126, 131, 275–282, 284, 361, 363
   time-slot interchanger (TSI), 277
Time-divisoin multiplexing (TDM), 5, 43, 275
Time-domain switching, 35–37
Timescales, 323
   timescales of information transfer, 9
Time-space interleaving priciple, 240
Time space time circuit switching, 35–38
Time-to-live (TTL), 308
Token(s), 155, 163–165, 223, 350–353, 373
Toll office, 4, 5
Trade-off, 16, 47, 249, 389, 390, 393–395
Traffic
   traffic demands, 323
   traffic flow, 325
   traffic load, 256–260, 343
   traffic matrix(s), 386, 410, 415
   traffic matrix decomposition, 389, 411, 415
   traffic policing, 326, 348
   traffic shaping, 324, 348
   traffic-versus-time, 327
Transmission
   transmission capacity, 43, 332, 354, 355, 358
   transmission delay, 282

transmission facilities, 2–6
transmission hierarchy, 4
transmission interface, 200
transmission medium, 2
transmission noise, 385, 386
transmission system(s), 387
transmitter-receiver, 6, 7
Truck number (TNs), 211
Truck-number translators, 211
Truncated banyan network, 123, 124
Truncated geometric distribution, 257
Truncated series, 130
Trunk(s), 15
T-S-T switch modules, 278

Unicasting, 206
Unidirectional shuffle-exchange network, 161, 173
Uniform-traffic distribution, 95
Unique-path, 31, 33, 51, 54, 152, 154
Unmarked packet, 151, 152
Unshared-buffer switch, 186
Unshuffle link, 166, 167, 174–178
Unshuffle plane(s), 167, 169, 174, 178
Unspecified bit rate(UBR), 325
Upper branch, 210
Upstream, 324, 325
User datagram protocol (UDP), 305
User-network interface (UNI), 303, 348

Variable bit-rate (VBR), 298, 299
Variable-length, 43, 283
VBR (variable bit-rate), 325, 380
VCI translation, 282
VCs, 287–295, 302–308, 323, 352
Video and speech coding algorithms, 12
Video conferencing, 2, 11
Virtual channel (VC), 287
   level, 324
Virtual circuit, 9
   virtual-circuit identifier (VCI), 44, 289
   virtual-circuit networks, 44
   virtual-circuit routing, 280
Virtual output queues (VOQs), 115
Virtual path (VP), 236, 287
   virtual-path identifier (VPI), 289
   virtual-path networking, 323

Virtual private networks (VPNs), 295
VLSI, 187
VP level, 323, 324

Waiting systems, 45, 47, 131
Waiting time, 105, 106, 109–113, 129
Weighted fair queueing (WFQ), 364, 419
WFQ scheduling algorithm, 420, 428
WF^{2}Q scheduling algorithm, 421

Wide-area computer networks, 1
Wide-area networks (WANs), 300
Wide-sense nonblocking, 17, 18
Window size, 249–254, 259
Winning packet, 73–75, 114, 120, 124, 135, 138, 160, 164
Work-conserving system, 102, 110

Zero–one principle, 61, 64, 71

WILEY SERIES IN TELECOMMUNICATIONS AND SIGNAL PROCESSING

John G. Proakis, Editor
Northeastern University

Introduction to Digital Mobile Communications
Yoshihiko Akaiwa

Digital Telephony, 3rd Edition
John Bellamy

ADSL, VDSL, and Multicarrier Modulation
John A. C. Bingham

Biomedical Signal Processing and Signal Modeling
Eugene N. Bruce

Policy-Driven Mobile Ad hoc Network Management
Ritu Chadha and Latha Kant

Elements of Information Theory, 2nd Edition
Thomas M. Cover and Joy A. Thomas

Erbium-Doped Fiber Amplifiers: Device and System Developments
Emmanuel Desurvire

Fiber-Optic Systems for Telecommunications
Roger L. Freeman

Fundamentals of Telecommunications, 2nd Edition
Roger L. Freeman

Practical Data Communications, 2nd Edition
Roger L. Freeman

Radio System Design for Telecommunications, 3rd Edition
Roger L. Freeman

Telecommunication System Engineering, 4th Edition
Roger L. Freeman

Telecommunications Transmission Handbook, 4th Edition
Roger L. Freeman

Introduction to Communications Engineering, 2nd Edition
Robert M. Gagliardi

Optical Communications, 2nd Edition
Robert M. Gagliardi and Sherman Karp

Efficient Algorithms for MPEG Video Compression
Dzung Tien Hoang and Jeffrey Scott Vitter

Advances in Multiuser Detection
Michael L. Honig, Editor

FTTX Concepts and Applications
Gerd Keiser

Active Noise Control Systems: Algorithms and DSP Implementations
Sen M. Kuo and Dennis R. Morgan

Mobile Communications Design Fundamentals, 2nd Edition
William C. Y. Lee

Robust Adaptive Beamforming
Jian Li and Petre Stoica, Editors

Expert System Applications for Telecommunications
Jay Liebowitz

Principles of Broadband Switching and Networking
Soung C. Liew and Tony Lee

Polynomial Signal Processing
V. John Mathews and Giovanni L. Sicuranza

Digital Signal Estimation
Robert J. Mammone, Editor

Digital Communication Receivers: Synchronization, Channel Estimation, and Signal Processing
Heinrich Meyr, Marc Moeneclaey, and Stefan A. Fechtel

Synchronization in Digital Communications, Volume I
Heinrich Meyr and Gerd Ascheid

Business Earth Stations for Telecommunications
Walter L. Morgan and Denis Rouffet

Wireless Information Networks, 2nd Edition
Kaveh Pahlavan and Allen H. Levesque

Satellite Communications: The First Quarter Century of Service
David W. E. Rees

Fundamentals of Telecommunication Networks
Tarek N. Saadawi, Mostafa Ammar, with Ahmed El Hakeem

Analogue and Digital Microwave Links: Theory and Design
Carlos Salema

Microwave Radio Links: From Theory to Design
Carlos Salema

Meteor Burst Communications: Theory and Practice
Donald L. Schilling, Editor

Digital Communication over Fading Channels, 2nd Edition
Marvin K. Simon and Mohamed-Slim Alouini

Ultra-Wideband Communications Systems: Multiband OFDM Approach
W. Pam Siriwongpairat and K. J. Ray Liu

Digital Signal Processing: A Computer Science Perspective
Jonathan (Y) Stein

Vector Space Projections: A Numerical Approach to Signal and Image Processing, Neural Nets, and Optics
Henry Stark and Yongyi Yang

Signaling in Telecommunication Networks, 2nd Edition
John G. van Bosse and Fabrizio Devetak

Telecommunication Circuit Design, 2nd Edition
Patrick D. van der Puije

Worldwide Telecommunications Guide for the Business Manager
Walter H. Vignault